ALLE ZEIT WACH
1842

Dianthus caryophyllus L., the clove pink. After a drawing in Schlechtendal DLF von, Langenthal LE, Schenk E (eds) Flora von Deutschland, 5th edn. Köhler, Gera-Untermhaus 1883

H.-D. Behnke T. J. Mabry (Eds.)

Caryophyllales

Evolution and Systematics

With 86 Figures

Springer-Verlag
Berlin Heidelberg New York London Paris
Tokyo Hong Kong Barcelona Budapest

Professor Dr. H.-Dietmar Behnke
Zellenlehre
Universität Heidelberg
Im Neuenheimer Feld 230
D-69120 Heidelberg, FRG

Professor Dr. Tom J. Mabry
Department of Botany
University of Texas
Austin, TX 78713-7640, USA

ISBN 978-3-642-78222-0 ISBN 978-3-642-78220-6 (eBook)
DOI 10.1007/978-3-642-78220-6

Library of Congress Cataloging-in-Publication Data. Caryophyllales: evolution and systematics / H.-D. Behnke, T. J. Mabry (eds.). p. cm. Papers originally presented at a symposium held at the Internationales Wissenschaftsforum der Universität Heidelberg, July 1–4, 1992. Includes bibliographical references and index. (Berlin: acid-free paper). – (New York: acid-free paper) 1. Caryophyllales – Classification – Congresses. 2. Caryophyllales – Evolution – Congresses. I. Behnke, H.-D. II. Mabry, T. J. (Tom J.) QK495.A12C38 1993 93 21309 583'.152–dc20

Softcover reprint of the hardcover 1st edition 1994

Typesetting: Best-set Typesetters Ltd., Hong Kong
31/3145-5 4 3 2 1 0 – Printed on acid-free paper

Preface

With about 10000 described species, the Caryophyllales is among the larger orders of dicotyledons and is presently one of the most well-defined groups of flowering plants. The order includes a number of families which contain economically important species (beets, spinach, quinoa, amaranths) and many common horticultural plants (cacti, carnations, four-o'clocks, bougainvillea). The ordinal name Caryophyllales is based on the former botanical name of the carnations, the genus *Caryophyllus* (Mill.), now conserved in the specific epithet of the clove pink, *Dianthus caryophyllus* L. (see frontispiece).

F.G. Bartling, in his *Ordines Naturalis Plantarum* (Dieterich, Göttingen 1830), was probably the first to use the term "caryophyll" in a taxon above the family level. His Caryophyllinae included five of the presently considered 11 core families of the order. However, for the next 150 years a more descriptive name for the order, the Centrospermae, found general acceptance.

In 1876 A.W. Eichler, in his *Syllabus der Vorlesungen über Phanerogamenkunde* (Schwer'sche Buchhandlung, Kiel), coined the term Centrospermae (meaning "central-seeded") for a group of families characterized by free-central or basal placentation. Even today, after the general usage of the ordinal name Caryophyllales in all recent systems of flowering plants, Eichler's name Centrospermae is still perhaps the better known.

Eichler's (1876) Centrospermae comprised eight families, all currently recognized as belonging to the Caryophyllales, and the Polygonaceae, a possible close associate. Subsequently, most workers retained these eight taxa in the order, but divided Portulacaceae and Aizoaceae into more narrowly defined families and separated several genera by raising them to monotypic or bigeneric families. Only one entirely new family has been added, the Didiereaceae (not known in Eichler's time).

Up to a dozen or so other families were occasionally suggested by taxonomists of that early period as being members of, or peripheral to, the Caryophyllales (see Chap. 1). Controversy regarding these taxa continued until the detection, in the 1960s, of both the betalains, a structurally distinct class of pigments belonging exclusively to the order (see Chaps. 10 and 11), and an order-specific subtype of sieve-element plastids (see Chap. 5). These characters provided the crucial arguments for decisions regarding ordinal affinity.

The present circumscription of the order was broadly agreed upon during the XIIth International Botanical Congress held in Leningrad (1975) at which the Caryophyllales were the subject of a symposium (published 1976 in *Plant Systematics and Evolution* Vol. 126, exactly 100 years after Eichler's Syllabus).

Unlike the characterization of the Caryophyllales and its delimitation against other dicotyledons, the circumscription and phylogeny within the order is far from settled. Therefore, some 30 specialists, from plant morphologists to organic chemists, gathered during 1992 in Heidelberg, Germany, for a symposium on the *Evolution and Systematics of the Caryophyllales.* Fifteen speakers presented new data and assessed the progress made in all fields since 1975. The organizers had hoped that research conducted during the intervening 17 years would allow a consensus phylogeny for the Caryophyllales, but the lively discussions indicated otherwise. However, an important result, in addition to this volume, was the establishment of a working group to exchange new ideas, future results and research material.

We want to emphasize the wide range of information presented in this volume, from papers on taxonomic history and systematics of the order (Chaps. 1 and 15), chromosome numbers (Chap. 2), vegetative (Chaps. 3–5) and reproductive (Chaps. 6–7 and 13) characters, chemical and molecular data (Chaps. 8–11 and 14), to cladistic analyses (Chap. 12) and the putative origin and relationship of the order (Chaps. 13 and 14). The two papers on cpDNA studies (Chaps. 8 and 9) highlighted the potential future of macromolecular work in the Caryophyllales.

The symposium and this volume are dedicated to the memory of the late Professor Arthur Cronquist (1922–1992), one of the great generalists of angiosperm taxonomy, who would have addressed the symposium participants on "Nomenclatural and Taxonomic History" of the Caryophyllales. The symposium opened with Professors Robert Thorne and Billie Turner offering personal glimpses from their long friendships with Arthur.

The symposium was held at the Internationales Wissenschaftsforum der Universität Heidelberg, July 1–4, 1992, and was supported by the Volkswagen-Stiftung and the Stiftung Universität Heidelberg.

September, 1993

H.-DIETMAR BEHNKE, Heidelberg
TOM J. MABRY, Austin, Texas

Contents

Contributors

WILHELM BARTHLOTT
Botanisches Institut
Universität Bonn
Meckenheimer Allee 170
D-53115 Bonn, FRG

H.-DIETMAR BEHNKE
Zellenlehre
Universität Heidelberg
Im Neuenheimer Feld 230
D-69120 Heidelberg, FRG

JOHN S. CLEMENT
Department of Botany
University of Texas
Austin, Texas 78713-7640
USA

ARTHUR CRONQUIST,
deceased
formerly Herbarium
New York Botanical Garden,
Bronx, New York 10458-5126,
USA

ANDRÉ S. DREIDING
Organisch-Chemisches Institut
Universität Zürich
Winterthurer Str. 90
CH-8057 Zürich, Switzerland

STEPHEN R. DOWNIE
Department of Plant Biology,
University of Illinois
505 South Goodwin Avenue
Urbana, Illinois 61801-3707
USA

CLAUDIA ERBAR
Institut für Systematische
Botanik
und Pflanzengeographie
Universität Heidelberg
Im Neuenheimer Feld 345
D-69120 Heidelberg, FRG

ARTHUR C. GIBSON
Department of Biology
and Laboratory of Biomedical
and Environmental Science
University of California
Los Angeles, California
90024-1606, USA

URSULA HOFMANN
Systematisch-Geobotanisches
Institut
Universität Göttingen
Untere Karspüle 2
D-37073 Göttingen, FRG

KLAUS KUBITZKI
Institut für Allgemeine
Botanik und Herbarium
Universität Hamburg
Ohnhorststr. 18
D-22609 Hamburg, FRG

PETER LEINS
Institut für Systematische
Botanik
und Pflanzengeographie
Universität Heidelberg
Im Neuenheimer Feld 345
D-69120 Heidelberg, FRG

ALICIA LOURTEIG
Muséum National d'Histoire Naturelle
Laboratoire de Phanérogamie
16, rue de Buffon
F-75005 Paris, France

TOM J. MABRY
Department of Botany
University of Texas
Austin, Texas 78713-7640
USA

JAMES R. MANHART
Department of Biology
Texas A & M University
College Station, Texas
77843-3258, USA

JOAN W. NOWICKE
Botany Department
National Museum
of Natural History
Smithsonian Institution
Washington DC 20560
USA

JEFFREY D. PALMER
Department of Biology
Indiana University
Bloomington, Indiana 47405
USA

JEFF M. RETTIG
Department of Biology
College of the Ozarks
Point Lookout, Missouri
65726, USA

JAMES E. RODMAN
Division of Environmental Biology
National Science Foundation
Washington DC 20550, USA

DIETER STRACK
Institut für Pflanzenbiochemie
Weinberg 3
D-06120 Halle, FRG

ROBERT F. THORNE
Rancho Santa Ana Botanic Garden
1500 North College Avenue
Claremont, California
91711-3157, USA

BILLIE L. TURNER
Department of Botany
University of Texas
Austin, Texas 78713-7640
USA

HUGO WYLER
Institute de Chimie Organique
Université de Lausanne
Rue de la Barre 2,
CH-1005 Lausanne
Switzerland

VICTOR WRAY
Gesellschaft für
Biotechnologische Forschung
(GBF)
Mascheroder Weg 1
D-38124 Braunschweig, FRG

Dedication to Arthur Cronquist

Robert F. Thorne

On 22 March 1992, the distinguished angiosperm phylogenist, Arthur Cronquist, a good friend to many of us, died suddenly of a heart attack in the herbarium of Brigham Young University at Provo, Utah. Art literally "died with his boots on" in the glorious western tradition, as he would have preferred, working on specimens of his beloved western plants, apparently at that time on members of the Loasaceae, in one of his favorite herbaria. The shock of his death was compounded a few days later by the death of his only son, Dr. John Cronquist, Professor of Philosophy at California State University, Fullerton, also of a sudden heart attack. Arthur's widow, Mabel Allred Cronquist, whom he had married in 1940, has borne these tragedies surprisingly well, no doubt much aided by the support of their daughter, Mrs. Robert Crowe, four grandchildren, Art's half-sister, Marian Alexander, and many, many friends and colleagues. An upbeat memorial service was held at the New York Botanical Garden, Bronx, New York on 5 May 1992. It was very well attended by his former students, botanical colleagues, relatives, and other friends coming from many of the United States.

Arthur was born 19 March 1919, in San Jose, California, and was reared in Oregon and Idaho. In 1938 he graduated from Utah State University with a BS and in 1940 with an MS. His PhD was earned at the University of Minnesota in 1944. In 1946–1948 Art was an Assistant Professor at the University of Georgia, Athens, then at the same rank at Washington State University from 1948–1951 and later he was a Research Associate from 1953–1966. He served the Belgian Government as a Technical Advisor, 1951–1952, and UN-FAO in Iran as a Technical Consultant, 1969.

Arthur had worked at the New York Botanical Garden as a Technical Assistant, 1943–1944, and as an Assistant Curator, 1944–1946. He returned there after his stint with the Belgian Government, in 1952–1957 as Associate Curator, 1957–1965 as Curator, and 1965–1971 as Senior Curator. He also served at the New York Botanical Garden as Administrator of Graduate Studies, 1967–1969 and 1976–1984, as Director of Botany, 1971–1974, and as Senior Scientist, 1974–1992. He was editor of *Botanical Review* from 1969 until his death. He held Adjunct Professorships at Columbia University, 1964–1992, and at City University of New York, 1968–1992.

Arthur's popularity among his fellow botanists is attested to by his several presidencies: of the American Society of Plant Taxonomists in 1962, the Botanical Society of America in 1973, and of the Torrey Botanical Club in 1976. He also served on the Council of the International Association for Plant Taxonomy, 1976–1992. Many deserved honors came to Arthur: the Leidy Medal from the Philadelphia Academy of Natural Sciences, 1970; Merit Award, Botanical Society of America, 1974; Honorary Vice President of the XII Botanical Congress, Leningrad, 1975; Distinguished Service Award, The New York Botanical Garden, 1981; Asa Gray Award, American Society of Plant Taxonomists, 1985; Linnean Medal for Botany, Linnean Society of London, 1986; Gleason Award, The New York Botanical Garden, 1982; and Honorary Doctorate, Utah State University, 1987. It is a national disgrace that Arthur was never voted to membership in the National Academy of Sciences (USA). That lapse is indicative of the old-boy network of the Academy and its disdain for the natural sciences, especially taxonomic biology.

The books authored or coauthored by Art Cronquist are numerous and greatly valued. Among the most significant of these are *An Integrated System of Classification of Flowering Plant*, 1981; *The Evolution and Classification of Flowering Plants*, two editions, 1968 and 1988; *Manual of Vascular Plants of Northeastern United States and Adjacent Canada*, two editions, 1963 and 1991; *Vascular Plants of the Pacific Northwest*, five volumes, 1955–1969; *Flora of the Pacific Northwest*, 1973; *Intermountain Flora. Vascular Plants of the Intermountain West*, six volumes, 1972-present; *The Natural Geography of Plants*, 1964; *Introductory Botany*, two editions, 1961 and 1971; and *Basic Botany*, two editions, 1973 and 1982. In addition, he authored important contributions for our expanded knowledge of his favorite family, the Asteraceae.

Art was a prodigious worker and a keen scholar. He was highly intelligent but somewhat traditional-minded and rather stubborn about his beliefs. He would listen carefully to his fellow phylogenists and agree that the arguments were cogent but state that he did not intend to change his mind. I believe the only time I succeeded in convincing him of anything phylogenetic was that *Forsellesia* belonged to the Crossosomataceae and that the family belonged in the Rosales rather than in the Dilleniales. I considered that a major victory. We greatly enjoyed our verbal jousting, and always remained close friends despite our phylogenetic differences. Arthur was a splendid, patient teacher, generous with his time and his knowledge. He encouraged many botanists in addition to his own graduate students. He was always friendly and good humored. He had a booming voice, consistent with his large stature, and he loved to sing and to tell jokes. He was a self-reliant field botanist who made excellent specimens, and was a most pleasant field companion. He travelled widely about the world and was well known in botanical circles, especially in the former Soviet Union. I never was able to spend much time with Art but can recall being with him in such varied places as Georgia, the Olympic Peninsula, Ottawa, Hamburg, Edinburgh, Leningrad, Sydney, St. Louis, Berlin, Reading, and numerous herbaria and university campuses throughout the United States, usually at botanical gatherings. Arthur will be greatly missed by his colleagues and other friends and admirers throughout the botanical world.

1 Nomenclatural and Taxonomic History

ARTHUR CRONQUIST (deceased) and ROBERT F. THORNE

1.1 Introduction

The ordinal names Chenopodiales (Lindley 1833), Caryophyllales (Braun 1864) and Centrospermae (Eichler 1878) are equally acceptable under the Code, which does not extend the principle of priority to the ordinal level. Thorne originally preferred the older name, Chenopodiales, but has agreed to conform to Cronquist's preference for the more familiar ordinal name Caryophyllales and superordinal name Caryophyllanae. However, in the following discussion it is sometimes convenient to resurrect the well-known irregular name Centrospermae.

Eckardt (1976) presented such a clear history of the concept of the Centrospermae ("centrospermous families") from 1864 to 1975 that we hesitate to plow over the same ground. Still, some things must be repeated in order for us to provide a coherent story. We begin our history with Linnaeus.

1.2 Early History

Linnaeus had no concept at all of the Caryophyllales, although he did not do badly on some of the component families. In his fragment of a natural system (part of *Philosophia Botanica*, 1751), he put *Phytolacca* into his group I, Piperatae, along with genera such as *Piper*, *Arum*, and *Calla*. His group 42, Caryophyllei, included most of his genera of the present Caryophyllaceae, along with some extraneous genera such as *Frankenia*. His group 46, Succulentae, included *Cactus*, *Mesembryanthemum*, *Tetragonia*, *Portulaca*, and *Claytonia*, along with some very different things such as *Sedum*, *Saxifraga*, and *Geranium*. His group 56, Holeraceae, included mainly genera of the present Chenopodiaceae and Amaranthaceae, along with the extraneous genus *Callitriche*. His group 58 included *Mollugo*, *Montia*, *Mirabilis*, and *Basella*, along with a number of other genera that he considered to be *incertae sedis*.

Jussieu (1789) scarcely improved on Linnaeus. Genera that we would now regard as centrospermous appear in five orders, within three classes (classes 6, 7, and 14). In each group a number of genera that we would now associate in the same or related families appear together, along with other genera that we would now regard as extraneous. No concept of the Caryophyllales as a group emerges here.

Robert Brown (1810, 1819, 1827) had to deal with a number of genera of Caryophyllales in his *Prodromus Florae Novae-Hollandiae et Insulae Van-Diemen*, but he got no farther than his contemporaries in approaching a concept of the Caryophyllales as a group. Neither de Candolle (1824, 1849) nor Endlicher (1836–40) had a unified concept of the Caryophyllales, although each of them managed to get many of the genera into two well-separated groups that also contained other genera.

Lindley (1830, 1833) made but little progress toward the modern concept of the Caryophyllales. In 1830 he had what we would now call the families Phytolaccaceae, Aizoaceae, Amaranthaceae, Chenopodiaceae, Portulacaceae, Basellaceae, Molluginaceae, and Caryophyllaceae in the same general part of the system, thoroughly intermingled with our taxa Frankeniaceae, Tamaricaceae, Elatinaceae, Fouquieriaceae, *Galax* (Diapensiaceae), Crassulaceae, *Nitraria* (Zygophyllaceae), Polygonaceae, and Begoniaceae. The cacti were well separated from these groups. In 1833 he grouped the families (as we would now call them) into nixi (orders) and cohorts (subclasses). The families of the modern Caryophyllales occur in several different nixi in two cohorts, each of which included also some things we now consider extraneous. He did manage to get the Phytolaccaceae, Amaranthaceae, Chenopodiaceae, Nyctaginaceae, and a portion of the Caryophyllaceae (*Scleranthus*) along with the Polygonaceae into adjacent nixi in his cohort Curvembryonae in the subclass Incompletae. Basically this is no improvement over the systems of Endlicher or de Candolle.

Bartling (1830) made some progress toward the modern concept. He put into his class Caryophyllinae eight orders, which we would now consider to constitute five families, the Chenopodiaceae, Amaranthaceae, Phytolaccaceae, Portulacaceae, and Caryophyllaceae. Note that this class includes taxa both with and without evident petals. The Aizoaceae and Cactaceae were still far distant in his system, in different classes, where they were associated with miscellaneous other groups.

Bentham (1862) was well aware that the major families we now put into the Caryophyllales all belong together, as shown by his comment: "The series of orders in which natural affinities are the most dissevered by the Candollean arrangement is undoubtedly that of the Curvembryonous group . . . necessarily dispersed in the three great classes of Thalamiflorae, Calyciflorae, and Monochlamydeae." Despite his qualms, Bentham could not bring himself to disrupt the broad Candollean arrangement to the degree necessary to put these several families all together, and in the Bentham and Hooker *Genera Plantarum* (1862–1883) they remain as the principal constituents of three widely separated cohorts (orders) assigned to the three "great classes" mentioned above. Furthermore, the cohort Caryophyllinae included the Frankeniaceae and Tamaricaceae as well as the Caryophyllaceae and Portulacaceae.

1.3 Refinement of the Definition from Alexander Braun (1864) to the Present

The Caryophyllales in the modern sense made their first appearance in Alexander Braun's synopsis in the introduction to Ascherson's *Flora der Provinz Brandenburg*, in 1864. There we have an order Caryophyllinae Bartling emend A. Braun, with the families Nyctaginaceae, Chenopodiaceae, Amaranthaceae, Caryophyllaceae, Phytolaccaceae, Portulacaceae, Aizoaceae, and Opuntiaceae (which we would now call Cactaceae). Only the Didiereaceae (which were then unknown), the Basellaceae, and some segregate families, usually subsumed in the Phytolaccaceae and Aizoaceae, are missing. We may suppose that *Basella* and its immediate allies would have been included in the Portulacaceae or another of the recognized families had he taken these genera into account.

In 1876 Eichler had a Gruppe Centrospermae, with three orders, the Caryophyllinae, Opuntiinae, and Oleraceae. These three orders included taxa that we would now call Caryophyllaceae, Cactaceae, Aizoaceae, Portulacaceae, Phytolaccaceae, Nyctaginaceae, Chenopodiaceae, Amaranthaceae, and Polygonaceae. Thus, the Centrospermae of Eichler in 1876 are essentially identical to Braun's Caryophyllinae of 1864, with the addition of the Polygonaceae. In 1878 Eichler modified his arrangement by having a Reihe (order) Centrospermae, with nine families, not including the Cactaceae, which were associated uncertainly with the Passifloraceae. Our present families Molluginaceae and Basellaceae were included in the Aizoaceae and Portulacaceae, respectively. Only the Polygonaceae, Theligonaceae, and Gyrostemonaceae (as part of the Phytolaccaceae) are extraneous from our present point of view.

The Caryophyllales in nearly their modern form appeared under the name Centrospermae in the first (1892) edition of the Engler *Syllabus*. The families Chenopodiaceae, Amaranthaceae, Nyctaginaceae, Phytolaccaceae, Aizoaceae, Portulacaceae, Basellaceae, and Caryophyllaceae, all still retained in the order, were included. Also included were the Cynocrambaceae (= Theligonaceae) and Gyrostemonaceae (as a subfamily of Phytolaccaceae), which are now generally excluded from the group. The Molluginaceae, now generally taken as a family, were included in the Aizoaceae. The Cactaceae were treated as an order Opuntiales, following the Parietales. The Didiereaceae had not yet been described as a family and were not provided for. From this time on, we can see that the order, under whatever name, has a continuous identity, but it was subjected to vicissitudes in the various general systems of classification.

Small changes were made in subsequent editions of the *Syllabus*. The fifth (1907) edition extracted the Batidaceae, as an order Batidales between the Juglandales and Julianales in the subclass Archichlamydeae. In the combined ninth and tenth editions (1924) the Didiereaceae made their first appearance in the *Syllabus*, as a family of the order Sapindales. In the 12th edition (Melchior 1964) the family Didiereaceae was discussed in a separate paragraph immediately following the Centrospermae, and the Cactales (Opuntiales) followed directly thereafter. The Molluginaceae, previously included in the Aizoaceae, appeared for the first time (in the *Syllabus*) as a full-fledged family. The Gyrostemonaceae still remained in the Centrospermae in the 12th edition.

Wettstein (1907, 1911) closely approached the modern concept of the Caryophyllales. In the first edition of his *Handbuch* he included in the Centrospermae all the families now referred there, except for the then only recently described Didiereaceae, which were not provided for. The only other two families beyond those now accepted were the Batidaceae and Theligonaceae, the latter only doubtfully. In the second edition the Batidaceae also became doubtful. Thus, if these two doubtful families are removed, and the Didiereaceae added, Wettstein's concept of the Centrospermae (Caryophyllales) is the modern concept as well.

Hallier's treatment (1912) was no real improvement over the past, except for the inclusion of the Didiereaceae. He included in his order Caryophyllinae all the traditional centrospermous families, the not so traditional Cactaceae and Didiereaceae, and also the Crassulaceae, Plumbaginaceae, and Polygonaceae. He did exclude the Bataceae and Theligonaceae.

Neither Bessey nor Hutchinson contributed significantly to the concept of the Caryophyllales. Bessey (1915) got the Amaranthaceae, Chenopodiaceae, Phytolaccaceae, Nyctaginaceae, Basellaceae, Portulacaceae, and Aizoaceae (including Molluginaceae) together into the order which also included the extraneous families Elatinaceae, Frankeniaceae, Tamaricaceae, Salicaceae, Podostemonaceae, Hydrostachyaceae, Cynocrambaceae, and Batidaceae. The cacti were treated as an order Cactales after the Myrtales and Loasales.

Hutchinson (1926) got most of the centrospermous families into two orders: Caryophyllales, with the Caryophyllaceae, Molluginaceae (here recognized as a family for the first time), Aizoaceae, and Portulacaceae, plus the extraneous Elatinaceae; and Chenopodiales, with the Phytolaccaceae, Chenopodiaceae, Amaranthaceae, and Basellaceae, plus the extraneous Cynocrambaceae, Batidaceae, and Gyrostemonaceae. Between the Caryophyllales and Chenopodiales were the Polygonales, with the Polygonaceae and Illecebraceae (the latter now generally treated as a part of the Caryophyllaceae). These three orders were included in the major group Herbaceae. The cacti and Didiereaceae were well removed from these families in the major group Lignosae. The Cactales formed an order following the

Passiflorales and Cucurbitales, whereas the Didiereaceae were referred to the Sapindales. The second (1959) and third (1973) editions of his *Families of Flowering Plants* had no significant changes in these regards, except for the removal of some segregate families (Barbeuiaceae, Agdestidaceae, and the extraneous Gyrostemonaceae) from the Phytolaccaceae.

In 1954 Takhtajan included in the Caryophyllales all the families now generally referred there, notably including the Cactaceae and Didiereaceae, but he also included the small families Bataceae, Gyrostemonaceae, and Theligonaceae, which are now generally excluded. His 1959 treatment was virtually the same, except for the doubtful addition of the Simmondsiaceae. By 1966 he had the Caryophyllales in almost the present form, differing only in the inclusion of the extraneous small families Bataceae and Gyrostemonaceae. In 1980 (1980a,b) he excluded these two families and referred them to the order Sapindales, thus holding precisely to the present concepts of the limits of the Caryophyllales. In 1987 he moved Bataceae and Gyrostemonaceae to a bifamilial order Batales, but that did not affect the delimitation of the Caryophyllales.

Cronquist (1957) got nearly all the essential centrospermous families together in the order Caryophyllales (Phytolaccaceae, Nyctaginaceae, Aizoaceae, Portulacaceae, Basellaceae, Caryophyllaceae, Chenopodiaceae, and Amaranthaceae). The Polygonaceae and Theligonaceae were also included. *Mollugo* was not mentioned, but may be presumed to be subsumed under Aizoaceae. The Batidales and Cactales were treated as separate, unifamilial orders following the Caryophyllales. By 1968 Cronquist had arrived at essentially the present definition of the order, except that he included the Gyrostemonaceae within the Phytolaccaceae. He excluded both the Bataceae and Theligonaceae from the order, but he still retained the Batales within the subclass Caryophyllidae. His 1981 and 1988 books are precisely in accord with present and recent concepts of the Caryophyllales.

Emberger's treatment (1960) is fairly ordinary in the context of the time, and brings nothing unusual except the doubtful inclusion of the Sphenocleaceae.

Buxbaum (1961) presented a detailed study of the limits and organization of the order. He forcefully argued for the inclusion of the Cactaceae, as he had also done in 1948. He excluded the Polygonaceae and Theligonaceae, but included the Bataceae and said nothing about the Didiereaceae, Gyrostemonaceae, or Molluginaceae. It may be presumed that the Gyrostemonaceae were included in the Phytolaccaceae, as was common at the time, and that the Molluginaceae were included in the Aizoaceae. It is curious that he did not mention the Didiereaceae, inasmuch as there was already some ferment at that time about the possible inclusion of this family in the Centrospermae.

Soó (1967) included all the families we now put into the Caryophyllales, except for the Cactaceae, which make up the next order in his system. He also included the Gyrostemonaceae, Theligonaceae, and Bataceae, which are now generally excluded. His treatment was no improvement over others of the time, and from our present viewpoint was not so good as that of Takhtajan in 1954. Takhtajan then included the same families as Soó, plus the Cactaceae.

In 1968 Thorne included all the present families of the order, plus the Gyrostemonaceae and Polygonaceae. In 1976 he had the order (as Chenopodiales) in its present form except for the inclusion of the small family Rhabdodendraceae, now generally handled elsewhere. His 1992 paper is precisely in accord with present and recent concepts of the definition of the order.

By 1971 Ehrendorfer had a completely modern list of 11 families in the Caryophyllales, but the Bataceae, Gyrostemonaceae, and Theligonaceae were not provided for in the overall scheme. Stebbins, in 1974, likewise had a completely modern list but did not provide for the Gyrostemonaceae and Theligonaceae. He did exclude the Bataceae.

Dahlgren (1975) may have been the first major systematist to define the Caryophyllales precisely as we would today, with regard to both inclusion and exclusion. He recognized more families in 1975 than in 1980, but that is purely a matter of splitting or lumping, not affecting the limits of the order. In both treatments he excludes the Gyrostemonaceae, Bataceae, and Theligonaceae.

As recently as 1979, Benson continued to include the Polygonaceae and Gyrostemonaceae (as a family) in the Caryophyllales, while excluding the Cactaceae (which were put into the immediately following unifamilial order Cactales). Benson excluded *Batis*, and put *Theligonum* on his list of *incertae sedis*.

With the sole exception of Goldberg (1986) competent opinion on the limits of the order appears to be unanimous from 1980 to the present. In the face of the evidence, Goldberg continues to include the Gyrostemonaceae and Theligonaceae in the order.

1.4 Use of Characters Other Than Classical Morphology in Defining the Order

Several sets of characters in addition to those of classical morphology have contributed to the progressive refinement of taxonomic opinion as to the delimitation of the Caryophyllales. Notable among these, in historical sequence, are the centrospermous embryological syndrome, the betalains, and the sieve-element plastids.

Drawing on the work of Rocén (1927) and others, Schnarf (1933) may have been the first to point out the centrospermous embryological syndrome. Schnarf apparently missed the facts that the ovule is campylotropous rather than anatropous, and that the food reserve in the seed is perisperm rather than endosperm. By 1939 Mauritzon used the embryology as a strong argument for the unity of the order, inclusion of the Cactaceae, and exclusion of the Theligonaceae. Maheshwari (1950) enumerated ten features of the syndrome. None of these features is confined to the Caryophyllales, and some of them do not occur in all members of the order, but collectively they provide a strong indication of membership.

By 1945 the betalains, as we now call them, were attracting attention as possible markers of the Caryophyllales. Gibbs (1945) noted that "nitrogenous anthocyanins" were known in only eight families: the Aizoaceae, Amaranthaceae, Basellaceae, Cactaceae, Chenopodiaceae, Nyctaginaceae, Phytolaccaceae, and Portulacaceae. All of these, except the Cactaceae, were then commonly included in the Caryophyllales (or Centrospermae), and some authors had included the Cactaceae as well. Reznik (1955) further emphasized the importance of nitrogenous anthocyanins as a marker of a taxonomically unified group. He suggested the possibility that the Caryophyllales (including the Molluginaceae and the Caryophyllaceae) might have to be treated as an order distinct from the Chenopodiales (embracing the rest of the Centrospermae), but he reached no conclusion. In 1963 Mabry et al. proposed to define the Centrospermae by the presence of betacyanins. The Caryophyllaceae were regarded as close to the betacyanin group, but necessarily in another order. Mabry (1964) gave a good history of the recognition of nitrogenous anthocyanins and their nomenclatural transformation into betacyanins after their structure was elucidated. By 1966 Mabry was leaving the Caryophyllaceae and Molluginaceae in limbo, suggesting that: "Betacyanins and betaxanthins apparently developed solely in the Centrospermae at a very early time, perhaps even before anthocyanins appeared generally in the angiosperms." In 1968 the fundamental similarity of betacyanins and betaxanthins was recognized, and the combined group was named betalains (Mabry and Dreiding 1968). Mabry and Dreiding defined the Centrospermae on the presence of betalains, continuing to leave the Caryophyllaceae and Molluginaceae twisting in the wind.

Exclusion of the Caryophyllaceae and Molluginaceae from the Centrospermae was not well received by students of the general system of classification of flowering plants (Takhtajan 1966; Cronquist 1968; Dahlgren 1975; Eckardt 1976; Thorne 1976). By 1973 Mabry had retreated to the position that the Centrospermae could properly be defined broadly to include both betalain and anthocyanin families, but he still maintained that "the 11 Centrospermae families were derived from a common ancestral line from the angiosperm ancestor; this major evolutionary line subsequently gave rise to two lines prior to the origin of floral pigments." He held to the same position in 1976. Later, he more whole-heartedly accepted the present concept of the order, and together with some coauthors (Behnke et al. 1983a) he suggested that *Macarthuria*, in the Molluginaceae, occupies a key "central position" in the Centrospermae. His position with regard to the taxonomic and evolutionary significance of betalains now appears to be mainstream.

Even as the controversy about defining the Centrospermae on the basis of pigmentation was raging, another even more esoteric character attracted attention. Behnke (1969) showed that the Caryophyllales characteristically have a unique kind of sieve-element plastid, with a subperipheral ring of proteinaceous filaments. Behnke and Turner (1971) proposed to restrict the subclass Caryophyllidae to families with such plastids. Within the subclass they recognized two orders, the Caryophyllales and Chenopodiales. The Caryophyllales contained the families Caryophyllaceae and Molluginaceae, with anthocyanins but without betalains, and the Chenopodiales contained the several betalain families, lacking anthocyanin. Behnke (1972, 1976a) further expounded the plastid character and in 1976 proposed to use it to define the Caryophyllales, including both pigment types. In the same symposium, and relying partly on Behnke's data, Mabry (1976) accepted Behnke's definition of the order, with the Caryophyllaceae and Molluginaceae forming a distinctive suborder, Caryophyllineae, in contrast to the suborder Chenopodiineae for the betalain families.

In summing up the 1975 symposium, Ehrendorfer (1976) emphasized the evolutionary and taxonomic unity of the order as now defined. He suggested that the ancestors of the Caryophyllales had anthocyanin, which is still retained in the Caryophyllaceae and Molluginaceae, and that the betalain families evolved from centrospermous ancestors that had lost anthocyanin. In 1988 (p. 246) Cronquist elaborated on an unpublished suggestion by Giannasi that blockage of the terminal step in anthocyanin synthesis may have promoted a shunt from phenylalanine to betalain instead of anthocyanin.

Several other characters of more or less limited distribution among angiosperms as a whole show up often enough in the Caryophyllales to have contributed to the often intuitive thinking leading to the present concept of the order. As long ago as 1912 Wernham pointed out that "critical tendencies are no less important than critical characters" in the perception of taxonomic groups. Such critical tendencies in the Caryophyllales include pantoporate pollen (Skvarla and Nowicke 1976; Nowicke and Skvarla 1977), anomalous secondary growth, the succulent habit, and either CAM or C_4 photosynthesis. None of these features appears to be plesiomorphic for the order as a whole, but they appear often enough, as parallel apomorphies, to attract taxonomic attention. Furthermore, the frequent presence of triterpenoid saponins and the complete absence of ellagic acid contribute to a chemical syndrome found in few groups outside the Caryophyllales (Gibbs 1974; Frohne and Jensen 1979). The Polygonaceae, for example, are tanniferous but not saponiferous.

The definition of the Caryophyllales that crystallized in association with the papers presented at a symposium at the International Botanical Congress in 1975 in Leningrad (Mabry and Behnke 1976) soon became widely accepted. The features of pigmentation and sieve-element plastids were soon used to exclude the Plumbaginaceae, Polygonaceae, Theligonaceae, Batidaceae, Gyrostemonaceae, Vivianiaceae, and Rhabdodendraceae, which had sometimes been attributed to the Centrospermae.

In 1981 Hartley and Harris reported that the Caryophyllales characteristically have bound ferulic acid in unlignified cell walls. Ferulic acid is present in the cell walls of several families of monocots but not, so far as known, in dicots other than the Caryophyllales. This discovery might have had some impact on taxonomic thinking had it come a decade or two sooner, but now it seems more like frosting on the cake.

1.5 Inclusion or Exclusion of some Particular Families

Aside from some splitting of some of the families, only two families have gained widely recognized admission to membership in the Caryophyllales since Braun's formal publication of the order in 1864. These are the Cactaceae and Didiereaceae. The Cactaceae were in fact included in the Centrospermae by Eichler in 1876, and in the same order (under whatever name) by a number of later authors (Wettstein 1907; Hallier 1912; Mauritzon 1939; Buxbaum 1948, 1961; Takhtajan 1959) before the combination of betalains and order-specific sieve-element plastids removed all doubts some two decades ago.

Didierea was described by Baillon (1880) and assigned to the Sapindaceae. The family Didiereaceae, together with a new genus *Alluaudia*, was described in 1903 by Drake del Castillo. Hallier (1912) was doubtless the first to include the family within the Caryophyllales. The Didiereaceae did not make their way into the Engler *Syllabus* until the combined ninth and tenth edition (1924), where they were assigned to the order Sapindales. The 12th (1964) edition of the *Syllabus* has the Didiereaceae as an appendix to the Centrospermae. Takhtajan (1959) included the Didiereaceae in the Caryophyllales, making reference to studies of the pollen by Erdtman (1948, 1952). Rauh and Reznik (1961) put some emphasis on the pigmentation in assigning the Didiereaceae to the Centrospermae. Jensen (1965) adduced serological reactions to support the inclusion of the Didiereaceae in the Centrospermae.

In addition to the bona fide members of the Caryophyllales by present standards, a number of other families have been included in the group by one or another (or many) botanists during the century since the group took shape. Notable among these are the Polygonaceae, Theligonaceae, Bataceae, Gyrostemonaceae, Vivianiaceae, Plumbaginaceae, and Rhabdodendraceae. None of these families has betalains (Rhabdodendraceae not yet investigated), none has the centrospermous type of sieve-element plastid, and none has the full complement of the centrospermous embryological syndrome.

The Polygonaceae have long been regarded as more or less closely allied to the families now included in the Caryophyllales. Sometimes they were included in the same order (Eichler 1876, 1878; Hallier 1912; Bessey 1915; Cronquist 1957; Thorne 1968; and Benson 1979). Other botanists excluded the Polygonaceae but held them nearby (Engler 1892 et seq.; Wettstein

1907, 1911; Takhtajan 1959 and later works; Cronquist 1968, 1981, 1988; Thorne 1976 and later works; Stebbins 1974; Frohne and Jensen 1979). Benson (1979) was probably the last major systematist to include the Polygonaceae in the Caryophyllales. The Polygonaceae do not share the full centrospermous embryological syndrome. Notably they have endosperm instead of perisperm in the seeds, and they have mostly anatropous rather than campylotropous ovules. Furthermore, they have a different set of secondary metabolites, with tannins and anthraquinone glycosides rather than triterpenoid saponins as prominent components. Discoveries about betalains and sieve-element plastids during the 1960s and 1970s fostered the present consensus that the Polygonaceae do not belong to the Caryophyllales.

The Bataceae and Gyrostemonaceae may well be considered together here. Until two or three decades ago the Bataceae were often included in the Caryophyllales or Centrospermae (see Mabry and Turner 1964, for diverse treatments) and the Gyrostemonaceae were more regularly included, either as a family or more often as a subfamily of the Phytolaccaceae. Since that time the evidence has accumulated against their inclusion in the Caryophyllales (Goldblatt et al. 1976; Carlquist 1978). They lack both perisperm and betacyanins. Their sieve elements have S-type plastids (Behnke and Turner 1971; Behnke 1977). They have mustard oils, otherwise unknown in the Caryophyllales (Jørgensen 1981). They do not have the full centrospermous embryological syndrome, notably in lacking perisperm. All hands now agree on the exclusion of these two families from the Caryophyllales, excepting only Goldberg (1986).

The position of the Bataceae and Gyrostemonaceae, once they are excluded from the Caryophyllales, is still debatable. They differ from the vast majority of angiosperms in their solid exine, and as long ago as 1965 Kuprianova emphasized external pollen morphology in assigning them to a bifamilial order of their own. The presence of mustard oils has led several taxonomists to associate them in one way or another with the Capparales. Very recently, Tobe and Raven (1991) have referred the Gyrostemonaceae to the Capparales, but excluded the Bataceae.

The status of the Theligonaceae (Cynocrambaceae) as a member of the Caryophyllales was long in doubt. A thorough study by Wunderlich (1971) established its affinity with the Rubiaceae, although opinion is still divided as to whether it should be included in that family or placed alongside it. Wunderlich's conclusion was anticipated by S.S. Nenyukov in an unpublished manuscript dated 1939. Nenyukov lost his life in the Second World War, and his manuscript came to attention at the Komarov Institute only during the 1970s. *Theligonum* is tanniferous and has unitegmic, tenuinucellate ovules, unlike the Caryophyllales, and its seeds have endosperm rather than perisperm. The demonstration by Behnke (1975) and by Mabry et al. (1975) that *Theligonum* lacks betalains and has S-type sieve-element plastids would appear to bolt the door against any possible alliance with the Caryophyllales.

Friedrich (1956) made what at the time seemed like a strong case for inclusion of the Plumbaginaceae in the Centrospermae. Hallier (1912) had long before included the Plumbaginaceae in the order, but his views were generally ignored. More detailed study militates strongly against inclusion of the Plumbaginaceae in the Caryophyllales. They do not have the full centrospermous embryological syndrome: notably, the ovules are mostly anatropous, and the seeds usually have endosperm, never perisperm. The production of anthocyanins instead of betalains by the Plumbaginaceae does not exclude the family from the Caryophyllales, but it does nothing to support their inclusion. Nowicke and Skvarla (1977) consider that the pollen of Plumbaginaceae is distinctive, unlike the Caryophyllales. Furthermore, the corolla of the Plumbaginaceae is sympetalous. Finally, the Plumbaginaceae have S-type sieve-element plastids, instead of the unique plastids of the Caryophyllales (Behnke 1976b). We do not believe that any systematist in the past three decades has included the Plumbaginaceae in the Caryophyllales.

Bortenschlager (1967) proposed on the basis of pollen morphology to include the Vivianiaceae in the Centrospermae, between the Caryophyllaceae and Amaranthaceae. Takhtajan (1973) accepted Bortenschlager's proposal, but in subsequent works (e.g., 1980a,b, et seq.) he returned the Vivianiaceae to the Geraniales. The Vivianiaceae have anatropous ovules and endospermous seeds (Lefor 1975), which would be extraordinary for the Caryophyllales. Behnke and Mabry (1977) have shown that the Vivianiaceae have anthocyanins and S-type sieve-element plastids. We know of no present-day support for inclusion of the Vivianiaceae in the Caryophyllales.

Rhabdodendron has a checkered nomenclatural and taxonomic history, which was elucidated by Prance (1968). It has been variously associated with the Rutaceae, Chrysobalanaceae, and Phytolaccaceae. Prance was led to consider it in connection with his monograph of the Chrysobalanaceae (eventually published in 1972). Influenced especially by its anomalous secondary growth, Prance assigned *Rhabdodendron* to the Centrospermae as a separate family. Subsequent studies, especially by Puff and Weber (1976), have shown *Rhabdodendron* to be so anomalous in the Centrospermae as to be necessarily excluded. It has a unitegmic ovule. The seeds lack both endosperm and perisperm. The sieve-element plastids are P-type (Behnke 1976a), but very different in structure from those of the Centrospermae. If the pigments have been analyzed, the fact has not come to our attention. The proper position of *Rhabdodendron* is still debatable. Thorne (1992) treats it as a family in the Rutales, whereas Cronquist (1981) has it in the Rosales. In either case it is far removed from the Caryophyllales.

1.6 Recent Developments

Attention to the major taxonomy of the Caryophyllales during the last decade or so has focused on relationships within and among the families, associated with intraordinal or intrafamilial classification. A cladistic approach to the families of the order by Rodman et al. (1984) has been vigorously criticized by Hershkovitz (1989). Rodman has responded (1990, and Chap. 12) and both authors evidently intend to continue their efforts. Carolin (1987) attempted a cladistic revision of generic limits in the Portulacaceae. Bedell (1980) undertook a general consideration of the small family Stegnospermataceae (a segregate from the Phytolaccaceae), and Narayana and Narayana (1986) considered especially the embryology as bearing on the status and position of the family. Bittrich and Hartman (1988) undertook a de novo approach to the Aizoaceae. None of these studies challenges the delimitation of the order that has been current since 1975.

Preliminary studies of the molecular structure of a chloroplast gene, the larger subunit (rbcL) of ribulose-1,5-biphosphate carboxylase, support a close relationship among the very few tested members of the Caryophyllales (*Atriplex*, *Spinacia*, *Amaranthus*, and *Dianthus*), and the rather wide separation of these genera from *Rheum* (Polygonaceae), *Plumbago* (Plumbaginaceae), and tested members of a number of other families (Giannasi et al. 1992). Although the sequence data here are in harmony with established concepts about the Caryophyllales, it would be premature to put much weight on them. We must await the accumulation of a more secure database.

The consensus on the definition of the Caryophyllales since about 1975 may or may not last. Most of us feel very comfortable with it, but taxonomy is never final. Next week, next year, or next decade someone may come up with a newly recognized character that will require a reconsideration of our ideas.

1.6.1 Families Now Generally Included in the Caryophyllales

For the purposes of this list, the families are defined narrowly. Names in parentheses indicate the families to which some of the items on the list might be referred by authors who prefer broader definitions.

Achatocarpaceae (Phytolaccaceae)
Agdestidaceae (Phytolaccaceae)
Aizoaceae
Alsinaceae (Caryophyllaceae)
Amaranthaceae
Barbeuiaceae (Phytolaccaceae)
Basellaceae
Cactaceae
Caryophyllaceae
Chenopodiaceae
Didiereaceae
Dysphaniaceae (Chenopodiaceae)
Ficoidaceae (Aizoaceae)
Gisekiaceae (Phytolaccaceae)
Halophytaceae (Chenopodiaceae)

Hectorellaceae (Portulacaceae)
Illecebraceae (Caryophyllaceae)
Mesembryanthemaceae (Aizoaceae)
Molluginaceae
Nyctaginaceae
Petiveriaceae (Phytolaccaceae)
Phytolaccaceae
Pisoniaceae (Nyctaginaceae)
Portulacaceae
Salicorniaceae (Chenopodiaceae)
Sesuviaceae (Aizoaceae)
Stegnospermataceae (Phytolaccaceae)
Tetragoniaceae (Aizoaceae)

1.6.2 Families Now Generally Excluded from the Caryophyllales, Although Sometimes Included by Past Authors

Bataceae
Gyrostemonaceae
Plumbaginaceae
Polygonaceae
Rhabdodendraceae
Theligonaceae (= Cynocrambaceae)
Vivianiaceae

1.6.3 Summary of Our Present Knowledge of the Caryophyllales

To summarize our present knowledge of the Caryophyllales, we can confidently state that there is no other dicot order more clearly defined than the centrosperms. Almost all interested current taxonomists will agree with our present definition of the order due to the overwhelming list of features that characterize its members. These features include the large syndrome of centrospermous embryological characteristics, such as the campylotropous, bitegmic, crassinucellar ovule; food reserve as perisperm rather than endosperm; and the peripheral embryo curved or spiralled around the central perisperm. Also definitive are the betalain pigmentation replacing anthocyanins in all but two centrosperm families; the unique sieve-element plastids with a subperipheral ring of proteinaceous filaments (see Chap. 5); bound ferulic and other plant acids in unlignified cell walls, restricted to the centrosperms among dicots (Hartley and Harris 1981); and absence of the chloroplast *rpl2* intron in all investigated centrosperms (see Chap. 8).

Also to be mentioned here is the suite of critical tendencies prominent among Caryophyllales: tendencies to pantoporate pollen grains with spinulose tectum and punctae or annular perforations (see Chap. 7), anomalous secondary growth (except in the Portulacineae) (see Chap. 3), either CAM or C_4 photosynthesis, the frequent presence of triterpenoid saponins and absence of ellagic acid, and the centrifugal emergence of stamens in pluristaminate taxa (see Chaps. 6 and 13).

Although there is little controversy regarding the limits of the Caryophyllales, there is still much diversity in the treatment of the ordinal contents. Listed below are the preferred classifications of the order by Cronquist and by Thorne.

1.6.4 Preferred Classification of Caryophyllales by Cronquist

1. Phytolaccaceae (including Agdestidaceae, Barbeuiaceae, Gisekiaceae, Petiveriaceae, Stegnospermataceae)
2. Achatocarpaceae
3. Nyctaginaceae
4. Aizoaceae (Ficoidaceae, Mesembryanthemaceae, Sesuviaceae, Tetragoniaceae)
5. Didiereaceae
6. Cactaceae
7. Chenopodiaceae
8. Amaranthaceae
9. Portulacaceae (Hectorellaceae)
10. Basellaceae
11. Molluginaceae
12. Caryophyllaceae (Alsinaceae, Illecebraceae).

1.6.5 Preferred Classification of Caryophyllales by Thorne

Numbers in parentheses indicate the number of genera and number of species accepted by various experts, often rounded off to the nearest five or ten for species.

Caryophyllanae (Chenopodianae, Centrospermae) (564/8585)
- Caryophyllales (Chenopodiales, Centrospermae) (564/8585)
  - Caryophyllineae (83/1840)
    - Caryophyllaceae (70/1750)
      - Alsinoideae (Alsinaceae)
      - Paronychioideae (Illecebraceae)
      - Caryophylloideae
    - Molluginaceae (13/90)
  - Achatocarpineae (2/10)
    - Achatocarpaceae (2/10)
  - Portulacineae (125/1955)
    - Portulacaceae (19/500)
    - Hectorellaceae (2/2)
    - Basellaceae (4/40)

Didiereaceae (4/11)
Cactaceae (93/1400)
Pereskioideae (2/18)
Opuntioideae (4/250)
Cactoideae (87/1130)
Phytolaccineae (185/2420)
Stegnospermataceae (1/3)
Phytolaccaceae (13/85)
Phytolaccoideae (3/30)
Gisekioideae (Gisekiaceae) (1/5)
Rivinoideae (Petiveriaceae, Rivinaceae) (including *Lophiocarpus*) (7/45)
Agdestidoideae (Agdestidaceae) (1/1)
Barbeuioideae (Barbeuiaceae) (1/1)
Nyctaginaceae (30/290)
Aizoaceae (140/2040)
Aizooideae (6/80)
Aptenioideae (Mesembryanthemaceae, p.p.) (9/90)
Ruschioideae (including Caryotophoroideae, Hymenogynoideae, Mesembryanthemaceae, p.p.)
Sesuvioideae (Sesuviaceae) (4/20)
Tetragonioideae (Tetragoniaceae) (2/50)
Halophytaceae (1/1)
Chenopodiineae (169/2360)
Chenopodiaceae (105/1510)
Chenopodioideae (including *Dysphania*, *Microtea*)
Salicornioideae
Salsoloideae (Salsolaceae)
Sarcobatoideae (1/2)
Amaranthaceae (65/850)
Amaranthoideae
Gomphrenoideae.

By way of explanation of the Thorne system of the Caryophyllales, devised in part in consultation with Arthur Gibson, much emphasis has been placed on the presence or absence of anthocyanins vs betalains, normal vs anomalous growth patterns, strong tendency toward succulence, and types of sieve-element plastids, pollen grains, and epicuticular wax crystalloids. Thus, the two families of the Caryophyllineae have anthocyanins, whereas the remaining families have betalains replacing and apparently performing the functions of anthocyanins. The Caryophyllaceae and some genera of the Molluginaceae, in addition, have form-Pcf sieve-element plastids with a polygonal protein crystal (Behnke 1976b; Behnke and Barthlott 1983; Behnke et al. 1983b; see Chap. 5) and rather similar epicuticular waxes (Engel and Barthlott 1988; see also Chap. 4). Downie and Palmer (Chap. 8) found that these two anthocyanin-containing families occur in the same portion of the trees developed from their survey of restriction site variation in chloroplast DNAs, though they do not consider them basal to the order.

The species of Achatocarpaceae investigated do not seem to fit well into any of the suborders currently accepted. Though often associated with the Phytolaccaceae, members of the American Achatocarpaceae have only normal secondary growth (see Chap. 3), and the compound, unilocular ovary is quite distinct from the loculate gynoecium or separate carpels of the Phytolaccaceae. Pollen grains of the Achatocarpaceae are rather distinctive within the Caryophyllales because they have four to seven irregular and poorly defined pores and a scabrate tectum (Skvarla and Nowicke 1982; see also Chap. 7). Their wax crystalloids of irregular platelets resemble those of the Phytolaccaceae (see Chap. 4), but the form-Pcf sieve-element plastids have polygonal crystals like those in the Caryophyllaceae, some showing transitions to globular crystals (see Chap. 5). It would seem best, therefore, to treat the Achatocarpaceae in a separate suborder near the Caryophyllineae, and possibly transitional to the Phytolaccineae.

The Portulacineae also have normal growth patterns but a strong tendency to succulence. The unity of the group is attested to by results of gene sequence data (see Chap. 9) and restriction site studies (see Chap. 8). Although he includes the Aizoaceae, otherwise Rodman (Chap. 12), in his strict consensus tree, considered the Cactaceae and Portulacaceae (including Basellaceae, the Madagascan Didiereaceae, and the New Zealand–Kerguelen Hectorellaceae) as a distinct assemblage.

The Phytolaccineae have anomalous growth patterns with secondary thickening and only Aizoaceae and the Patagonian Halophytaceae show a strong tendency to succulence. Because of their similarities in pollen grains, epicuticular waxes, and sieve-element plastids, several of the families segregated from the Phytolaccaceae, the Mexican–Central American Agdestidaceae, Petiveriaceae, and the Madagascan Barbeuiaceae are returned to subfamily status in Phytolaccaceae as suggested by Nowicke (Chap. 7). Excepting Stegnospermataceae, the results of restriction site studies in chloroplast DNA (see Chap. 8) and gene sequence data, also from chloroplast DNA (see Chap. 9), treat the Nyctaginaceae, Aizoaceae, and Phytolaccaceae (with Petiveriaceae) as a major clade. The North and Central American *Stegnosperma*, though it does have anomalous growth, fits marginally at best into the Phytolaccineae. The form-Pcf sieve-element plastids have a central polygonal crystal like those of the Caryophyllaceae (see Chap. 5), yet the pollen grains (see Chap. 7) and epicuticular wax crystalloids (see Chap. 4) show resemblances to the Phytolaccaceae. It seems best to treat the Stegnospermataceae at the beginning of the Phytolaccineae as transitional to the Caryophyllineae.

The Chenopodiineae, with the closely related subcosmopolitan Chenopodiaceae and Amaranthaceae, show strong tendencies also to succulence but are disintinguished by their lack of 3-colpate pollen grains (type I of Nowicke

1975; also Skvarla and Nowicke 1976; see also Chap. 7) and presence of the form-Pf or Pfs sieve-element plastids (excluding *Sarcobatus*), characterized by a peripheral ring of protein filaments but no central protein crystal (Behnke 1976b). *Sarcobatus* is distinctive in the Chenopodiaceae in having form-Pcf plastids with a central globular crystal (see Chap. 5), and probably does deserve subfamilial treatment.

1.6.6 Relationships of the Caryophyllales

The relationships of the Caryophyllales, unlike the general acceptance of their contents, are a source of much speculation and argument. Cronquist (1981) places the Caryophyllales along with the Polygonales and Plumbaginales in the subclass Caryophyllidae, and suggests the Ranunculaceae as at least collateral ancestors. Thorne (1992), on the other hand, treats the Caryophyllales as a single order in his superorder Caryophyllanae, and considers the group more closely related to the Theanae than to the Magnolianae. The Plumbaginaceae are treated as a suborder Plumbagininеae of the Primulales, and the Polygonaceae as a separate but related order Polygonales, also in the Theanae. The affinities of the Nepenthaceae, Droseraceae, and Dilleniaceae to the Caryophyllales, reported lately by some molecular taxonomists (see Chap. 8), are really to be expected since they are all considered members of the Theales currently by Thorne.

Because of their present disjunct distribution about the world and their highly distinctive characteristics, the Caryophyllales must be an ancient group despite their relatively short paleobotanical record (back to the Late Cretaceous Maestrichtian) and rather specialized anatomy. They may have been an early spin-off from protodicotyledonous ancestors. They have certain characteristics in common with the Commelinanae and Arecanae in being the only tested dicots that characteristically have bound ferulic and other plant acids in unlignified cell walls (Harris and Hartley 1980; Hartley and Harris 1981). But whatever their relationships the Caryophyllales are one of the most well-defined groups among the angiosperms.

References

Baillon H (1880) Sur le *Didierea*. Bull Mens Soc Linn Paris 1:258–259

Bartling FG (1830) Ordines naturales plantarum eorumque characteres et affinitates adjecta generum enumeratione. Dieterich, Göttingen

Bedell HG (1980) A taxonomic and morphological re-evaluation of Stegnospermaceae (Caryophyllales). Syst Bot 5:419–431

Behnke H-D (1969) Über Siebröhren-Plastiden und Plastidenfilamente der Caryophyllales. Planta (Berl) 89:275–283

Behnke H-D (1972) Sieve-tube plastids in relation to angiosperm systematics – an attempt towards a classification by ultrastructural analysis. Bot Rev 38:155–197

Behnke H-D (1975) Elektronenmikroskopische Untersuchungen zur Frage der verwandtschaftlichen Beziehungen zwischen *Theligonum* und Rubiaceae: Feinbau der

Siebelement-Plastiden und Anmerkungen zur Struktur der Pollenexine. Plant Syst Evol 123:317–326
Behnke H-D (1976a) Sieve-element plastids of *Fouquieria*, *Frankenia* (Tamaricales), and *Rhabdodendron* (Rutaceae), taxa sometimes allied with the Centrospermae (Caryophyllales) Taxon 25:265–268
Behnke H-D (1976b) Ultrastructure of sieve-element plastids in Caryophyllales (Centrospermae), evidence for the delimitation and classification of the order. Plant Syst Evol 126:31–54
Behnke H-D (1977) Phloem ultrastructure and systematic position of Gyrostemonaceae. Bot Not 130:255–260
Behnke H-D, Barthlott W (1983) New evidence from the ultrastructural and micromorphological fields in angiosperm classification. Nord J Bot 3:43–66
Behnke H-D, Mabry TJ (1977) S-type sieve-element plastids and anthocyanins in Vivianiaceae: Evidence against its inclusion into Centrospermae. Plant Syst Evol 126:371–375
Behnke H-D, Turner BL (1971) On specific sieve-tube plastids in Caryophyllales. Further investigations with special reference to the Bataceae. Taxon 20:731–737
Behnke H-D, Mabry TJ, Neuman P, Barthlott W (1983a) Ultrastructural, micromorphological and phytochemical evidence for a "central position" of *Macarthuria* (Molluginaceae) within the Caryophyllales. Plant Syst Evol 143:151–161
Behnke H-D, Pop L, Sivarajan VV (1983b) Sieve-element plastids of Caryophyllales: Additional investigations with special reference to the Caryophyllaceae and Molluginaceae. Plant Syst Evol 142:109–115
Benson L (1979) Plant classification, 2nd edn. DC Heath, Lexington, MA
Bentham G (1862) Notes on the Caryophylleae, Portulaceae, and some allied orders. J Proc Linn Soc Bot 6:55–77
Bentham G, Hooker JD (1862–1883) Genera plantarum, 3 vols. Reeve, London
Bessey CE (1915) The phylogenetic taxonomy of flowering plants. Ann Mo Bot Gard 2:239–254
Bittrich V, Hartmann HEK (1988) The Aizoaceae – a new approach. Bot J Linn Soc 97:239–254
Bortenschlager S (1967) Vorläufige Mitteilungen zur Pollenmorphologie in der Familie der Geraniaceen und ihre systematische Bedeutung. Grana Palynol 7:400–468
Braun A (1864) Übersicht des natürlichen Systems nach der Anordnung desselben. In: Ascherson P (ed) Flora der Provinz Brandenburg, der Altmark und des Herzogthums Magdeburg, vol 1. Hirschwald, Berlin, pp 22–67
Brown R (1810, 1819, 1827) Prodromus florae Novae-Hollandiae et Insulae Van-Diemen. Johnson, London
Buxbaum F (1948) Zur Klärung der phylogenetischen Stellung der Aizoaceae und Cactaceae im Pflanzenreich. Jahrb Schweiz Kakt Ges 2:3–16
Buxbaum F (1961) Vorläufige Untersuchungen über Umfang, systematische Stellung und Gliederung der Caryophyllales (Centrospermae). Beitr Biol Pflanz 36:1–56
Candolle AP de (1824, 1849) Prodromus systematis naturalis regni vegetabilis, vol 1, 1824; vol 13(2), 1849. Masson, Paris
Carlquist S (1978) Wood anatomy and relationships of Bataceae, Gyrostemonaceae and Stylobasiaceae. Allertonia 1:297–330
Carolin RC (1987) A review of the family Portulacaceae. Aust J Bot 35:383–412
Cronquist A (1957) Outline of a new system of families and orders of dicotyledons. Bull Jard Bot Brux 27:13–40
Cronquist A (1968) The evolution and classification of flowering plants. Houghton Mifflin, Boston
Cronquist A (1981) An integrated system of classification of flowering plants. Columbia University Press, New York
Cronquist A (1988) The evolution and classification of flowering plants, 2nd edn. New York Bot Gard, New York
Dahlgren RMT (1975) A system of classification of the angiosperms to be used to demonstrate the distribution of characters. Bot Not 128:119–147

Dahlgren RMT (1980) A revised system of classification of the angiosperms. Bot J Linn Soc 80:91–124
Drake del Castillo ME (1903) Note sur les plantes recueillies par M. Guillaume Grandider, dans le Sud de Madagascar, en 1898 et 1901. Bull Mus Hist Nat (Paris) 9:35–46
Eckardt T (1976) Classical morphological features of centrospermous families. Plant Syst Evol 126:5–25
Ehrendorfer F (1971) Spermatophyta. In: Denffer D von, Schumacher W, Mägdefrau K, Ehrendorfer F (eds) Strasburger's Lehrbuch der Botanik, 30th edn. Fischer, Stuttgart, pp 586–741
Ehrendorfer F (1976) Closing remarks: Systematics and evolution of centrospermous families. Plant Syst Evol 126:99–106
Eichler AW (1876) Syllabus der Vorlesungen über Phanerogamenkunde. Schwer'sche Buchhandlung, Kiel
Eichler AW (1878) Blüthendiagramme. Zweiter Teil. Engelmann, Leipzig
Emberger L (1960) Les végétaux vasculaires. In: Chadefaud M, Emberger L (eds) Traité de botanique, vol 2. Systématique. Masson & Cie, Paris
Endlicher S (1836–40) Genera plantarum secundum ordines naturales disposita. Beck, Wien
Engel T, Barthlott W (1988) Micromorphology of epicuticular waxes in centrosperms. Plant Syst Evol 161:71–85
Engler A (1892) Syllabus der Vorlesungen über specielle und medicinisch-pharmaceutische Botanik. Borntraeger, Berlin
Engler A (1907) Syllabus der Pflanzenfamilien, 5th edn. Borntraeger, Berlin
Engler A (1924) Syllabus der Pflanzenfamilien. Combined 9th and 10th edn. Borntraeger, Berlin
Erdtman G (1948) Pollen morphology and plant taxonomy. VIII. Didiereaceae. Bull Mus Hist Nat (Paris) 20:387–394
Erdtman G (1952) Pollen morphology and plant taxonomy. Angiosperms. Almquist & Wiksell, Stockholm
Friedrich H-C (1956) Studien über die natürliche Verwandtschaft der Plumbaginales und Centrospermae. Phyton (Horn) 6:220–263
Frohne D, Jensen U (1979) Systematik des Pflanzenreichs. Fischer, Stuttgart
Giannasi DE, Zurawski G, Learn G, Clegg MT (1992) Evolutionary relationships of the Caryophyllidae based on comparative rbcL sequences. Syst Bot 17:1–15
Gibbs RD (1945) Comparative chemistry as an aid to the solution of problems in systematic botany. Trans Soc Can III 39(5):71–103
Gibbs RD (1974) Chemotaxonomy of flowering plants, vol 2. McGill-Queens University Press, Montreal
Goldberg A (1986) Classification, evolution, and phylogeny of the families of dicotyledons. Smithson Contrib Bot 58:1–314
Goldblatt P, Nowicke JW, Mabry TJ, Behnke H-D (1976) Gyrostemonaceae: Status and affinity. Bot Not 129:201–206
Hallier H (1912) L'origine et la système phylétique des Angiospermes exposés à l'aide de leur arbre généalogique. Arch Néerl Sci Exact Nat Sér 3B 1:146–234
Harris PJ, Hartley RD (1980) Phenolic constituents of the cell walls of monocotyledons. Biochem Syst Ecol 8:153–160
Hartley RD, Harris PJ (1981) Phenolic constituents of the cell walls of dicotyledons. Biochem Syst Ecol 9:189–203
Hershkovitz MA (1989) Phylogenetic studies in Centrospermae: a brief appraisal. Taxon 38:602–610
Hutchinson J (1926) The families of flowering plants, vol 1. Dicotyledons. Macmillan, London
Hutchinson J (1959) The families of flowering plants, 2nd edn. Clarendon Press, Oxford
Hutchinson J (1973) The families of flowering plants, 3rd edn. Clarendon Press, Oxford

Jensen U (1965) Serologische Untersuchungen zur Frage der systematischen Einordnung der Didiereaceae. Bot Jahrb Syst 84:233–253
Jørgensen LB (1981) Myrosin cells and dilated cisternae of the endoplasmic reticulum in the order Capparales. Nord J Bot 1:433–445
Jussieu AL de (1789) Genera plantarum. Herrisat and Barrois, Paris
Kuprianova LA (1965) Palinologia serezhkotsvetnykh. Nauka, Moscow
Lefor MW (1975) A taxonomic revision of the Vivianiaceae. Univ Conn Occas Pap Biol Sci Ser 2(15):225–255
Lindley J (1830) Introduction to the natural system of botany. Longman et al., London
Lindley J (1833) Nixus plantarum. Ridgway and Sons, London
Linnaeus C (1751) Philosophia botanica. Kiesewetter, Stockholm & Chatelain, Amsterdam
Mabry TJ (1964) The betacyanins, a new class of red-violet pigments, and their phylogenetic significance. In: Leone CA (ed) Taxonomic biochemistry and serology. Ronald Press, New York, pp 239–254
Mabry TJ (1966) The betacyanins and betaxanthins. In: Swain T (ed) Comparative phytochemistry. Academic Press, London, pp 131–144
Mabry TJ (1973) Is the order Centrospermae monophyletic? In: Bendz G, Santesson J (eds) Chemistry in botanical classification. Nobel Foundation, Stockholm and Academic Press, London, pp 275–285
Mabry TJ (1976) Pigment dichotomy and DNA-RNA hybridization data for centrospermous families. Plant Syst Evol 126:79–94
Mabry TJ, Behnke H-D (eds) (1976) Evolution of centrospermous families. Plant Syst Evol 126:1–106
Mabry TJ, Dreiding AS (1968) The betalains. Recent Adv Phytochem 1:145–160
Mabry TJ, Turner BL (1964) Chemical investigations of the Batidaceae. Betaxanthins and their systematic implications. Taxon 13:197–200
Mabry TJ, Taylor A, Turner BL (1963) The betacyanins and their distribution. Phytochemistry 2:61–64
Mabry TJ, Eifert IJ, Chang C, Mabry H, Kidd C, Behnke H-D (1975) Theligonaceae: pigment and ultrastructural evidence which excludes it from the order Centrospermae. Biochem Syst Ecol 3:53–55
Maheshwari P (1950) An introduction to the embryology of angiosperms. McGraw-Hill, New York
Mauritzon J (1939) Die Bedeutung der embryolgischen Forschung für das natürliche System der Pflanzen. Lunds Univ Årsskr NF II 35(15):1–70
Melchior H (ed) (1964) A. Engler's Syllabus der Pflanzenfamilien, 12th edn, vol 2. Centrospermae by Th. Eckardt, with Didiereaceae as an "Anhang" by H Melchior. Borntraeger, Berlin, pp 79–102
Narayana PS, Narayana LL (1986) The embryology of Stegnospermataceae, with a discussion on its status, affinities and systematic position. Plant Syst Evol 154:137–146
Nowicke JW (1975) Pollen morphology in the order Centrospermae. Grana 15:51–77
Nowicke JW, Skvarla JJ (1977) Pollen morphology and the relationship of the Plumbaginaceae, Polygonaceae, and Primulaceae to the order Centrospermae. Smithson Contrib Bot 37:1–64
Prance GT (1968) The systematic position of *Rhabdodendron* Gilg & Pilg. Bull Jard Bot Nat Belg 38:127–146
Prance GT (1972) A monograph of the Rhabdodendraceae. Fl Neotropica Monog 11. Hafner, New York
Puff C, Weber A (1976) Contributions to the morphology, anatomy, and karyology of *Rhabdodendron*, and a reconsideration of the systematic position of the Rhabdodendraceae. Plant Syst Evol 125:195–222
Rauh W, Reznik H (1961) Zur Frage der systematischen Stellung der Didiereaceen. Bot Jahrb 81:94–105
Reznik H (1955) Die Pigmente der Centrospermen als systematisches Element. Z Bot 43:499–530
Rocén T (1927) Zur Embryologie der Centrospermen. Diss Uppsala

Rodman JE (1990) Centospermae revisited, part I. Taxon 39:383–393
Rodman JE, Oliver MK, Nakamura RR, McClammer JU, Bledsoe, AH (1984) A taxonomic analysis and revised classification of Centrospermae. Syst Bot 9:297–323
Schnarf K (1933) Die Bedeutung der embryologischen Forschung für das natürliche System der Pflanzen. Biol Gen 9:271–288
Skvarla JJ, Nowicke JW (1976) Ultrastructure of pollen exine in centrospermous families. Plant Syst Evol 126:55–78
Skvarla JJ, Nowicke JW (1982) Pollen fine structure and relationships of *Achatocarpus* Triana and *Phaulothamnus* A. Gray. Taxon 31:244–249
Soó R (1967) Die modernen Systeme der Angiospermen. Acta Bot Acad Sci Hung 13:201–233
Stebbins GL (1974) Flowering plants. Evolution above the species level. Belknap Press of Harvard University Press, Cambridge, Mass
Takhtajan AL (1954) Proiskhozhdenie pokrytosemennykh rasteniy. Nauka, Moscow. Translated by Gankin OL and published in 1958 by AIBS, Washington, as Origins of angiospermous plants
Takhtajan AL (1959) Die Evolution der Angiospermen. Fischer, Jena
Takhtajan AL (1966) Sistema i filogenia tsvetkovykh rasteniy. Nauka, Moscow (in Russian)
Takhtajan AL (1973) Evolution und Ausbreitung der Blütenpflanzen. Fischer, Stuttgart
Takhtajan AL (1980a) Outline of the classification of flowering plants (Magnoliophyta). Bot Rev 46:225–359
Takhtajan AL (1980b) Systema Magnoliophytorum. Nauka, Leningrad (in Russian)
Takhtajan AL (1987) Sistema Magnoliophytov. (Systema Magnoliophytorum). Nauka, Leningrad (in Russian)
Thorne RF (1968) Synopsis of a putatively phylogenetic classification of the flowering plants. Aliso 6(4):57–66
Thorne RF (1976) A phylogenetic classification of the angiosperms. Evol Biol 9:35–106
Thorne RF (1992) Classification and geography of the flowering plants. Bot Rev 58:225–348
Tobe H, Raven PH (1991) The embryology and relationships of Gyrostemonaceae. Aust Syst Bot 4:407–420
Wernham HF (1912) Floral evolution: With particular reference to the sympetalous dicotyledons. New Phytol 11:373–397
Wettstein R (1907) Handbuch der systematischen Botanik, 2 vols, 1901–1908. The part containing the Centrospermae, vol 2. Deuticke, Leipzig, pp 250–262
Wettstein R (1911) Handbuch der systematischen Botanik, 2 vols, 1910–1911. The part containing the Centrospermae, vol 2. Deuticke, Leipzig, pp 532–546
Wunderlich R (1971) Die systematische Stellung von *Theligonum*. Oesterr Bot Z 119:329–394

2 Chromosome Numbers and Their Phyletic Interpretation

BILLIE L. TURNER

2.1 Introduction

Ehrendorfer (1976a) briefly reviewed and critically evaluated chromosome numbers and their patterns of variation as indicators of phylogeny in the Caryophyllales. He concluded that the probable ancestral base number for the group was $\underline{x} = 9$, with radiation and advancement in several lines through subsequent dysploidy and polyploidy. At the time of his study relatively few chromosome counts were available for most of the genera. I have reviewed in a cursory manner all chromosome counts of the order Caryophyllales up to 1989 (the last published account of the *Index to Plant Chromosome Numbers*; Goldblatt and Johnson 1991). In a few instances I have included data for genera published after 1989. It should be emphasized that no attempt has been made to verify the *exact* number of species counted for the various genera, this being a most difficult task for the larger genera. Rather, the figures posited are believed to be very reliable, good estimates based on the data listed in the references given. All this information is summarized in Table 2.1. My interpretation of the chromosomal data is discussed in the account that follows and is summarized pictorially in Fig. 2.1.

2.2 Chromosome Numbers of Caryophyllales

Aizoaceae/Molluginaceae ($\underline{x} = 8$). Pax and Hoffmann (1934a) treated this family as comprised of some 1100 species in 23 genera distributed among six tribes. At the time of their study, most of the species were positioned in the large genus *Mesembryanthemum*. Recent workers have tended to segregate out of the latter numerous small genera, and this has been followed in my construction of Table 2.1.

Chromosome counts (up to 1989) are available for about 350 species and 80 genera of the Aizoaceae (s.l.), representing all the tribes recognized by Pax and Hoffmann (1934a) except for the Orygieae (two genera) and Limeeae (five genera). The degree of polyploidy among the taxa investigated

Table 2.1. Chromosome numbers in Caryophyllales (Taxa arranged according to families recognized by Thorne, see Chap. 1)

Genus	No. taxa	No. counted	No. polyploids	Base chromosome numbers (x)[a] 7	8	9	10	11	12	13	14	15	17 (8+9)
Achatocarpaceae													
Achatocarpus	No counts available												
Aizoaceae													
Acrodon	2	1	–			9							
Aizoanthemum	1	1	–		8								
Aizoon	15	3	1		8								
Antegibbaeum	1	1	–			9							
Apatesia	3	1	–			9							
Aptenia	2	1	–			9							
Argyroderma	10	10	–			9							
Aridaria	40	10	7			9							
Astridia	1	1	–			9							
Bergeranthus	5	4	2			9							
Bijlia	1	1	–			9							
Braunsia		1	–			9							
Brownanthus		8	4			9							
Carpanthea	1	1	–			9							
Cephalophyllum	30	25	10			9							
Cerochlamys	1	1	–			9							
Cheiridopsis	25	12	8			9							
Conophytum	80	6	2			9							
Corpuscularia	2	1	–			9							
Cylindrophyllum	3	1	–			9							
Dactylopsis	2	2	–			9							
Delosperma	30	11	4			9							
Dicrocaulon	10	7	3			9							
Didymaotis	1	1	–			9							
Dinteranthus	4	1	–			9							
Diplosoma	1	1	–			9							
Disphyma	5	4	4			9							
Dorotheanthus	2	2	–			9							
Drosanthemum	10	3	3			9							

Table 2.1. *Continued*

Genus	No. taxa	No. counted	No. polyploids	Base chromosome numbers (x)[a] 7	8	9	10	11	12	13	14	15	17 (8+9)
Eberlanzia	1	1	1			9							
Echinus	2	1	–			9							
Erepsia	1	1	–			9							
Faucaria	4	8	1			9							
Fenestraria	1	1	–			9							
Gibbaeum	30	25	8			9							
Giselina		1	1			9							
Glottiphyllum	36	4	–			9							
Hereroa	10	6	1			9							
Hymenocyclus	6	5	3			9							
Hymenogyne	2	1	–			9							
Imitaria	1	1	–			9							
Jacobsenia	4	2	–			9							
Jordaaniella	10	4	1			9							
Lampranthus		3	–			9							
Lapidaria	1	1	–			9							
Leipoldtia	10	10	2			9							
Lithops	20	10	–			9							
Machairophyllum	2	1	–			9							
Malephora	1	1	1			9							
Mesembryanthemum (s.l.)	800	100	20			9							
Maughaniella	1	1	–			9							
Meyerophytum	1	1	–			9							
Mitrophyllum	20	10	–			9							
Monilaria		6	1			9							
Muiria	1	1	–			9							
Nananthus	5	4	–			9							
Nelia	1	1	–			9							
Odontophorus	10	4	–			9							
Oophytum	4	3	–			9							
Ophthalmophyllum	10	4	–			9							
Oscularia	1					9							

Table 2.1. *Continued*

Genus	No. taxa	No. counted	No. polyploids	Base chromosome numbers (x)[a] 7	8	9	10	11	12	13	14	15	17 (8+9)
Pleiospilos	4	4	–			9							
Prenia	10	6	1			9							
Psilocaulon	40	2	1			9							
Rhinephyllum	4	2	–			9							
Ruschia	5	3	–			9							
Schwantesia	2	1	–			9							
Sesuvium	3	1	1		8								
Stomatium	10	6	1			9							
Tetragonia	50	5	5		8								
Titanopsis	5	1	–			9							
Trianthema	15	6	2		8								
Trichodiadema	4	3	2			9							
Vanheerdia	1	1	1			9							
Vanzijlia	3	2	1			9							
Amaranthaceae													
Achyranthes	10	6	5	7	8	9							17
Acnida	8	3	3		8								
Aerva	10	5	3	7	8	9							
Allmania	2	2	1		8								
Alternanthera	170	10	9	7	8	9							17
Amaranthus	100	80	20		8								17
Banalia	1	1	–		8								
Bosea	5	1	–			9							
Celosia	60	7	6		8	9							
Charpentiera	1	1	1							13			17
Cyathula	27	6	5	7	8								17
Deeringia	8	3	–		8	9	10						
Digera	3	3	1			9							
Froelichia	20	1	1		8	9							17
Gomphrena	100	10	5		8	9	10	11		13			
Iresine	70	4	4		8	9	10					15	17
Nothosaerva	1	1	–			9							
Pleuropetalum	4	1	–		8	9							17
Psilotrichum	15	1	–		8								

Table 2.1. *Continued*

Genus	No. taxa	No. counted	No. polyploids	Base chromosome numbers (x)[a] 7	8	9	10	11	12	13	14	15	17 (8+9)
Ptilotus	100	15	14			9							
Pupalia	1	1	1				10						
Sericostachys	2	1	1					11					
Telanthera	3	1	1		8	9							17
Tidestromia	5	1	–				10						
Basellaceae													
Anredera	1	1	1		8								
Basella	3	2	2		8		11						
Boussingaultia	14	3	3		8								
Ullucus	1	1	1		8								
Cactaceae													
[Ca. 84 genera and ca. 800 species have been counted all with base numbers of x = 11]													
Caryophyllaceae													
Acanthophyllum	50	15	7									15	
Agrostemma	2	2	2						12				
Alsinanthe	2	2	1									15	
Alsinidendron	1	1	–				10					15	
Ammodenia	1	1	–										17
Ankyropetalum	3	1	–							13			
Arenaria	200	165	135	7	8	9	10	11	12	13			
Buffonia	20	6	4		8	9							17
Cerastium	150	120	115			9						15	17
Chaetonychia	1	1	–	7									
Conosilene	1	1	–				10						
Corrigiola	10	2	2		8	9							
Cucubalus	1	1	–						12				
Cyathophylla	1	1	1									15	
Dianthus	300	220	70									15	
Dicheranthus	1	1	1		8								
Drymaria	70	5	5					11	12				
Drypis	2	2	2									15	
Eremogone								11					

Table 2.1. *Continued*

Genus	No. taxa	No. counted	No. polyploids	Base chromosome numbers (x)[a] 7	8	9	10	11	12	13	14	15	17 (8+9)
Eudianthe	5	3	–						<u>12</u>				
Gastrolychnis	10	8	6						<u>12</u>				
Gymnocarpos	2	1	–									<u>15</u>	
Gypsophila	125	50	8						12	13		15	<u>17</u>
Heliosperma	7	4	–						<u>12</u>				
Herniaria	20	16	12			<u>9</u>							
Holosteum	6	3	–				<u>10</u>						
Honkenya	2	2	2										<u>17</u>
Illecebrum	1	1	–				<u>10</u>						
Kohlrauschia	5	4	2				<u>10</u>						
Krauseola	1	1	–		<u>8</u>								
Lepyrodiclis	3	1	–										<u>17</u>
Loeflingia	7	1	–						<u>12</u>				
Lychnis	40	35	3						<u>12</u>				
Malachium	1	1	–								<u>14</u>		
Melandrium	50	40	11						<u>12</u>				
Merckia	1	1	1					<u>11</u>	<u>12</u>				
Minuartia	140	90	40			9	<u>10</u>	11	<u>12</u>	<u>13</u>	14	15	
Moehringia	25	18	4						<u>12</u>				
Moenchia	6	3	3						<u>12</u>				
Myosoton	1	1	–								<u>14</u>		
Oberna	1	1	–						<u>12</u>				
Ortegia	1	1	–			<u>9</u>							
Otites	1	1	–						<u>12</u>				
Paronychia	50	24	14		8	<u>9</u>			<u>12</u>				
Petrocoma	2	1	–						<u>12</u>				
Petrocoptis	7	5	–						<u>12</u>				
Petrorhagia	20	10	5							13		<u>15</u>	
Pollichia	1	1	–			<u>9</u>							
Polycarpaea	50	6	2			<u>9</u>							
Polycarpon	16	5	2		<u>8</u>	9							
Pseudostellaria	16	4	1						<u>12</u>		14		
Pteranthus	2	1	–		<u>8</u>								
Pycnophyllum	17	1	–								14	15	
Queria	2	1	–									15	

Table 2.1. *Continued*

Genus	No. taxa	No. counted	No. polyploids	Base chromosome numbers (x)[a] 7	8	9	10	11	12	13	14	15	17 (8+9)
Sagina	30	25	6			9	10	11	12				
Saponaria	30	10	–								14	15	
Schiedea	18	1	1									15	
Scleranthus	10	6	5					11	12				
Silene	400	240	50			9	10		12				
Spergula	15	13	6			9							
Spergularia	50	42	35			9							
Stellaria	150	80	70				10	11	12	13	14		
Telephium	5	3	–			9							
Tryphane	1	1	–						12				
Tunica	5	3	2									15	
Uebelinia	2	2	1						12				
Vaccaria	5	4	4				10					15	
Velezia	6	1	–								14		
Viscaria	10	6	–						12				
Wilhelmsia	1	1	1				10	11					
Willwebera	1	1	–								14		
Chenopodiaceae													
Acroglochin	2	2	2			9							
Aellenia	6	1	1			9							
Allenrolfea	3	1				9							
Anabasis	30	2	–			9							
Archiatriplex	1	1	–			9							
Arthrocnemum	4	4	3			9							
Atriplex	140	100	35			9	10						
Axyris	2	2	–			9							
Bassia	26	7	–			9							
Beta	10	9	6			9							
Bienertia	1	1	–			9							
Blitum	1	1	–			9							
Camphorosma	2	2	–	(x = 6)									
Ceratocarpus	1	1	–			9							
Ceratodes	3	3	–			9							

Table 2.1. *Continued*

Genus	No. taxa	No. counted	No. polyploids	Base chromosome numbers (x)[a] 7	8	9	10	11	12	13	14	15	17 (8+9)
Chenopodium	200	140	80		8	9							
Climacoptera	23	4	3			9							
Corispermum	60	6	–			9							
Cornulaca	7	1	–			9							
Cremnophyton		1					10						
Cycloloma	1	1	1			9							
Eurotia	8	2	1			9							
Gamanthus	7	1	–			9							
Girgensohnia	5	1	–			9							
Grayia	2	2	2			9							
Hablitzia	1	1	–			9							
Halimione	3	1	1			9							
Halocharis	12	1	–			9							
Halocnemum	1	1	1			9							
Halogeton	3	2	1			9							
Halothamnus	1	1	1			9							
Haloxylon	10	5	–										
Hammada	12	2	1			9							
Iljinia	1	1	1			9							
Kirilowia	3	1	–			9							
Kochia	10	6	1			9							
Londesia		1				9							
Maireana	58	2	–			9							
Microcnemum	1	1	–			9							
Monolepis	4	4	–			9							
Nitrophila	8	2	–			9							
Obione	100	4	1			9							
Ofaiston	2	1	1			9							
Panderia	1	1	–			9							
Petrosimonia	12	2	–			9							
Polycnemum	7	1	–			8							
Rhagodia	11	2	1			9							
Salicornia	20	15	10			9							

Table 2.1. *Continued*

Genus	No. taxa	No. counted	No. polyploids	Base chromosome numbers (x)[a] 7	8	9	10	11	12	13	14	15	17 (8+9)
Salsola	150	40	20			9							
Sarcobatus	1	1	–			9							
Schanginia	1	1	–			9							
Sclerolaena	77	1	–			9							
Seidlitzia	8	1	–			9							
Spinacia	3	3	–	(x = 6)									
Suaeda	120	25	15			9							
Suckleya	1	1	1			9							
Didiereaceae													
Alluaudia	6	6	6		8								
Alluaudiopsis	2	2	2		8								
Decaryia	1	1	1		8								
Didierea	2	2	2		8								
Halophytaceae													
Halophytum	No counts available												
Hectorellaceae													
Hectorella	2	1	1		8								
Molluginaceae													
Glinus	10	3	–			9							
Hypertelis	1	1	–		8								
Mollugo	15	8	5			9							
Nyctaginaceae													
Abronia	20	3	3					11		13			
Allionia	1	1	1				10						
Boerhavia	40	4	2						12	13			
Bougainvillea	16	10	6		8	9	10						17
Commicarpus	16	6	6				10						[x = 20,21]
Mirabilis	10	7	7				10						
Oxybaphus	80	3	3				10						
Pisonia	20	2	2		8	9							17
Quamoclidium	4	2	2										

Table 2.1. *Continued*

Genus	No. taxa	No. counted	No. polyploids	Base chromosome numbers (x)[a] 7	8	9	10	11	12	13	14	15	17 (8+9)
Phytolaccaceae													
Gallesia	2	1	1			9							17
Gisekia	1	1	1			9							
Hilleria	4	1	1			9							
Petiveria	1	1	1			9							
Phytolacca		8	7			9							
Rivina	1	2	2			9							
Trichostigma	3	1	1			9							
Portulacaceae													
Anacampseros	50	5	2			9							
Calandrinia (s.st.)[b]	150	12	6		8				12				
Calyptridium	6	6	–					11					
Ceraria	4	3	3		8								
Cistanthe[b]							10	11					
Claytonia	40	36	22		8		10						
Grahamia	1	1	–			9							
Lewisia	20	2	2			9							
Montia	50	10	6		8	9	10		12				
Montiastrum	3	2	1										
Montiopsis[b]							10						
Naiocrene	3	2	1					11					
Neopaxia	1	1	1		8								
Oreobroma	2	1	1		8	9							17
Portulaca	100	1	9		8	9	10						
Portulacaria	2	1	1					11					
Talinaria[b]						9							
Talinum	50	14			8				12				
Xenia						9							
Stegnospremataceae													
Stegnosperma	No counts available												

[a] The numbers underlined are the most common for the genera concerned
[b] Information provided by Hershkovitz (pers. comm.)

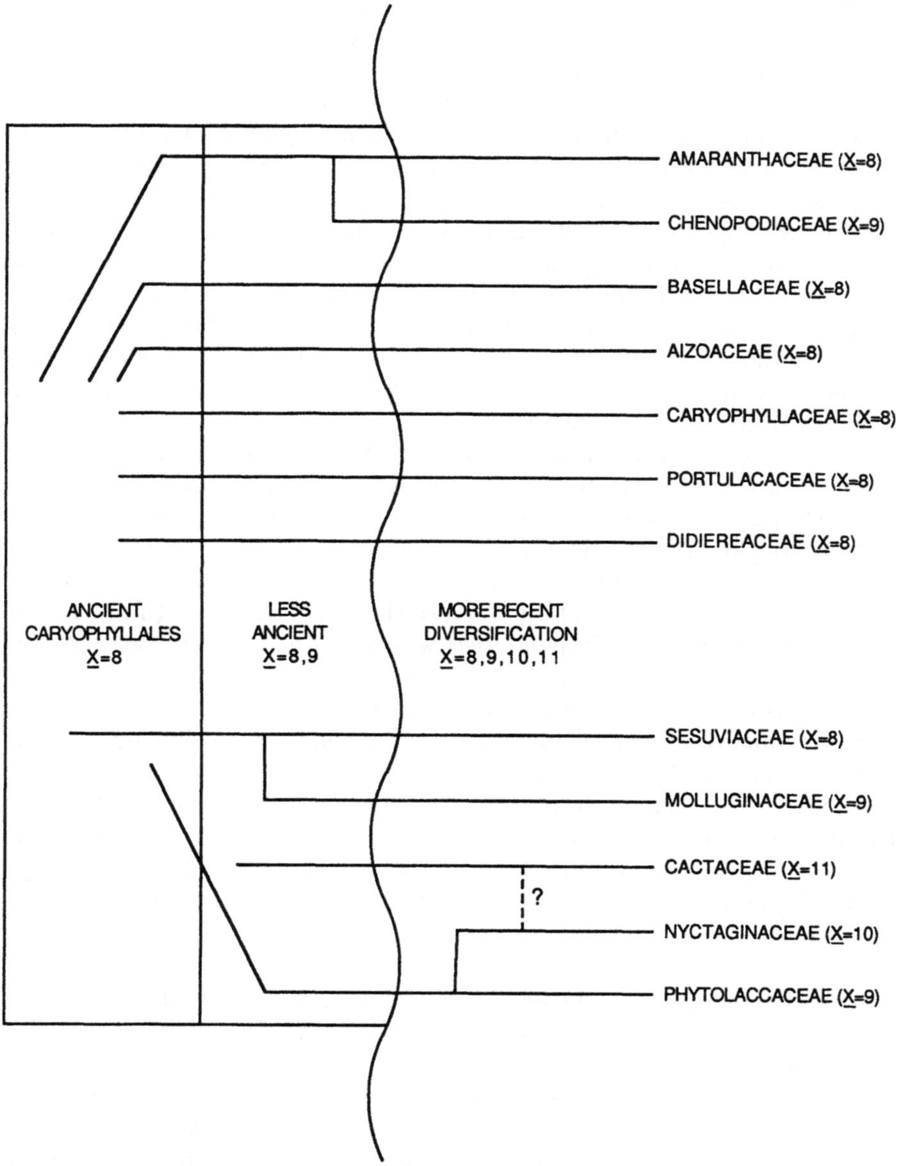

Fig. 2.1 Hypothetical phylogeny of Caryophyllales, based upon a base chromosome number of x = 8

is relatively low (ca. 30%). As can be ascertained from Table 2.1, Aizoaceae has base numbers of x = 8 and 9, the latter being much more common, occurring in all the tribes recognized by Pax and Hoffmann (1934a) counted to date, except for the Tetragonieae which has a base of x = 8. The base number x = 8 also characterizes the subtribe Sesuviinae of the tribe

Mollugineae (sensu Pax and Hoffmann), but the subtribe Molluginінae has a base number of x = 9. This perhaps lends credence to the treatment of the cohorts of *Sesuvium* and *Tetragonia* as a distinct phyletic line, either within a separate family Sesuviaceae, as suggested by Ehrendorfer (1976b), or positioned within an expanded Portulacaceae in which the base number x = 8 predominates. Bittrich and Hartmann (1988), however, in their circumscription of the Aizoaceae, which includes five subfamilies, excluded the subtribe Molluginінae (x = 9) but retained within Aizoaceae both *Sesuvium* (positioned along with *Cypselea*, *Trianthema*, and *Zaleya* in the subfamily Sesuvioideae) and *Tetragonia* (positioned along with *Tribulocarpus* in the subfamily Tetragonioideae). Additionally, Bittrich and Hartmann (1988) note that taxa with x = 8 characterize the subfamily Aizooideae (e.g., *Aizoon* and *Aizoanthemum*); thus three of their five subfamilies have base numbers of x = 8, while two (Aptenioideae and Ruschioideae) have base numbers of x = 9. Bittrich (1990) has provided an excellent account of the problems associated with the circumscription of the Aizoaceae.

Following the views of Ehrendorfer (1976a), Bittrich and Hartmann (1988) thought the base number x = 8 was a dysploid reduction from an ancestral base number of x = 9. My own view regarding the above is that the ancestral base number of the Caryophyllales (s.l.) is x = 8 (if not x = 4) and that early on major dysploid lines with x = 7 and 9 arose, some of the latter giving rise to x = 10 and 11 through more recent dysploidy. I consider the fairly commonly encountered base number x = 17 to be derived through ancient amphiploidy between lines with x = 8 and 9, or possibly by dysploid reduction or addition from old polyploid lines on a base of x = 9 or 8, respectively. Looked at from this perspective it seems conceivable that the Aizoaceae might be best circumscribed by those groups having base numbers of x = 8, those on a base of x = 9 being construed as the family Mesembryanthemaceae (cf. Bittrich and Hartmann 1988; Bittrich 1990 for discussion).

Amaranthaceae (x = 8). As treated by Schinz (1934), the Amaranthaceae has approximately 850 species distributed among 64 genera. These are divided among two subfamilies and three tribes. The family is relatively poorly studied in chromosomal terms. Counts are available for about 165 species and 24 genera, including taxa from both the subfamilies and all the tribes. About two-thirds of the genera remain to be counted. Polyploidy is very common in the family, occurring in about 50% of the taxa counted to date.

As can be seen from Table 2.1, diploid counts on a base of x = 8 and 9 are commonly found in the family, these sometimes occurring within genera which also possess x = 17. For this reason, as discussed above, I have considered the latter counts to be ancestral amphidiploids (8 + 9). So construed, the most commonly encountered base number is x = 8 (occurring in 16 genera) followed by x = 9 (occurring in 14 genera). Interestingly, genera on a base of $\underline{x} = 10$ and 11 also occur in the Amaranthaceae, but these appear to be mostly higher dysploid derivatives from the fairly widespread base number of $\underline{x} = 9$. Certainly the base numbers of $\underline{x} = 11$ (occurring in the genera *Sericostachys* and *Gomphrena*) have little to do with the base number of $\underline{x} = 11$ found in the Cactaceae. However, the array of diploid counts on a base of $\underline{x} = 7, 8, 9, 10$, and 11 found in this rather natural family strongly suggests that, over time, both dysploidy ($7 \leftarrow 8 \rightarrow 9, 10, 11$) and amphiploidy ($17 = 8 + 9$) have been a factor, both in the evolution of the family Amaranthaceae and of the order Caryophyllales in general. Taxa with base numbers of $\underline{x} = 13$ (e.g., species of *Charpentiera* and *Gomphrena*) are probably dysploid derivatives from ancestral relatives with base numbers of $\underline{x} = 12$, although diploid counts of $2^n = 24$ are currently unknown in the family. Alternatively, taxa with $\underline{x} = 13$ might be equally well explained by dysploid loss from taxa with $\underline{n} = 14$, base numbers of $\underline{x} = 7$ presumably occurring in several genera, as shown in Table 2.1.

Basellaceae ($\underline{x}$ = 8). As treated by Ulbrich (1934a), the Basellaceae contains five genera distributed among two tribes. Chromosome counts are available for four of the genera and both tribes, as indicated in Table 2.1.

All the counts for taxa of this family are presumably polyploid: *Anredera* ($2^n = 24$); *Basella* ($2^n = 44, 48$); *Boussingaultia* ($2^n = 24, 36, 48$); *Ullucus* ($2^n = 24, 36$). In keeping with my treatment of base numbers for similar polyploids in yet other families of the Caryophyllales, I consider the ancestral base number of Basellaceae to be $\underline{x} = 8$, although it is perhaps likely that the family as it is known today evolved from a base of $\underline{x} = 12$. This is suggested by the fact that four of the five genera are based on $\underline{x} = 12$, the anomalous number $2^n = 44$ found in *Basella rubra* (Huang et al. 1986) presumably derived by aneuploid loss ($n = 12 - 1 = 11$).

Cactaceae ($\underline{x}$ = 11). This is a large and diverse family variously estimated to have as many as 2000 species distributed among 100 or more genera. Species counts are available for 800 or more species distributed among some 84 genera.

Regardless, except for the occasional anomaly, all the taxa are based on $\underline{x} = 11$. Because of this, the various genera for which counts are available are not shown in Table 2.1. While a base number of $\underline{x} = 11$ clearly holds for the Cactaceae as it is known today, it is possible that this was derived from an ancestral base number of $\underline{x} = 12$ via aneuploid loss ($12 - 1 = 11$) and that the number $\underline{x} = 12$ probably arose from a paleobasic number of $\underline{x} = 8$, much as assumed for the Basellaceae, Didiereaceae and yet other families of the Caryophyllales where numbers on a base of $\underline{x} = 12$ are encountered. It is also possible, however, that it was derived by dysploid gain from the $\underline{x} = 10$ line postulated for the family Nyctaginaceae.

Hershkovitz (unpubl.) discussed in some detail the possible relationships of *Pereskia* ($\underline{x} = 11$) with the genus *Grahamia* of the family Portulacaceae ("strictly on the basis of overall similarity"). Diploid chromosome counts of

Grahamia (2^n = 18) by Bernardello (1989) clearly suggest that the proposed character similarities are mostly convergent, as Hershkovitz suspected might be the case.

Caryophyllaceae ($\underline{x}$ = 8). As treated by Pax and Hoffmann (1934b), the Caryophyllaceae has about 2300 species distributed among approximately 100 genera. These are divided among ten tribes arrayed in three subfamilies. The family is very well studied chromosomally. As of 1989, chromosome numbers are available for approximately 1360 species and about 80 genera (Table 2.1). About 60% of the taxa counted to date are tetraploid or have tetraploid derivatives. Counts are available for all the subfamilies, nine of the ten tribes, and most of the subtribes of the family (counts not available for the subtribes Xerotieae and Habrosieae).

As can be seen from Table 2.1, a wide array of base numbers occurs in the family, both at the diploid level and as inferred from likely polyploid derivatives. Thus, all polyploid taxa with numbers of 2^n = 24, 28, 30 (or polyploids thereof) are taken to be stabilized polyploids from ancestral taxa with base numbers of $\underline{x}$ = 8, 7, and 10, respectively. Likewise, polyploid taxa clearly on a base of $\underline{x}$ = 9 (e.g., 2^n = 18, 27, 36, etc.) are listed as such, and taxa with $\underline{x}$ = 17 are taken to be aneuploid derivatives of ancestral polyploid relatives with $\underline{x}$ = 8 (16 + 1 = 17) or $\underline{x}$ = 9 (18 − 1 = 17). Regardless, it seems likely that the very remote ancestral base number or numbers is $\underline{x}$ = 7, 8, or 9 (if not $\underline{x}$ = 4). Indeed, a number of closely related taxa in the family are known to have diploid counts of 2^n = 14, 16, 18, 20 or 22, often within the same genus. Thus, the capacity of this family to form an array of varying numbers within closely related species, or even *within* a species, is remarkable, as is well documented for *Arenaria* and yet other groups.

Assuming my inferences are correct as regards ancestral base numbers, it would appear that the most common and widespread base number is $\underline{x}$ = 8, which occurs in about 33 of the approximately 80 genera counted to date, this number occurring in all the subfamilies, including all the tribes except for the Sperguleae, which is based upon $\underline{x}$ = 9. The second most common and widely encountered base number is $\underline{x}$ = 9, which also occurs, albeit sporadically, in all the subfamilies, but is absent from three of the subtribes for which counts are available.

Chenopodiaceae ($\underline{x}$ = 9). As treated by Ulbrich (1934b), the Chenopodiaceae has approximately 1400 species distributed among 102 genera. These are divided among eight subfamilies and 14 tribes. The family is fairly well studied chromosomally, counts being available for about 400 species and 53 genera. About 50% of the taxa are polyploids or have polyploid races. Counts are available for all the subfamilies and 12 of the 14 tribes (absent in the tribe Nucularieae), as treated by Ulbrich.

All the genera counted to date (Table 2.1) have base numbers of $\underline{x}$ = 9, except for *Camphorosma* and *Spinacea* which have $\underline{x}$ = 6 (based upon 2^n =

12 for all the taxa within these groups counted to date). These two genera are positioned within two closely adjacent tribes of the subfamily Atripliceae and the counts appear to be independently derived from a base number of $\underline{x}$ = 9. It is possible, however, that $\underline{x}$ = 6 is ancestral and that numbers which appear to be on a base of $\underline{x}$ = 9 (e.g., 2^n = 18, 36, 72, etc.) are old derived hexaploids or higher polyploids.

Didiereaceae ($\underline{x}$ = 8). This small family of only 11 species distributed among four genera is confined to Madagascar. It is fairly well known chromosomally with counts for all species and all the genera. The counts are mostly high polyploids on a hypothetical base of $\underline{x}$ = 8 (Table 2.1), ranging from 2^n = 24 to 240. Hershkovitz (pers. comm.), however, thought the base number might be $\underline{x}$ = 12 [presumably following the suggestion of Ehrendorfer (1976a)]. In keeping with my reasoning throughout the present contribution, I believe the base number for the Didiereaceae is $\underline{x}$ = 8, this being the lowest divisible number for the polyploid series reported to date (ignoring the lower possible numbers of $\underline{x}$ = 2, 4, 6).

In short, chromosomal data for this family, as I interpret it, suggests that the family is a relic from the $\underline{x}$ = 8 ancestral plexus of the Caryophyllales, presumably not especially close to the Nyctaginaceae ($\underline{x}$ = 10), which appears to be an ancestral higher dysploid derivative from the $\underline{x}$ = 9 plexus (Fig. 2.1).

Some workers link the Didiereaceae to the Cactaceae and it is possible that the latter arose out of an ancestral didiereoid phylad on a base of $\underline{x}$ = 12 via dysploid reduction (12 − 1 = 11).

Nyctaginaceae ($\underline{x}$ = 10). Heimerl (1934a) recognized the Nyctaginaceae as having about 300 species arranged in 30 genera and five tribes. The family is poorly known chromosomally, largely because the chromosomes are small and difficult to count, especially from meiotic material. Counts are available for about 40 species from only eight or nine genera, this representing only two of the five tribes, most of these polyploids, seemingly on base numbers of $\underline{x}$ = 10. In Table 2.1, as for other families, ancestral base numbers of $\underline{x}$ = 8 and 9 have been inferred where counts of 2^n = 34 ($\underline{x}$ = 17) have been reported. Many of the specific counts are uncertain, however; thus closely related taxa have been reported as 2^n = 44, 46 (*Abronia*); 2^n = 40, 42 (*Boerhavia*); 2^n = 40, 42 (*Commicarpus*); 2^n = 26, 33, 58 (*Mirabilis*); and 2^n = 52, 58 (*Oxybaphus*).

Because of the above, any hypothesized ancestral base number for the Nyctaginaceae based upon present chromosomal data must be highly speculative. According to Ehrendorfer (1976b), the woody condition and yet other characters suggest that the family is best considered as derivatives of the largely neotropical Phytolaccaceae. If so, the commonly encountered base number of $\underline{x}$ = 10 for the Nyctaginaceae was probably derived from an ancestral base of $\underline{x}$ = 9 (9 + 1 = 10), which predominates in the Phytolaccaceae. This speculation is reflected in the arrangement of Fig. 2.1.

Phytolaccaceae ($\underline{x}$ = 9). Heimerl (1934b) recognized this family as having 110 species in 17 genera, distributed among five tribes. Counts are available for about 15 species from six genera, these occurring in only two of the five tribes. While poorly sampled, the family appears to be rather uniform as regards the base chromosome number, all the taxa for which counts are available possessing base numbers of $\underline{x}$ = 9, as indicated in Table 2.1.

Portulacaceae ($\underline{x}$ = 8). Pax and Hoffmann (1934c) recognized in this family ca. 500 species distributed among 19 or 20 genera. These were distributed among two subfamilies and three tribes. Chromosome counts are available for ca. 80 species distributed among 19 genera (including segregates since the treatment of Pax and Hoffmann). All the subfamilies and tribes have been examined and the family appears to be multibasic with base numbers of $\underline{x}$ = 7, 8, 9, 10, 11, and 12. Assuming that taxa with base numbers of $\underline{x}$ = 12 are actually polyploids on a base of $\underline{x}$ = 8 (and even without this assumption), the most commonly encountered and widespread base number is $\underline{x}$ = 8, as indicated in Table 2.1.

2.3 Discussion and Conclusions

The most commonly encountered, widely occurring (phyletically speaking) base chromosome number in the Caryophyllales is $\underline{x}$ = 8. This number is found, or can be inferred to have once occurred, in all the families except for the Chenopodiaceae ($\underline{x}$ = 9), Phytolaccaceae ($\underline{x}$ = 9), Nyctaginaceae ($\underline{x}$ = 10), and Cactaceae ($\underline{x}$ = 11). The lowest chromosome number reported to date ($\underline{n}$ = 6 pairs or 2^n = 12) occurs in two genera of the Chenopodiaceae (*Camphorosma* and *Spinacia*) and appears to be derived. The highest numbers are found in the Didiereaceae (2^n = ca. 240). Indeed, polyploidy is relatively common in the order and found in nearly all the families. Chromosome size varies widely both within and between families, but the Nyctaginaceae is especially noteworthy for having small chromosomes which are difficult to count.

Because of its widespread occurrence in the order, the base chromosome number of $\underline{x}$ = 8 is also considered primitive in the Caryophyllales. This fits nicely with the views of Grant (1982a,b) who postulated a base of $\underline{x}$ = 8 for primitive families generally. Nevertheless, there must have been a very early dichotomy in the Caryophyllales between groups with $\underline{x}$ = 8 and $\underline{x}$ = 9, the latter being also quite widespread in the order and almost the sole number found in the families Chenopodiaceae and Phytolaccaceae. Indeed, Ehrendorfer (1976a), based on data available up to 1975, thought $\underline{x}$ = 9 might be the ancestral base number for the Caryophyllales.

Families with base numbers of $\underline{x}$ = 10 (Nyctaginaceae), if current counts are correct, and $\underline{x}$ = 11 (Cactaceae) might possibly be derived from relatively recent dysploid derivatives from ancestral groups with a base of $\underline{x}$ = 9, although I now favor the view, based upon evidence presented at this symposium, that the Cactaceae probably arose as a dysploid derivative of an ancestral group with $\underline{x}$ = 12, the latter itself being perhaps a polyploid derivative of 2^n = 24 (3$\underline{x}$ = 24) or $\underline{x}$ = 8.

References

Bernardello LM (1989) The chromosomes of *Grahamia* (Portulacaceae). Plant Syst Evol 163:127–131

Bittrich V (1990) Systematic studies in Aizoaceae. Mitt Inst Allg Bot Hamb 23b:491–507

Bittrich V, Hartmann HEK (1988) The Aizoaceae–a new approach. Bot J Linn Soc 97:239–254

Ehrendorfer F (1976a) Chromosome numbers and differentiation of centrospermous families. Plant Syst Evol 126:27–30

Ehrendorfer F (1976b) Closing remarks: Systematics and evolution of centrospermous families, Plant Syst Evol 126:99–106

Goldblatt P, Johnson DE (1991) Index to plant chromosome numbers 1988–1989. Monogr Syst Bot Mo Bot Gard 30:1–243

Grant V (1982a) Periodicities in the chromosome numbers of the angiosperms. Bot Gaz 143:379–389

Grant V (1982b) Chromosome patterns in primitive angiosperms. Bot Gaz 143:390–394

Heimerl A (1934a) Nyctaginaceae. In: Engler A (ed) Die natürlichen Pflanzenfamilien, 2nd edn, vol 16c. Engelmann, Leipzig, pp 86–134

Heimerl A (1934b) Phytolaccaceae. In: Engler A (ed) Die natürlichen Pflanzenfamilien, 2nd edn, vol 16c. Engelmann, Leipzig, pp 135–164

Huang S-F, Chen S-J, Q: Q-Y, Shi X-H (1986) Plant chromosome count(2). Subtrop For Sci Tech 3:41–47

Pax F, Hoffmann K (1934a) Aizoaceae. In: Engler A (ed) Die natürlichen Pflanzenfamilien, 2nd edn, vol 16c. Engelmann, Leipzig, pp 79–123

Pax F, Hoffmann K (1934b) Caryophyllaceae. In: Engler A (ed) Die natürlichen Pflanzenfamilien, 2nd edn, vol 16c. Engelmann, Leipzig, pp 275–364

Pax F, Hoffmann K (1934c) Portulacaceae. In: Engler A (ed) Die natürlichen Pflanzenfamilien, 2nd edn, vol 16c. Engelmann, Leipzig, pp 234–262

Schinz H (1934) Amaranthaceae. In: Engler A (ed) Die natürlichen Pflanzenfamilien, 2nd edn, vol 16c. Engelmann, Leipzig, pp 7–85

Ulbrich E (1934a) Basellaceae. In: Engler A (ed) Die natürlichen Pflanzenfamilien, 2nd edn, vol 16c. Engelmann, Leipzig, pp 263–271

Ulbrich E (1934b) Chenopodiaceae. In: Engler A (ed) Die natürlichen Pflanzenfamilien, 2nd edn, vol 16c. Engelmann, Leipzig, pp 379–584

3 Vascular Tissues

ARTHUR C. GIBSON

3.1 Introduction

Vascular anatomy of the Caryophyllales provides an interesting database that has potential for phylogenetic analyses, given that several peculiar apomorphic features have been documented within the order. Chief among the structural oddities is anomalous secondary growth of roots and stems, wherein most secondary thickening of vascular tissues is produced by arcs of lateral meristems that are formed to the outside of the initial vascular cambium of the organ (Pfeiffer 1926; Esau 1965a). Even during the last century anatomists recognized that anomalous secondary thickening characterizes several closely related centrospermous families (Sanio 1863; de Bary 1884; Georghieff 1887; Solereder 1908), and this feature has been used for evaluating familial relationships (Eckardt 1976; Cronquist 1981; Thorne 1983; Gibson and Nobel 1986). Nonetheless, reporting on anomalous secondary thickening in stems and roots of different ages has been incomplete. Primary shoot vasculature of this order also includes some fascinating variations, e.g., the apparent occurrence of vascular bundles within pith (Wilson 1924; Dastur 1925; Joshi 1934; Metcalfe and Chalk 1950; Pant and Mehra 1961) and open and closed systems within the same family (Wilson 1924; Bisalputra 1962; Gibson 1976; Beck et al. 1982).

Using computer analyses Rodman and coworkers (1984) evaluated the number of leaf gaps per leaf (multilacunar or trilacunar versus unilacunar), occurrence of anomalous secondary thickening, and occurrence of pit borders on wood fibers; of these the unilacunar node was identified as one of six defining synapomorphies for Centrospermae. Brown and Varadarajan (1985) analyzed the presence or absence of anomalous secondary thickening in Phytolaccaceae s.l., but they were incorrect about *Stegnosperma* (Bedell 1980; Horak 1981a,b) and may have improperly scored taxa that were not thoroughly examined.

To date, generalizations about the vascular tissues of Caryophyllales have been unobtainable because observations are uneven, being plentiful in certain subfamilies but mostly lacking in others. Leaf venation has been used to evaluate relationships of particular species within the Cactaceae (Bailey 1960, 1968) and Portulacaceae (Hershkovitz 1991) but not in other families. Numerous important studies have been published on the ontogeny of secondary vascular tissues in roots (Artschwager 1926; Millner 1934; Joshi 1937; Esau and Cheadle 1969; Stevenson and Popham 1973; Mikesell and Popham 1976; Mikesell 1979; Webster and Wilson 1980), but these have not been followed by broader systematic surveys and analyses. Descriptions of phloem cell types, other than features of sieve-element plastids (see Chap. 5), have been scanty, and thorough analyses of cells in secondary xylem are available only for several families, particularly in the Cactaceae (Gibson 1973, 1975, 1977a,b, 1978b) and a few genera of special phylogenetic significance (Gibson 1978a; Bedell 1980; Horak 1981a).

To identify homologous characters of primary and secondary vascular systems, ideally workers should employ detailed ontogenetic studies on: (1) growth and differentiation of primary shoot vasculature, as it relates primordia and phyllotaxis; (2) origin of the initial vascular cambium in stems and roots and the nature of its derivatives; (3) inception of the first anomalous lateral meristem and its course of axial growth; (4) ontogeny of vascular tissues formed by anomalous lateral meristems; and (5) initiation of subsequent supernumerary arcs and their relationships to other vascular layers and phyllotaxis or branching. Even *Bougainvillea*, probably the most thoroughly studied centrosperm with anomalous vascular structure (Balfour 1965; Esau and Cheadle 1969; Pulawska 1973; Stevenson and Popham 1973; Zamski 1980), is incompletely known for several key ontogenetic details of vascularization.

The importance of evaluating vascular tissues for the centrosperms is two-fold: (1) to search for synapomorphies that may clarify the phylogenetic relationships of the subfamilies and families within the order, and (2) via new comparative studies, to determine the suite of vegetative features expected in the hypothetical ancestor of the order. All previous attempts to reconstruct the phylogeny of centrosperms have been limited by the lack of a defensible outgroup, which is needed to define plesiomorphic versus apomorphic character states and estimate polarity for the evolution of each character. In Rodman et al. (1984) for example, a consensus outgroup was constructed using seven candidates that had been suggested in earlier literature; however, it is possible that none was the real outgroup and, instead, researchers must look for a yet unidentified taxon to be the sister taxon.

3.2 Materials and Methods

To supplement published accounts on vascular tissues of centrosperms, representative species from many of the families and subfamilies were selected from the cultivated and wild plants in Los Angeles County, California. Emphasis was given to species having nonsucculent leaves and for which the primary shoot vasculature was previously unknown or poorly described. Genera that have been excluded from the order by biochemical and ultrastructural analyses, e.g., *Batis*, *Theligonum*, *Rhabdodendron*, Polygonaceae, and Gyrostemonaceae (Mabry 1977; Carlquist 1978), were not sampled. Several materials were obtained from herbarium materials and from colleagues.

Rapidly growing shoot tips, mature leaves, stems of different ages, and, where available, roots were liquid-preserved in a 50–70% mixture of formalin, acetic acid, and alcohol (FAA) and embedded in Paraplast. Only vegetative shoot tips were used, and samples were chosen that tended to lack sylleptic branching, so that formation of branch trace bundles would not have to be included in the analysis. New developmental observations were based largely on the following taxa: *Aptenia cordifolia* (L.f.) N.E. Brown 'Red Apple' and *Tetragonia tetragonioides* (Pallas) Kuntze (Aizoaceae); *Bosea amherstiana* Hooker and *Pleuropetalum darwinii* Hooker (Amaranthaceae); *Anredera cordifolia* (Tenore) Steen. (Basellaceae); *Atriplex lentiformis* (Torrey) S. Watson (Chenopodiaceae); *Alluaudia procera* Drake (Didiereaceae); *Halophytum ameghinoi* (Speg.) Speg. (Halophytaceae; poor materials available); *Bougainvillea spectabilis* Willd. and *Mirabilis oligantha* (Standl.) MacBr. (Nyctaginaceae); *Phytolacca americana* L. and *P. dioica* L. (Phytolaccaceae); *Calyptrotheca somalensis* Gilg (poor material) and *Ceraria namaquensis* Pears. & Steph. (Portulacaceae); and *Stegnosperma cubense* A. Rich. (Stegnospermataceae). Typically, two shoot tips per species were serially sectioned at 10 µm, stained with safranin and fast green, and studied with transmission and polarized light (Gibson 1976).

Liquid-preserved and several herbarium materials of older stems and roots of numerous Caryophyllales were sectioned either in paraffin or with a sliding microtome and macerated with a 5% macerating fluid (Gibson 1973, 1977b). Leaves of the sampled species were also cleared using 5% NaOH and stained with safranin. Some stem materials were either dehydrated and critical-point dried or air-dried between sheets of paper before being coated with 200 nm of gold-palladium and examined with an ETEC Autoscan at 10 kV.

3.3 Primary Vascular Systems

3.3.1 Procambial Differentiation in the Shoot

As in typical dicotyledons and demonstrated previously in Cactaceae (Gibson 1976), the procambium of centrosperms is a sympodial system that differentiates acropetally and continuously through residual meristem as discrete strands in relationship to primordia that flank the shoot apex.

In most species the median and lateral leaf-trace bundles for a leaf occurred adjacent to each other in a provascular ring that forms from the residual meristem. However, particularly in Nyctaginaceae and *Phytolacca* the median leaf-trace bundle was separated from the provascular ring by parenchymatization of cells just outside the differentiating protophloem, resulting in the appearance that these bundles were located within a pith. The course of each median bundle remained direct to the primordium, and the provascular ring was actually displaced outward slightly by cell enlargement of parenchyma. Maheshwari (1930) referred to the median leaf-trace bundles as "large central bundles," but many reports have termed these medullary bundles (Dastur 1925; Joshi 1931a,b, 1934; Metcalfe and Chalk 1950; Pant and Mehra 1961; Kirchoff and Fahn 1984). Other treatments have simply referred to these large bundles as the "inner ring of vascular bundles" of the primary vascular system (Stevenson and Popham 1973) or "interior vascular bundles" (Mikesell and Popham 1976). Because these bundles do not arise from ground tissue, i.e., the pith-rib meristem, authors should be cautious not to call these medullary bundles, and pith should be defined as a narrow region interior to these bundles. The paired central bundles in stems of certain Nyctaginaceae, particularly *Achyranthes* (Maheshwari 1930; Joshi 1934; Pant and Mehra 1961) are derived from two large strands of the initial two axial bundles of the seedling and certainly are apomorphic (Raj and Nagar 1980, 1989). Medullary bundles in the Caryophyllales also occur within the subfamily Cactoideae of the cacti (Boke 1951, 1957). Separate cortical systems have also evolved independently in fleshy halophytes of Chenopodiaceae (Fahn and Arzee 1959) and Cactoideae.

Given that many authors have compared certain features of centrosperms to those of monocotyledons, it is worth noting that the ontogeny of seedling vasculature in Chenopodiaceae (Fron 1899; Bisalputra 1961; Esau 1965a, 1977) has been used as a dicotyledonous model for describing transition in plants between the primary root and cotyledons, origins of epicotyl leaves, and patterns of differentiation of primary phloem and xylem.

3.3.2 Sympodial Nature of Primary Shoot Vasculature

Numerous workers have demonstrated the desirability to understand the developmental and physiological integrity of primary vascular systems of shoots based on detailed analyses of serial transections (Esau 1965b; Dormer 1972; Howard 1974; Larson 1975; Beck et al. 1982). For the Caryophyllales relatively few taxa have been sampled, producing extensive knowledge on Chenopodiaceae (Wilson 1924; Joshi 1937; Fahn and Arzee 1959; Fahn and Broido 1963; Zamski and Azenkot 1981; Fahn and Zimmermann 1982) and Cactaceae (Gibson 1976; Gibson and Nobel 1986) but partial data on other families (Dastur 1925; Maheshwari 1930; Joshi 1931a,b, 1934; Inouye 1956; Pant and Mehra 1961; Balfour and Philipson 1962; Philipson and Balfour 1963; Stevenson and Popham 1973; Kirchoff and Fahn 1984).

A basic type of open vascular system has been identified in taxa of betalain-containing families having helically alternate phyllotaxis. Its simplest form, but not necessarily the primitive condition, has been illustrated for Chenopodiaceae (Wilson 1924; Bisalputra 1962), a system with five vascular sympodia (with 2/5 phyllotaxis) having unilacunar nodes with three major traces per leaf, two from one axial bundle and one from the other. In some cases these have been drawn as three separate divergences. Published examples of separate divergence of three trace bundles are *Chenopodium glaucum* (Fig. 3.1A) and *Celosia cristata* (Wilson 1924) and *Rhagodia spinescens* and *Atriplex spongiosa* (Bisalputra 1962). In other species or specimens the median and one lateral leaf-trace bundle may have a common origin (Fig. 3.1B), i.e., one trace is bipartite, as was observed in *Bougainvillea* (Balfour 1965). Minor variations of the three-trace pattern include more subdivisions of lateral traces to form five or more bundles entering the leaf, as in *Phytolacca dioica* (Kirchoff and Fahn 1984), *Chenopodium album* (Wilson 1924), which may also produce more than three traces from the axial bundles, and *Pereskia grandifolia* (Fig. 3.1C), which possessed seven or more bundles from the bipartite and lateral traces. In this study *Atriplex lentiformis* also exhibited the bipartite form. Also in the current study, *Anredera cordifolia* and *Calyptrotheca somalensis* had open systems with three traces per leaf, probably having the bipartite form, but materials were not sufficient to observe all aspects of those vascular sympodia. In *A. cordifolia*, the median leaf-trace bundle was less developed at the time of divergence than were the two laterals.

Among other open vascular systems in the betalain families, published illustrations show that in some species venation for each leaf may diverge from only one axial bundle, becoming an arc with five or more bundles entering the petiole, as in *Pereskia humboldtii* (Fig. 3.1D), one bundle forming a broad leaf trace, as in *Salsola australis* (Bisalputra 1962), to one bundle forming a single leaf-trace bundle, as in *Maireana* (Fig. 3.1E), *Babbagia*, and *Bassia* (Bisalputra 1962). A single bundle that forms three leaf-trace bundles also occurs in *Tetragonia tetragonioides* (Fig. 3.1F). *Halophytum* typically has four bundles per primordium, two abaxial and two adaxial ones, apparently all arising as branches of a single trace bundle.

Also similar are those systems that become closed at a prescribed distance from the apex. In *Pereskia aculeata* (Gibson 1976) a bridge bundle formed at the inception of a bipartite leaf-trace bundle and fused with the adjacent axial bundle before the other leaf-trace bundle could diverge into the leaf. In *Atriplex* bridges may also form, producing a closed system (Bisalputra 1962). More advanced types of closed vasculature have been recorded in *Hectorella* (Skipworth 1962), Opuntioideae (Gibson 1976), and *Amaranthus* (Wilson 1924) and are clearly derived types.

In taxa having opposite decussate phyllotaxis, relatively few open systems have been observed, excepting samples of *Arthrocnemum* (Bisalputra 1962). Caryophyllaceae which have unilacunar nodes need to be carefully

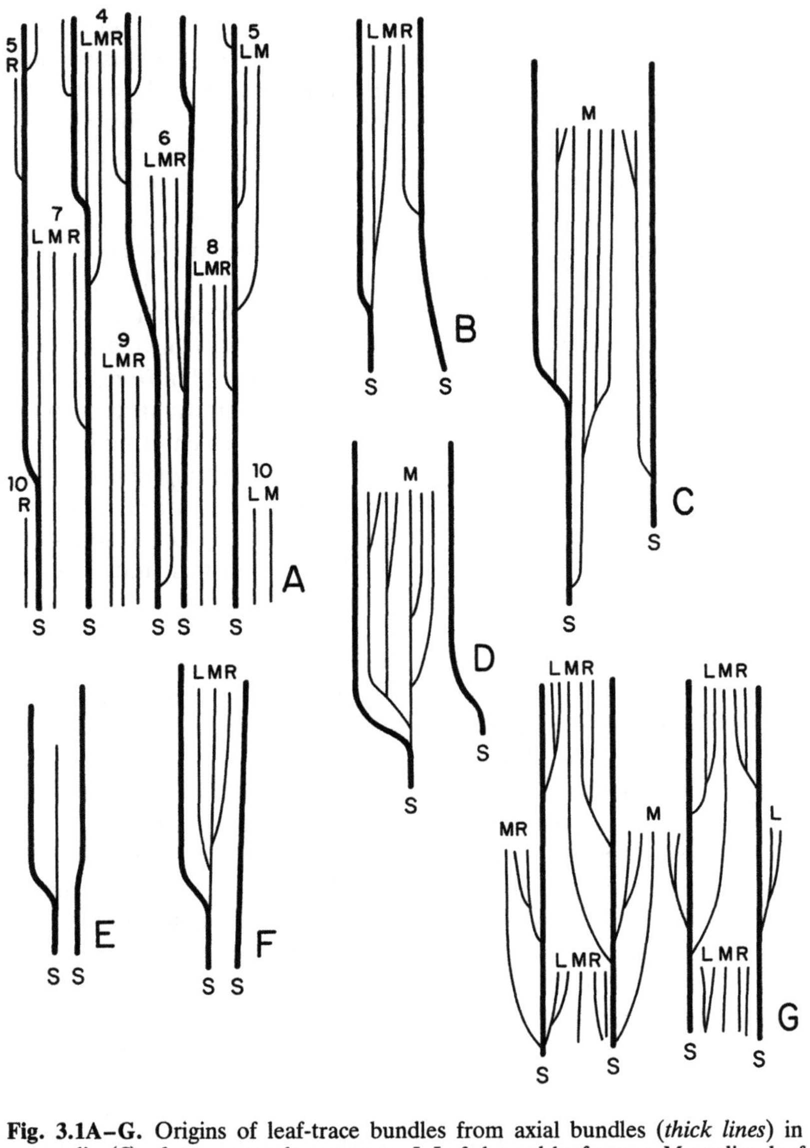

Fig. 3.1A–G. Origins of leaf-trace bundles from axial bundles (*thick lines*) in vascular sympodia (*S*) of open vascular systems. *L* Left lateral leaf trace; *M* median leaf trace; *R* right lateral leaf trace. **A** *Chenopodium glaucum* (after Wilson 1924), having five vascular sympodia and three traces per primordium arising from adjacent axial bundles. *4–10* Successive leaf traces. **B** *Bougainvillea spectabilis* (after Balfour 1965), showing divergence of a bipartite bundle that forms two of the three principal bundles to the young primordium. **C** *Pereskia grandifolia* (after Gibson 1976). **D** *Pereskia humboldtii* (after Gibson 1976). **E** *Maireana astrotricha* (after Bisalputra 1962). **F** *Tetragonia tetragonioides*. **G** *Aptenia cordifolia*, a species with opposite decussate phyllotaxis and four vascular sympodia that give rise to three principal traces for each primordium

surveyed. An open system was diagnosed in *Aptenia cordifolia* (Fig. 3.1G) and perhaps occurs in many Aizoaceae; this is an interesting example, reminiscent of the leaf-trace origin of species with helically alternate phyllotaxis, having three diverging bundles, two from one axial bundle and one from the other, usually forming five traces. In *Aptenia*, the median leaf-trace bundle diverged two nodes beneath the leaf primordium, but the two laterals originated only one node beneath. In the majority of Chenopodiaceae with opposite decussate phyllotaxis, systems are closed by forming a network between adjacent axial bundles (Fahn and Arzee 1959; Bisalputra 1962).

When proposing probable trends of evolution for the origin of leaf traces in Chenopodiaceae, Bisalputra (1962) assigned as primitive an open system having a unilacunar node with two equal and discrete traces. This type was once considered by some nodal anatomists to be the primitive state for dicotyledons (Beck et al. 1982). From that hypothetical type Bisalputra derived those having three traces from two axial bundles, as in *Chenopodium glaucum*, and independently the variant with a single leaf trace supplied by two axial bundles. Other types of nodal arrangement were considered to be later derivatives. However, given a sample of betalain-containing families, the parsimonious model would begin with a leaf that receives three or more bundles arising from two axial bundles (e.g., either Fig. 3.1A or B), then deriving closed systems, those with more or fewer traces, and those forming traces from only one axial bundle. Without details of primary vasculature for anthocyanin-containing families, a more robust model cannot be hypothesized for the Caryophyllales.

3.3.3 Differentiation Patterns of Bundles

3.3.3.1 *Phytolacca dioica*

A detailed description of vascular differentiation has been published for the basal taxa of Cactaceae (Gibson 1976), showing that cacti possess typical dicotyledonous ontogenetic patterns. Likewise, xylem differentiation in *P. dioica*, representative of species having anomalous secondary thickening, was also typical for dicotyledons. Protoxylem formation was bidirectional from where the median trace entered the primordium and discontinuous from the axial bundle supplying it. On one shoot tip, P6-P1 (the six youngest primordia) had no mature protoxylem; xylem only occurred in median leaf-trace bundles for P7 and P8 and was discontinuous beginning about 400 µm below the primordium; xylem in lateral bundles 12R (right lateral leaf trace for P12) and 11L (left lateral leaf trace for P11) was also discontinuous for 2500 µm below those primordia. In this species, as in all Caryophyllales, protoxylem was composed of vessels with simple perforation plates, which Bierhorst and Zamora (1965) showed as characteristic of highly derived angiosperms, and sieve-element members had advanced simple plates (Esau 1969).

Phytolacca dioica with 3/8 phyllotaxis was also used to study the origin of "medullary bundles". As illustrated in Kirchoff and Fahn (1984) each leaf received a median and weakly differentiated pairs of left and right lateral leaf-trace bundles. The median and two pairs of lateral leaf-trace bundles always arose within the provascular ring, but the median leaf-trace bundles were displaced to the inside of the ring. At each stem level there were eight median leaf-trace bundles in the inner ring, e.g., those for P15-P8, appearing there one orthostichal level before entering a leaf. Subdivided or undivided left and right lateral leaf-trace bundles for those primordia resided within the provascular ring, positioned approximately opposite the respective median trace. For the median leaf-trace bundles of P15-P8, P11-P14 had amphivasal organization, especially 12M (median leaf trace for P12), but collateral organization where each median bundle entered and exited the parenchyma core.

In much older portions of stem, regions of median leaf-trace bundles having amphivasal organization formed a vascular cambium between the core of phloem and centrifugal poles of xylem. Files of secondary xylem consisted of vessel elements with paratracheal axial parenchyma, xylary fibers, and ray cells; secondary phloem consisted of a few sieve tubes with companion cells. Lateral bundles and segments of median bundles residing within the original provascular ring initiated a continuous ring of true secondary thickening via the establishment of fascicular and interfascicular vascular cambium, this far from the apex.

Comparing *Phytolacca* with *Bougainvillea* (Stevenson and Popham 1973; Kirchoff and Fahn 1984), one observes that the vasculature of the Nyctaginaceae was more complicated because an outer ring of bundles, these mostly lateral traces, also became displaced inward from the provascular ring. This must be regarded as a derived state which has also been accompanied by substantial reduction in normal cambial activity before the onset of the supernumerary cambia. Precocious development of axillary buds and thorns also adds vascular bundles and more connections to the original sympodial system (Zamski 1980).

3.3.3.2 Notable Variations in Other Centrosperms

As expected in a set of plants with numerous differences in leaf size and phyllotaxis, species-specific differences were observed for the timing of differentiation. In *Atriplex lentiformis*, for example, with a 2/5 phyllotaxis, a substantial amount of mature xylem was present in the upper 500 µm of the shoot tip, and for this species discontinuous strands of xylem were less protracted and occurred at 140 µm for 4M, 340 µm for 5M, 480 µm for 6M, and 510 µm for 10R. In this species the median leaf trace was positioned slightly inside the provascular ring one orthostichal level below its divergence into a primordium, but medium traces never appear to be medullary and did not possess regions with amphivasal organization. In a shoot tip of *Stegnosperma cubense* having 2/5 phyllotaxis, which developed up to nine bundles per primordium, the protoxylem for 7M was present at leaf insertion and then was discontinuous for 5140 μm.

Many interesting taxa were sectioned but not reported on in detail because the growth forms are highly complex. For example, the Didiereaceae have sylleptic development of axillary spines, thus introducing branch traces into a primary vascular system that otherwise is an open system with unilacunar nodes and many leaf-trace bundles. All examples of sylleptic development, including cactus areoles, are presumed to be autapomorphies that evolved within families and should not be used to elucidate broad phylogenetic relationships of the families.

3.3.4 Leaf Venation

So few genera have been examined for venation patterns that any use of existing data for phylogenetic analysis would be unwarranted. A majority of the species have entire margins and hence no teeth; consequently, an important feature of tooth vasculature, found extremely useful for phylogenetic studies in other orders (Hickey and Wolfe 1975), is not available, and there appear to be no distinctive venation designs that clearly define lineages within the order.

Pate and Gunning (1969) sampled the families in Caryophyllales and found no vascular transfer cells in the leaves. Moreover, no transfer cells were observed in any of the families or orders that are serious candidates for being the sister taxon of the centrosperms.

3.4 Secondary Thickening

3.4.1 Occurrence of Normal and Anomalous Types

Secondary growth via a single vascular cambium has been observed in perennial axes of all Cactaceae, Didiereaceae, Portulacaceae, and Basellaceae that have been examined, although in certain cactus stems regions of interfascicular cambium never become established. In the Cactaceae, the least succulent, leaf-bearing species of *Pereskia* possess a solid cylinder of secondary xylem (Bailey 1962, 1963b; Gibson 1975; Gibson and Nobel 1986), documenting that normal cambial growth is the primitive state for the family, whereas species having much stem or root succulence tend to have zones within wood where cells lack lignified secondary cell walls (Bailey 1963a,c,d, 1964). In most cacti woody cylinders contain large primary rays of thin-walled parenchyma (Gibson 1978c). Other cacti, particularly small terrestrial forms, produce increments of secondary xylem lacking wood fibers but which are not anomalous in an ontogenetic sense (Gibson 1973, 1977a,b, 1978b). Early reports of anomalous structures in cacti were misinterpreted and actually were patches of nonfibrous wood or xylem parenchyma derived from true cambial activity and not included phloem tissue (Gibson 1973). All the Didiereaceae form solid cylinders of secondary xylem (Fig. 3.2A) via normal cambial activity (Heimsch 1942; Rauh 1961). Most Portulacaceae have thin, fleshy stems with little secondary growth, and this secondary xylem, like that of small cacti, may consist of weak fibers and much thin-walled parenchyma (Carlquist 1962; Prabhakar and Ramayya 1979). A relatively thick but soft woody cylinder occurs in *Portulacaria afra*, *Talinum guadalupense*, and *Talinella* (Hershkovitz, pers. comm.). Solid wood of *Calyptrotheca somalensis* was not available, but 1-year xylem resembled that of certain cacti. Woods of *Hectorella* and *Lyallia* have apparently not been described. The Basellaceae typically form little secondary xylem in their herbaceous stems, but only cambial secondary growth has been observed, and major primary vascular bundles were bicollateral.

Anomalous secondary thickening by successive cambia probably characterizes all the Aizoaceae sensu Bittrich and Hartmann (1988), Amaranthaeae, Chenopodiaceae (including *Dysphania*), Nyctaginaceae, Phytolaccoideae s.s., and the segregate unigeneric phytolaccaceous families Agdestidaceae, Barbeuiaceae, and Stegnospermataceae (Fig. 3.2B–H). In these, anomalous structures can be missed if only young stem axes or small individuals are examined, e.g., if plants are very small at maturity. *Phytolacca* and *Stegnosperma* (Fig. 3.3A–D) are examples of plants forming a normal cambium from the initial ring of primary vascular bundles, but that meristem eventually is covered by vascular tissues from supernumerary cambia which are initiated on the outer face of the phloem (Wheat 1977; Horak 1981a). Contrasting this, in Nyctaginaceae formation of supernumerary cambia occurs precociously and usually without any appreciable amount of normal cambial activity (Studholme and Philipson 1966; Esau and Cheadle 1969; Stevenson and Popham 1973). Betalain families exhibit a broad array of intermediary forms but typically have a narrow increment of normal cambial activity before anomalies are initiated.

Reports on the diverse forms of the phytolaccaceous alliance indicate that certain genera have normal secondary growth, whereas others exhibit anomalous structure, and the only anatomical summary (Metcalfe and Chalk 1950) provided insufficient data on which to base systematic conclusions. In the Petiveriaceae, anomalies have been reported in the literature or personally observed in *Gallesia*, *Petiveria*, *Rivina*, and *Seguieria* but are alleged not to occur in the remaining genera and *Microtea*; however, many of the poorly known species are herbaceous perennials that may have anomalous thickening in the rootstock. *Achatocarpus* and *Phaulothamnus* have only normal secondary growth. Stems of *Gisekia africana* did not have anomalous growth, but old roots need to be examined.

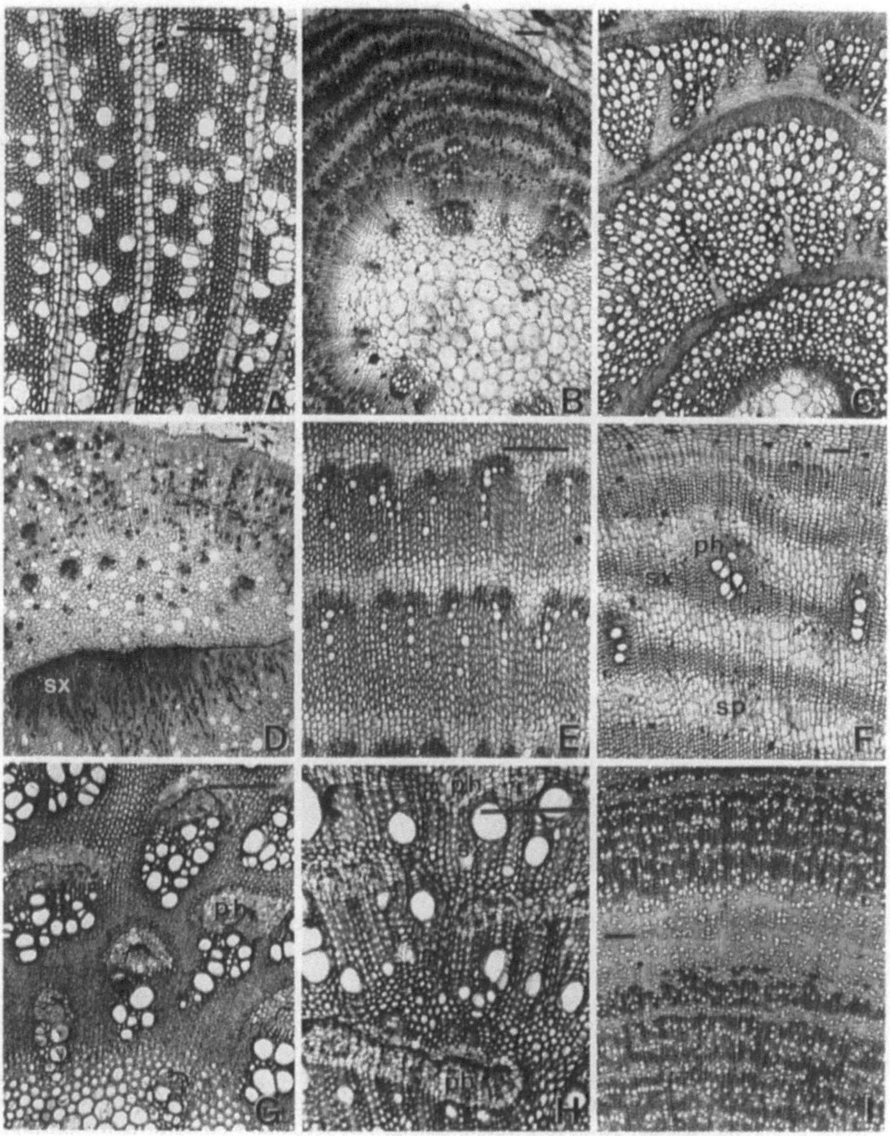

Fig. 3.2A–I. Secondary growth of stems in Caryophyllales. **A** *Didierea procera* has a solid woody cylinder and normal cambial activity. **B** *Aptenia cordifolia*, basal stem with numerous increments formed by supernumerary cambia. **C** *Tetragonia tetragonioides*, base of a vigorously growing stem, showing high densities of vessels in the first-formed secondary xylem and successive increments. **D** *Faucaria tigrinum*, a succulent species with much parenchyma between increments of secondary thickening. **E** *Lampranthus amoenus*, a fruticose specimen having numerous xylary fibers formed from each supernumerary cambium. **F** *Pisonia umbellifera*, an arborescent specimen having prominent rings of secondary parenchyma between increments of secondary xylem and phloem. **G** *Mirabilis oligantha*, a shrub with fascicular zones of vessels and phloem surrounded mostly by xylary fibers. **H** *Barbeuia madagascarensis* (Keating and Miller 2250), showing arcs of phloem between increments of secondary xylem. **I** *Dianthus plumarius*, a large subterranean stem having normal cambial activity but some zones in secondary xylem lack libriform fibers. *ph* Phloem formed by a supernumerary cambium; *sp* secondary parenchyma; *sx* secondary xylem. *Bars* are 100 μm

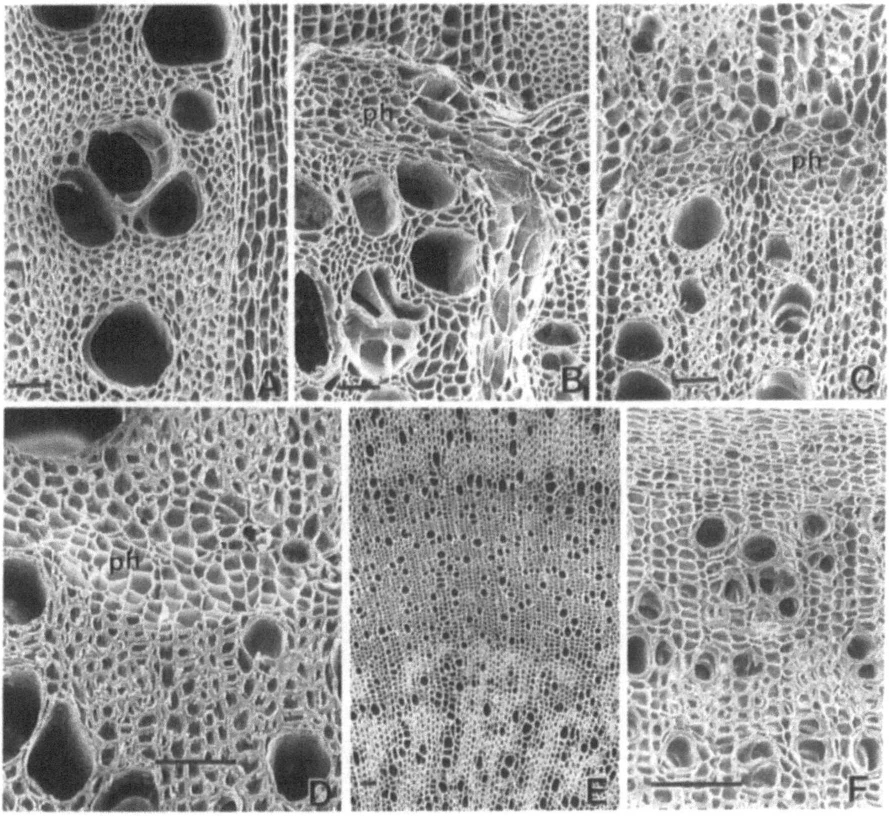

Fig. 3.3A–F. Scanning electron photomicrographs of stem vascular tissues. **A** and **B** *Phytolacca americana*, a plant with a basal woody axis of 5 cm: **A** region in which secondary xylem appears normal and has multiseriate vascular rays, thin-walled libriform fibers, and relatively wide vessels; **B** occurrence of phloem and parenchyma between successive increments of secondary xylem. **C** *Stegnosperma cubense*, lower stem with secondary phloem between increments of secondary xylem. **D** *Barbeuia madagascarensis* (Keating and Miller 2250), which has short arcs of included phloem. **E** and **F** *Dianthus plumarius*: **E** normal secondary thickening having high density of extremely narrow vessels and **F** no vascular rays; in **F** radial lineages of secondary phloem (*upper portion*) are easy to observe. *ph* Phloem formed by a supernumerary cambium. *Bars* are 50 μm

For the families with anthocyanins, Metcalfe and Chalk (1950) have listed both fully normal and anomalous types of secondary growth in the Caryophyllaceae (Figs. 3.2I and 3.3E–F) and Molluginaceae. Once again, modern surveys are needed to clarify the systematic distribution of each type, and especially old roots must be examined. In the Caryophyllaceae, anomalous secondary thickening appears commonly in the Paronychioideae, as in *Polycarpaea* and *Corrigiola*, but among subfamilies with normal secondary growth one can observe supernumerary cambia in roots of other taxa, e.g., *Spergularia marina*. The Molluginaceae has many species with relatively thin stems and roots, so that detecting anomalous structure is problematic, but in examined specimens the woody *Psammotropha apetala* clearly was normal, whereas that of *Macarthuria apetala* and *M. australis* exhibited anomalies. Materials of *Limeum sulcatum* were too young to diagnose.

To date, only normal cambial secondary growth has been reported from relatively old stems and roots of the fleshy annual *Halophytum* (Gibson 1978a), and *Gunniopsis* (Aizoaceae) may also have this feature (Bittrich, pers. comm.).

3.4.2 Nature of Anomalous Vascular Tissues

Vascular cylinders with anomalous secondary thickening (Fig. 3.2B–H) are complex, having included patches of phloem, discontinuous radial lineages of cells, units of conducting phloem and xylem cells arranged as if they are collateral bundles, and often tangential or radial bands of parenchyma between successive or adjacent increments of growth. Xylem commonly consists of short to extremely short libriform fibers and generally lacks vascular rays. Phloem sclerenchyma is absent within the xylem matrix.

Parenchymatization appears to be greater in roots than in stems and is most pronounced in storage roots, e.g., *Beta vulgaris* (Chenopodiaceae), *Mirabilis jalapa* and *Abronia* spp. (Nyctaginaceae), and *Mestoklema tuberosa* (Aizoaceae), where normally fibrous xylem differentiates instead as parenchyma with unlignified primary walls (Artschwager 1926; Metcalfe and Chalk 1950; Mikesell and Popham 1976; Webster and Wilson 1980). Large unlignified tangential bands of parenchyma also characterize anomalous thickening in stems of many succulent and fleshy Aizoaceae (Bhambie et al. 1977), as in *Aptenia cordifolia* (Fig. 3.2B), *Lithops localis*, *Sesuvium verrucosum*, *Tetragonia tetragonioides* (Fig. 3.2C), and *Faucaria tigrinus* (Fig. 3.2D), but shrubs tend to have phloem surrounded by cells with thin secondary cell walls, as in *Lampranthus amoenus* (Fig. 3.2E) and *Drosanthemum speciosum*, or rarely by thick-walled fibers, as in shrubby *Ruschia* spp. Stems of other families show similar ranges; for example, in Nyctaginaceae the tree *Pisonia umbellifera* (Fig. 3.2F) has rings of unlignified parenchyma between zones of vascular elements, the liana *Bougainvillea spectabilis* tends to have short bands of parenchyma with thin secondary cell walls, and the shrub *Mirabilis oligantha* (Fig. 3.2G) has relatively little parenchyma and strong development of xylary fibers. The Phytolaccoideae and Stegnospermataceae (Fig. 3.3A–C) also have nearly cylindrical bands of parenchyma with phloem, whereas parenchyma development is scanty in *Barbeuia* (Figs. 3.2H and 3.3D) and many shrubby Chenopodiaceae, such as in *Atriplex*, *Grayia*, and *Salicornia*. Consequently, no broad systematic patterns can be identified for either parenchyma or xylary fibers. The Nyctaginaceae, Ruschioideae and *Aptenia* (Aizoaceae), and *Anisomeria* and *Phytolacca*

(Phytolaccaceae) have calcium oxalate raphides in vascular parenchyma, whereas in other taxa, if present, crystals are either solitary prismatics, styloids, or druses. *Gisekia africana* had raphides in the cortex but not in young vascular tissues (Hofmann 1973). Several Aizoaceae also possessed mucilage cells (Metcalfe and Chalk 1950), but otherwise chemical idioblasts are absent.

Typical descriptions of anomalous secondary thickenings have observed that centrospermous axes tend to have fascicular organization (Fig. 3.2E–G), wherein vessels only occur opposite sieve tubes (de Bary 1884; Metcalfe and Chalk 1950). Several taxa did not show that striking pattern, e.g., in *Phytolacca*, *Stegnosperma*, *Seguieria* (Phytolaccaceae s.l.) and some woody Amaranthaceae, all of which possess vascular rays (Fig. 3.3A), and anomalous axes in the anthocyanin-containing Caryophyllaceae, e.g., *Corrigiola squamosa* and *Spergularia marina*, which are essentially rayless. In all these cases, parenchyma comprises a continuous cylinder between growth layers. It is clearly unwarranted to speculate that this represents the primitive type of anomalous secondary thickening for the order. Joshi (1937) proposed that because the root is a "conservative organ," occurrence of much parenchyma in roots is more primitive than having mostly fibrous cells. No evidence in this study supports that hypothesis.

3.4.2.1 Bidirectionally-Dividing Supernumerary Cambium

Most students of anomalous wood structure in centrosperms have concluded that these axes possess successive bidirectional cambia; however, several workers reinterpreted these as unidirectional meristems, which produce only centripetal derivatives (Stevenson and Popham 1973), or as a meristematic zone of which the outermost initials remain cambial (Philipson and Ward 1965; Balfour 1965). Esau and Cheadle (1969) conducted an ontogenetic study of secondary vascular tissues in stems and roots of *Bougainvillea* and provided convincing photographic documentation that phloem and xylem formed in radial lineages from a bidirectionally dividing lateral meristem, hence a vascular cambium in the conventional sense. For *Bougainvillea* Esau and Cheadle identified three stages in divisions of an anomalous cambium: (1) formation of some xylem fibers to the inside and secondary parenchyma to the outside; (2) production of secondary phloem opposite xylem fibers; and (3) to the inside initiation of vessels opposite where phloem sieve tubes were differentiating. Bidirectional cambium was confirmed for zones of sieve tubes and vessels in stems of *Phytolacca dioica* (Wheat 1977) and appeared to be the correct interpretation for all materials in the current investigation because phloem always differentiated in radial lineages, traceable as outer derivatives of a cambium. Hence, at least the conducting portions of each anomalous layer exhibit characteristic cambial activity. Much research is needed to determine whether nonconducting portions of supernumerary cambium ever function as unidirectional meristems, forming only centripetal derivatives.

Use of the term cambium received another temporary setback when several authors elected to reclassify these unusual regions as primary thickening meristems (Stevenson and Popham 1973; Mikesell and Popham 1976; Mikesell 1979; Yarrow and Popham 1981), until then generally regarded as unidirectional meristems of monocotyledons (Esau 1965a; DeMason 1983). From the outset this approach appeared flawed because primary thickening meristem characteristically applies to a layer with centripetal cell divisions immediately subtending a shoot apical meristem. Contrary to the situation, in centrosperms the thickening meristem is never very close to the shoot apex and may even never occur in a given stem.

When Stevenson and Popham accepted the terminology of primary thickening meristem for *Bougainvillea*, they were obligated to adopt corresponding terms for products of that meristem; they designated desmogen strand for each fascicular phloem and xylem unit that differentiates within the conjunctive tissue of these anomalies and suggested that each fascicular unit was formed as a meristematic entity and only afterwards developed a "desmogen cambium" forming "tertiary"–rather than secondary–xylem and phloem. Of course, this treatment was not a compatible interpretation with earlier observations on the same species (Esau and Cheadle 1969), showing that the cambial initials (desmogen cambium sensu Stevenson and Popham) actually produced conducting phloem and xylem elements, albeit that phloem differentiation was precocious. Likewise, treating such fascicular regions as discrete strands or bundles is flawed because radial and tangential anastomoses were observed between xylem vessels between adjacent strands without phloem involvement (Zamski 1980) or between phloem strands without xylem involvement (Horak 1981b).

If the concept of primary thickening meristem is abandoned for the Caryophyllales, then researchers should also refrain from using other terms associated with that type of meristem. For example, conjunctive tissue, which has been a common label especially for the parenchymatous tissue, should be replaced by secondary parenchyma, i.e., cambial derivatives formed to the outside, and xylary fibers, sclerenchyma formed to the inside.

3.4.2.2 Initiation and Progression of Vascular Cambia in Seedlings

Events in the early establishment of lateral meristems were described in considerable detail for *Bougainvillea spectabilis* (Stevenson and Popham 1973), *Mirabilis jalapa* (Mikesell and Popham 1976), *Phytolacca americana* (Mikesell 1979), and *Atriplex hortensis* (Yarrow and Popham, 1981).

In *Bougainvillea*, a typical stelar cambium became established in the radicle between primary phloem and xylem but not through pericycle opposite the two xylem poles, and by day 6 another lateral meristem was initiated "apparently at random" in root pericycle unassociated with that stelar cambium. Within several days that second cambium comprised a continuous cylinder, and regions of that cylinder developed acropetally as the primary root grew in length and somewhat later entered the hypocotyl (day 33) and the first node of the shoot (day 60). In shoots the meristem was initiated inside of the cortex, presumably from cells in the zone of residual meristem, and just outside conducting protophloem of leaf-trace bundles of the primary vascular system.

In the related *Mirabilis*, the same meristem was initiated just inside the cortex at the top of the cotyledonary node (day 18) and several days later had developed into the primary root and acropetally into the stem. New supernumerary cambia later developed external to conducting phloem of leaf-trace bundles. In *Phytolacca*, the meristem was initiated at the hypocotyl–primary root junction as two arcs parallel to the sides of the diarch vascular core and caused early thickening of that region (Mikesell 1979), whereas in *Atriplex* anomalous thickening commenced opposite xylem poles (Yarrow and Popham 1981). In the rosette-forming *Beta vulgaris*, the first supernumerary cambium presumably was induced along traces of leaves 1–4, but sampling was not refined enough to record the earliest initiation events (Zamski and Azenkot 1981). Detailed observations have not yet attempted to relate growth and differentiation of primordia on shoot tips of seedlings to initiation and growth of the first supernumerary cambium.

Evidence that supernumerary cambia develop acropetally in stems and roots can be seen from studies that have determined the lateral extent and thickness of secondary tissues at sampled points along the axis (Mikesell 1979; Horak 1981b). In some species the first anomaly may be detected several centimeters from the shoot apex, but in others anomalous secondary thickening does not occur in stems of the upper canopy. In *Stegnosperma*, Horak found that acropetal development of supernumerary cambia was present to the level of a major branch, so that upper stems had only normal secondary xylem. In herbaceous perennials acropetal development from rootstock may be so delayed that anomalies do not appear when typical stems have been examined (Joshi 1937). This may be the situation for the Petiveriaceae, in which anomalous secondary thickening has been reported in some stems but not others (Metcalfe and Chalk 1950; Rogers 1985). Development of supernumerary cambia from a dormant rootstock into new shoots has not been described.

3.4.2.3 Origin of Additional Supernumerary Cambia and Nature of Connections

The long list of labels used to identify the successive cambia of centrosperms (Mikesell and Popham 1976) partly reflects a desire to define their origins from undifferentiated parenchyma instead of procambium for vascular cambium. This is a hypocritical viewpoint given what occurs in typical periderm development of dicotyledons. Phellogen, a cambium, typically arises as a cylinder from a layer of outermost cortex or the epidermis but is succeeded by deeper sheets of phellogen arising first from arcs of cortical cells or phelloderm parenchyma from an older phellogen. In the same manner successive cambia of centrosperms arise progressively away from the original vascular cambium or where that cambium would have developed in a typical dicotyledonous plant. Each new cambium forms on the outer face of phloem in roots and stems from secondary parenchyma, outer derivatives of previous supernumerary cambia (Artschwager 1920; Esau and Cheadle 1969). As in the rhytidome of periderm, for vascular tissues the localized meristematic sheets become continuous with the older adjacent segments, resulting in the reticulate nature of this tissue. In roots the tangential extent of cambia is greater than in stems, thus producing longer arcs for each segment of growth.

Little is known about the three-dimensional nature of phloem and xylem in either the initial thickening layer of any species or the successive segments produced by supernumerary cambia. Laborious serial methods have been employed to visualize segments of these reticulate anomalies in several species (Fahn and Shchori 1968; Zamski 1980; Zamski and Azenkot 1981; Horak 1981b; Fahn and Zimmermann 1982). All workers have shown degrees of connections between conducting cells in different increments of secondary growth, and they have demonstrated that phloem and xylem of a given strand may have different connections within and between arcs of vascular tissues. This may result because phloem generally differentiates much before xylem in a given growth increment, i.e., each tissue tracks separate stimulating systems.

3.4.3 Structure and Cell Types of Secondary Xylem

3.4.3.1 Portulacineae (sensu Thorne 1983)

For the Cactaceae, Bailey (1966; see also Gibson 1975) determined that leafy species of *Pereskia* had woods that showed a high level of anatomical specialization. Vessel elements were moderately short (mean length 150–250 µm) and had slightly oblique to nearly transverse simple perforation plates and fully bordered, multiseriate alternate pits on the lateral walls. Xylary fibers were relatively short and libriform, often nucleated and sometimes septate. Axial wood parenchyma was scanty paratracheal, and vascular rays were multiseriate. Some specimens displayed diagonally storied axial elements, indicating that cacti tend to have a highly specialized stratified vascular cambium (Bailey and Srivastava 1962). Likewise, sieve elements in the phloem were also highly specialized for dicotyledons as a whole.

Gibson (1973, 1977a,b, 1978b) documented that many interesting features have evolved in the woods of Cactaceae, mostly in relation to differences in growth form and plant size. Extremely tall specialized growth habits had vessel elements (mean 450 μm) and libriform fibers (mean 900 μm) which were the longest in the Cactaceae. The degree of plant erectness was correlated with length, width, and wall thickness of xylem elements. Within each subfamily, low-growing succulent species had xylem lacking libriform fibers and instead possessed vessel elements with scanty helical to reticulate thickenings on lateral walls and unique vascular tracheids having either helical or annular secondary thickenings. Evolutionary change in the cell types comprising secondary xylem in succulent cacti was interpreted as wood paedomorphosis (Carlquist 1962), in which features that typically occur in primary xylem are protracted into secondary growth. Vascular tracheids, first described by Schleiden (1845), were interpreted as highly specialized homologues of wood fibers (Gibson 1977a). The wood anatomy of Didiereaceae has been described briefly (Heimsch 1942; Rauh 1961), but has never been compared with closely related families. As with cacti, secondary xylem of Didiereaceae consisted of short vessel elements with simple perforation plates and multiseriate alternate pitting, libriform fibers, scanty paratracheal axial wood parenchyma, and multiseriate vascular rays. In fact, wood transections of *Alluaudia comosa*, *A. procera* (Fig. 3.2A), and *Didierea trollii* qualitatively and quantitatively were extremely similar to those of columnar cacti of Mexico (Gibson 1973). Stellate druses occasionally occurred in ray cells lacking secondary walls, and this was a feature commonly found in pereskias (Bailey 1963a,b,c).

In general, the Portulacaceae are small plants with relatively little secondary growth and appear to have only normal cambial activity (Prabhakar and Ramayya 1979). *Portulacaria afra*, which is the most widely available woody species, had a number of features indicating evolutionary specialization. These included vessels with slitlike pits and narrow borders; storied nucleate libriform fibers with weakly lignified cell walls; paratracheal axial wood parenchyma, typically vasicentric to aliform, with no secondary walls; and wide multiseriate vascular rays composed of large cells with unlignified walls and containing many druses and pigmented mucilage idioblasts. Such pigmented mucilage cells also occurred in ground tissues here and in *Ceraria namaquensis*, and they appear elsewhere in the Caryophyllales in the stem cortex and lateral primordia of Didiereaceae. Adjacent to pith the vascular rays of *Portulacaria* were dilated by tangential elongation of the cells.

Talinum guadalupense (Portulacaceae), a stem and leaf shrub, had nonfibrous secondary xylem. Its extremely short vessel elements had helical secondary thickenings, xylary fibers were absent, and parenchyma was thin-walled and diffuse. This species was chosen by Carlquist (1962) as a prime representative of wood paedomorphosis and therefore cannot be used to estimate the most primitive type of wood for the family. Only a young stem of *Calyptrotheca somalensis* was available for study; this specimen had very thin-walled libriform fibers but otherwise most resembled the woods of certain nonmucilaginous stems in the leaf-bearing cacti, and as such probably can be used as similar to the original type of wood. Crystal patterns in the woods of Portulacaceae were unremarkable, but *Ceraria*, *Calyptrotheca*, and *Portulacaria*, like Didiereaceae, all Opuntioideae, some Pereskioideae, and many of the primitive-looking Cactoideae, all have calcium oxalate crystals in each epidermal cell. In fact, both leaf-bearing Cactaceae and *Calyptrotheca somalensis* have druses. Calcium oxalate crystals in stem epidermis have not been reported or observed personally in any other Caryophyllales and is likely a synapomorphy, just as the pigmented mucilage cells of certain African Portulacaceae are apparently synapomorphic with Didiereaceae.

Vascular anatomy of the Basellaceae was studied from transections of only two genera, and secondary xylem in the herbaceous stems was thin. Vessels are very wide (up to 200 μm), which is expected from its viney habit (Carlquist 1975); consequently, these axes showed no similarities with other Portulacineae. Stems showed no successive cambia, but this should be tested for in future studies because a possible relationship with lianous Agdestidaceae or Petiveriaceae should be considered.

3.4.3.2 Phytolaccaceous Alliance

So little is known about the xylem structure of Phytolaccaceae s.l. that generalizations at this time would be unwarranted. Moreover, a very confusing wood diagnosis was presented by Metcalfe and Chalk (1950), which lavishly employed data from the normal woods of *Gyrostemon* and anomalous woods of *Rhabdodendron*, so that specific features of most phytolaccaceous species are obscure.

There are at least four different patterns of wood architecture among the 18–20 genera of centrosperms (excluding Gyrostemonaceae and Rhabdodendraceae) that, by various authors, are still related to Phytolaccaceae (Brown and Varadarajan 1985), but all specimens have vessel elements with simple perforation plates, scanty paratracheal axial wood parenchyma, and multiseriate alternate pitting on the lateral vessel walls.

Both genera of Achatocarpaceae had only normal cambial activity. Vessels were round to elliptic and narrow, 25–40 μm in diameter, solitary and mostly radial chains of two to four (or six) members. Vessel elements tended to be 300–400 μm in length and were covered with minute bordered pits. Libriform fibers and axial parenchyma cells were filled with prominent starch grains. Vascular rays were uniseriate to biseriate and also had abundant starch but no crystals.

Phytolacca, *Anisomeria*, and *Ercilla* (Phytolaccoideae) have prominently developed anomalous secondary thickening, and the characteristics of *P. dioica* have been carefully described (Wheat 1977). Vessels tend to be relatively wide (Fig. 3.3A and B), and well-developed multiseriate vascular rays occur in the anomalous arcs. Unlignified secondary parenchyma and phloem generally form nearly complete rings separating successive increments of growth. These plants form an initial vascular cambium from the primary vascular system and therefore have a thin cylinder of secondary xylem surrounding pith. A sample stem of *Seguieria rigida* from a herbarium specimen was similar to Phytolaccoideae in showing a fairly even distribution of vessels in transection, whereas a published illustration of *S. paraguayensis* (Metcalfe and Chalk 1950) indicates a strongly fascicular pattern. *Seguieria rigida* possessed abundant calcium oxalate prismatics, including very large styloids in the stem cortex. Other Petiveriaceae and *Agdestis* probably have fascicular organization (Pfeiffer 1926) but need to be restudied.

Secondary growth of *Stegnosperma* is noteworthy because these plants produce 5–15 mm normal cambial growth before successive cambia are established (Bedell 1980; Horak 1981a). Vessels were mostly solitary, occasionally in radial pairs or short chains. Mean vessel length varied from 278 to 320 μm, and in outer xylem arcs vessel diameters may exceed 100 μm. Fiber tracheids were present, and multiseriate vascular rays were frequent and comprised half of the wood.

A 4-mm-diameter stem of *Barbeuia madagascarensis*, examined in transection, showed secondary xylem that was roughly comparable with that of *Stegnosperma*, because they both lacked pronounced fascicular organization and possessed solitary vessels, lignified fibers, and vascular rays within xylem arcs. However, there were no clear indications that relationships exist between the Madagascan monotype and other phytolacs in the study. *Barbeuia* had relatively short arcs of phloem surrounded by lignified tissues, quite dissimilar from Stegnospermataceae and Phytolaccoideae.

A stem sample of *Gisekia africana* was too young to detect anomalous secondary thickening, but the first-formed xylem was dissimilar from other phytolacs in the study and more similar to Molluginaceae, e.g., *Macarthuria australis*.

3.4.3.3 Other Betalain-Containing Families

Secondary xylem has been difficult to describe for axes having anomalous secondary thickening, because even on a single plant stems above ground often have a fundamentally different composition from that of below-ground stems and roots, and growth habit and plant stature undoubtedly are correlated characteristics of xylary fibers. Moreover, no objective method has been devised to sample vessel diameters in transections that have a strongly fascicular organization, including a gradient of diameters opposite each strand of secondary phloem.

Typical secondary xylem of Aizoaceae, Amaranthaceae, Chenopodiaceae, and Nyctaginaceae is composed of short to extremely short vessel elements, short libriform fibers that may be long-lived, scanty paratracheal axial wood parenchyma, and few or no rays. Xylary fibers and secondary parenchyma sometimes exhibit storying, especially in the Nyctaginaceae (Esau and Cheadle 1969) and Aizoaceae. Measurements from stem macerations in this study found mean vessel-element lengths to be less than 120 µm and mean fiber lengths to be mostly less than 400 µm. Metcalfe and Chalk (1950) reported mean vessel-element lengths to 300 µm in Amaranthaceae but extremely short ones in other families; published values for Chenopodiaceae are less than 100 µm (Baird and Blackwell 1980; Carlquist and Hoekman 1985). In Chenopodiaceae from deserts, vessels often have tertiary helical thickenings. Xylem of most species are rayless or have inconspicuous, very low and narrow vascular rays.

Xylem anatomy in these families probably gives no systematic insight but serves as a useful subject for ecological studies of evolution. For example, in Aizoaceae all species apparently form an increment of normal secondary xylem and then form supernumerary cambia in perennial axes (Bhambie et al. 1977), but the fast-growing herb *Tetragonia tetragonioides* was observed to have numerous relatively wide arcs of anomalous growth within stem bases (Fig. 3.2C). Old trailing stems of *Aptenia cordifolia* (Fig. 3.2B) and the halophyte *Sesuvium verrucosum* had very thin accumulation of anomalous growth, whereas the upright, rigid stems of woody shrubs such as *Ruschia spinescens* had fibrous increments and heavily lignified fibers with cell walls more than 4 µm thick. However, vessels of *R. spinescens* were only 15 µm in diameter and distributed in a sharply fascicular pattern, whereas the rapidly growing *T. tetragonioides* had moderately wide vessels to over 100 µm in diameter, at high densities, and this axis had no fibrous zones in outer increments. Species of *Carpobrotus* and *Lampranthus* (Fig. 3.2E) had axes composed largely of xylary fibers with relatively thin walls, whereas *Faucaria* (Fig. 3.2D), *Glottiphyllum*, and *Lithops*, very succulent low plants, possessed abundant thin-walled parenchyma, and *Lithops* had the paedomorphic feature of short vessel elements with only helical secondary thickenings. Because growth form is strongly coupled to the nature of vascular tissues, it would be difficult to choose a primitive condition of either for phylogenetic analysis.

Important for phylogenetic analysis is the observation that Nyctaginaceae is the only family in which normal cambial activity is extremely scant or absent before supernumerary cambia form secondary tissues. Nevertheless, much ontogenetic research has been conducted on the highly specialized liana *Bougainvillea spectabilis*, which should not be scored as unspecialized for any of its characters; *Bougainvillea*, *Reichenbachia*, and *Salpianthus* are also unusual in the family for having helically alternate phyllotaxis.

The xylem anatomy of *Halophytum* was carefully described (Gibson 1978a). This Argentine monotype only showed normal cambium activity in basal stems and main roots, so workers must assume that this annual does not form supernumerary cambia. Qualitative and quantitative comparisons reveal that the rayless wood sectors of *H. ameghinoi* are very similar to normal xylem sectors of *Tetragonia tetragonioides* (Fig. 3.2C) and probably many other members of that genus; an anatomical survey of

Tetragonia or related Aizoaceae may aid in clarifying the systematic position of Halophytaceae.

3.4.3.4 Anthocyanin-Containing Families

The Caryophyllaceae and Molluginaceae exhibit very similar xylem design. Published descriptions of these axes are few (Payne 1933; Metcalfe and Chalk 1950) and have done little to determine the systematic patterns of anomalous secondary thickening within the Caryophyllaceae and possibly Molluginaceae. Vessel elements were extremely short and narrow (Figs. 3.2I and 3.3F), with multiseriate alternate pitting and simple perforation plates, and woods had short and extremely narrow libriform fibers or diffuse parenchyma, and scanty paratracheal axial wood parenchyma. Most woods of the Caryophyllineae appear to be rayless (Fig. 3.3F), whether produced by normal or supernumerary cambia. Hence, the xylem anatomy most closely resembles that of the betalain-containing families Amaranthaceae and Aizoaceae, particularly the low-growing and nonsucculent forms.

3.5 Extraxylary Sclerenchyma of Stems

Sclerenchyma has been observed on the outer periphery of the vascular system in stems of all centrospermous families, but many succulents lack extraxylary sclerenchyma entirely (Gibson 1982). Aizoaceae appear to have none, given that *Orygia*, reported to have stone cells in the "pericycle" (Metcalfe and Chalk 1950), and *Macarthuria*, with a cylinder of fibers around phloem, are now considered as Molluginaceae. Even the generally nonsucculent Nyctaginaceae, including woody species such as *Pisonia umbellifera* and *Mirabilis oligantha*, generally have little or no extraxylary sclerenchyma in vegetative shoots.

Literature accounts have produced much confusion about fibers at the periphery of the primary phloem. When authors designated these as "pericyclic fibers", they generally did not discriminate between fibers derived from protophloem versus perivascular fibers, which originate inside the starch sheath of cortex but not from phloem tissue (Esau 1965a, 1969). Both types are known to occur within the order, and developmental studies with long serial sections would be required to demonstrate that a cluster of fibers in one species was perivascular, i.e., they differentiated from cells of the outer residual meristem that were not procambium of primary vascular bundles.

Among centrosperms the phloic origin of fibers has been demonstrated for the Cactaceae (Bailey 1961; Srivastava and Bailey 1962), and in taxa having normal secondary thickening, i.e., Cactaceae, Didiereaceae, and

Portulacaceae, such cells are easily recognized as primary phloem fibers, although they may not differentiate until after one or more years of growth. Strands of these fibers occur in the relatively unspecialized leaf-bearing cacti of *Pereskia* (subfamily Pereskioideae), *Pereskiopsis* and *Quiabentia* (Opuntioideae), all columnar cacti with putatively primitive features (Cactoideae), and nearly all species of epiphytes (Cactoideae), but never appear in the small terrestrial succulents, e.g., subtribes Cactinae and Echinocactinae of Mexico, and *Opuntia*. In some cactus epiphytes, differentiation of phloem fibers may rapidly follow the onset of cambial activity. For the Portulacaceae, genera such as *Calyptrotheca*, *Ceraria*, *Lenzia*, and *Portulacaria* have phloem fibers, whereas fibers are lacking in genera with very succulent and short-lived species. Although phloem fibers generally occur in distinct strands outside conducting protophloem, sometimes fibers form a continuous cylinder when interfascicular distances are short, as in the epiphytes *Disocactus* and *Rhipsalis*, or when parenchyma cells of rays differentiate as sclereid idioblasts, e.g., in *Epipyllum*. Primary phloem fibers with connecting sclereids produce a solid lignified ring around the vascular tissues in both genera of Achatocarpaceae. *Halophytum* also has flat strands of phloem fibers (Gibson 1978a).

In some taxa having anomalous secondary thickening, fibers that differentiate on the periphery of the vascular tissues are often termed perivascular (Esau and Cheadle 1969; Wheat 1977), whether or not ontogenetic studies have traced fiber appearance. Investigations have not elucidated a systematic pattern. In the Amaranthaceae and Chenopodiaceae, species having such fibers tend to differentiate in strands and appear to have a phloic origin. Species such as *Phytolacca dioica* and *Stegnosperma cubense* tend to have a nearly continuous cylinder of fibers, probably phloic, with relatively thick, lignified walls and connecting sclereids.

In another set of taxa a solid cylinder of sclerenchyma, often five to ten layers thick, forms precociously around the vascular tissues (Metcalfe and Chalk 1950). These cells have wide cell lumina and relatively thin, uniform cell walls, hence appearing to undergo mostly symplastic growth. Such cells may be perivascular fibers. Examples are found in the Basellaceae, Caryophyllaceae, Molluginaceae (e.g., *Limeum sulcatum*), and in *Gisekia africana* and *Macarthuria*, but other related taxa may instead possess strands of fibers. In situations when sclerenchyma differentiates as a solid ring, the stem diameter tends to be narrow, and the xylem generally is rayless and has little or no projection of protophloem strands.

To date there have been few observations on sclerenchyma in secondary phloem. In *Pereskia*, which has brachysclereids and fiberlike sclereids in secondary phloem, Bailey (1961) used sclereid types to delimit three clades within the genus. Also in the Cactaceae, narrow fibers have been observed in the outermost secondary phloem in stems of the epiphyte *Disocactus alatus*. Both genera of Achatocarpaceae form idioblastic secondary phloem fibers.

Sclereid idioblasts are mostly absent from ground tissue in stems, but sclerified parenchyma has been observed in several columnar cacti from North and South America, epiphytic cacti in the tribe Hylocereeae and subtribe Rhipsalinae, and the fleshy cortex of *Halophytum*. *Halophytum* also has a uniseriate hypodermis consisting of lignified fibers, also present around the bony fruit. *Barbeuia madagascarensis* is unique among the Caryophyllales in having idioblastic cortical fibers.

3.6 Phylogenetic Analysis

For betalain-containing families the hypothetical ancestor probably was a woody perennial with helically alternate, exstipulate, simple leaves, undoubtedly also petiolate and having entire margins. Normal secondary growth via a single vascular cambium must have characterized roots and stems, because the developmental pathway for producing supernumerary cambia is a highly derived feature in seed plants and apparently irreversible for perennial axes. Even when anomalous secondary thickening does not extend into the shoot, the pathway occurs early in roots of seedlings. Hence, treating the Phytolaccaceae s.l. as a basal taxon becomes awkward if citing those genera that produce anomalous secondary thickening, but the hypothesis of a basal phytolaccaceous ancestry would not be falsified here if species with normal cambial activity are used. It is possible, even highly likely, that anomalous growth arose more than once during the evolution of Caryophyllales, because included phloem has evolved independently numerous times in dicotyledons (Metcalfe and Chalk 1983). However, at least the ancestor of the Cactaceae, Didiereaceae, and Portulacaceae (Portulacineae), which have only normal cambial activity, must only have been an organism possessing a single cambium, and for this reason researchers should anticipate cladograms that depict Portulacineae diverging before Aizoaceae, Nyctaginaceae, Amaranthaceae, Chenopodiaceae, and any phytolaccaceous genera with anomalous secondary thickening. It is worth noting that anomalous secondary thickening has appeared sporadically within the Theales (Metcalfe and Chalk 1950), one of the proposed sister taxa of the Caryophyllales.

Given that basal taxa for each of the betalain-containing families have open primary vascular systems, unilacunar nodes, and several, often three, traces arising unequally from two axial bundles, these characters can be hypothesized as necessary plesiomorphic character states for the putative ancestor. Choosing a particular version of that sympodial system may be unrealistic given the diversity of shoot designs for the order. Those genera having closed systems or bundles seemingly scattered within the parenchymatous core of the stem must be regarded as derived dicotyledonous forms, even though such vascular architecture may resemble the complex vascular systems of monocotyledons (Beck et al. 1982).

Cells composing vascular tissues of these families are highly specialized for dicotyledons, including only vessel elements with simple perforation plates, and in secondary tissues mostly short to extremely short axial elements with advanced types of wall features appear in xylem and phloem. Some of the longest elements in the phytolaccaceous alliance seem to occur in *Stegnosperma* and Achatocarpaceae, and this may indicate that these are early derivatives of the centrospermous radiation. For the betalain families, scanty paratracheal parenchyma is typical in normal and anomalous secondary thickening and should be anticipated in the ancestor. Although libriform fibers characterize the Caryophyllales, at least *Stegnosperma* has fiber tracheids and therefore may be the expected ancestral form. Some of these features occur in Schisandraceae (Metcalfe 1987) and Theaceae, e.g., Ternstroemioideae (Metcalfe and Chalk 1950), which do not have some of the prerequisite embryological, biochemical, or plastid features that researchers hope to find in the ancestor (Behnke 1976; Mabry 1977).

Betalain families appear to lack demonstrable characters in vascular tissues that can be used as synapomorphies for cladistic analysis at the family level. In xylem crystal idioblasts, e.g., stellate druses in Portulacineae and raphides in some of the families with anomalous secondary thickening, are still untested characters, risky to use until researchers understand the structural and physiochemical aspects of crystal formation. *Pereskia* (Cactaceae) and Achatocarpaceae have idioblastic sclerenchyma in secondary phloem, but the woods of those taxa bear no similarities. However, the presence of pigmented mucilage cells in African Portulacaceae and Didiereaceae, coupled with calcium oxalate crystals in the stem epidermis, provides interesting new clues that these are sister taxa. The epidermal crystals also characterize many basal groups of Cactaceae and should move investigators to search for crystals in stem epidermis and mucilage cells (and no tannin cells) among potential candidates of ancestors.

Phylogeneticists must appreciate that the Aizoaceae do not share vegetative features of the Portulacineae and appear to fail as the sister taxon of Cactaceae. The origin of CAM in the Aizoaceae and Cactaceae is certainly a case of physiological convergence and evolved independently within both families (Gibson and Nobel 1986). Detailed comparisons are needed between *Pereskia* and Portulacaceae, particularly *Calyptrotheca* and *Talinella* in Africa and *Grahamia* in western South America, and a West Gondwanaland explanation may therefore reemerge for the evolution of the Portulacineae as well as for the taxa with anomalous secondary thickening. Similarities in the ontogeny of flowers in *Pereskia* and Aizoaceae (see Chaps. 6 and 13) should be regarded as symplesiomorphy.

For anthocyanin-containing families so little is known about the vascular tissues that any conclusions are premature, including how many times anomalous secondary thickening evolved in these families.

Probably the greatest mystery of the phylogenetic puzzle concerns why supernumerary vascular cambia have evolved. Comparatively little is known about the relationship between growth substances and the initiation of cambial activity (Shininger 1979; Rubery 1981; Aloni 1987), and certainly the topic is unexplored for the Caryophyllales. Changes in hormone transport patterns probably caused the dramatic shifts in how secondary tissues are produced, and simplistic hypotheses could be generated to point an accusing finger at auxins, which appear to be transported in phloem or on the outer edge of protophloem (Morris and Thomas 1978; Jacobs and Gilbert 1983). Recent knowledge suggests that flavonoids are involved in auxin transport (Jacobs and Rubery 1988). There is considerable uncertainty how to use many properties of vascular tissues as discrete and scoreable characters for cladistic analyses. Likewise, a detailed study of the Basellaceae, with its bicollateral bundles, needs to determine how internal phloem evolved and to which genus this family is most closely related.

Acknowledgements. I wish to thank undergraduates Jeffrey Low and Monica Varsanyi for great assistance in preparing many of the microslides used for this analysis. Selected research materials were generously contributed by Professors Richard Keating, Ursula Hofmann, and H.-D. Behnke.

References

Aloni R (1987) Differentiation of vascular tissues. Annu Rev Plant Phys 38:179–204

Artschwager EF (1920) On the anatomy of *Chenopodium album* L. Am J Bot 7:252–260

Artschwager EF (1926) Anatomy of the vegetative organs of the sugar beet. J Agric Res 33:143–176

Bailey IW (1960) Comparative anatomy of the leaf-bearing Cactaceae. I. Foliar vasculature of *Pereskia*, *Pereskiopsis* and *Quiabentia*. J Arnold Arbor 41:341–356

Bailey IW (1961) Comparative anatomy of the leaf-bearing Cactaceae. II. Structure and distribution of sclerenchyma in the phloem of *Pereskia*, *Pereskiopsis* and *Quiabentia*. J Arnold Arbor 42:144–156

Bailey IW (1962) Comparative anatomy of the leaf-bearing Cactaceae. VI. The xylem of *Pereskia sacharosa* and *Pereskia aculeata*. J Arnold Arbor 43:376–388

Bailey IW (1963a) Comparative anatomy of the leaf-bearing Cactaceae. VII. The xylem of pereskias from Peru and Bolivia. J Arnold Arbor 44:127–137

Bailey IW (1963b) Comparative anatomy of the leaf-bearing Cactaceae. VIII. The xylem of pereskias from southern Mexico and Central America. J Arnold Arbor 44:211–221

Bailey IW (1963c) Comparative anatomy of the leaf-bearing Cactaceae. IX. The xylem of *Pereskia grandifolia* and *Pereskia bleo*. J Arnold Arbor 44:222–231

Bailey IW (1963d) Comparative anatomy of the leaf-bearing Cactaceae. X. The xylem of *Pereskia colombiana*, *Pereskia guamacho*, *Pereskia cubensis*, and *Pereskia portulacifolia*. J Arnold Arbor 44:390–401

Bailey IW (1964) Comparative anatomy of the leaf-bearing Cactaceae. XI. The xylem of *Pereskiopsis* and *Quiabentia*. J Arnold Arbor 45:140–157

Bailey IW (1966) The significance of the reduction of vessels in the Cactaceae. J Arnold Arbor 47:288–292

Bailey IW (1968) Comparative anatomy of the leaf-bearing Cactaceae. XVII. Preliminary observations on the problem of transitions from broad to terete leaves. J Arnold Arbor 49:370–376

Bailey IW, Srivastava LM (1962) Comparative anatomy of the leaf-bearing Cactaceae. IV. The fusiform initials of the vascular cambium and the form and structure of their derivatives. J Arnold Arbor 43:187–202
Baird WV, Blackwell WH (1980) Secondary growth in the axis of *Halogeton glomeratus* (Bieb) Meyer (Chenopodiaceae). Bot Gaz 141:269–276
Balfour E (1965) Anomalous secondary thickening in Chenopodiaceae, Nyctaginaceae and Amaranthaceae. Phytomorphology 15:111–122
Balfour EE, Philipson WR (1962) The development of the primary vascular system of certain dicotyledons. Phytomorphology 12:110–143
Beck CB, Schmid R, Rothwell GW (1982) Stelar morphology and the primary vascular system of seed plants. Bot Rev 48:691–815
Bedell HG (1980) A taxonomic and morphological re-evaluation of Stegnospermaceae (Caryophyllales). Syst Bot 5:419–431
Behnke H-D (1976) A tabulated survey of some characters of systematic importance in centrospermous families. Plant Syst Evol 126:95–98
Bhambie S, Joshi ML, Gupta ML (1977) Anatomical studies in certain members of the Aizoaceae. Proc Ind Acad Sci B 85:399–406
Bierhorst DW, Zamora PM (1965) Primary xylem elements and element associations of angiosperms. Am J Bot 52:657–710
Bisalputra T (1961) Anatomical and morphological studies in the Chenopodiaceae. II. Vascularization of the seedling. Aust J Bot 9:1–19
Bisalputra T (1962) Anatomical and morphological studies in the Chenopodiaceae. III. The primary vascular system and nodal anatomy. Aust J Bot 10:13–24
Bittrich V, Hartmann HEK (1988) The Aizoaceae – a new approach. Bot J Linn Soc 97:239–254
Boke NH (1951) Histogenesis of the vegetative shoot in *Echinocereus*. Am J Bot 38:23–38
Boke NH (1957) Structure and development of the shoot in *Toumeya*. Am J Bot 44:888–896
Brown GK, Varadarajan GS (1985) Studies in Caryophyllales. I. Re-evaluation of classification of Phytolaccaceae s.l. Syst Bot 10:49–63
Carlquist S (1962) A theory of paedomorphosis in dicotyledonous woods. Phytomorphology 12:30–45
Carlquist S (1975) Ecological strategies of xylem evolution. University of California Press, Berkeley
Carlquist S (1978) Wood anatomy and relationships of Bataceae, Gyrostemonaceae, and Stylobasiaceae. Allertonia 1:297–330
Carlquist S, Hoekman DA (1985) Ecological wood anatomy of the woody southern California flora. IAWA Bull NS 6:319–347
Cronquist A (1981) An integrated system of classification of flowering plants. Columbia University Press, New York
Dastur RH (1925) The origin and course of vascular bundles in *Achyranthes aspera* L. Ann Bot 39:539–545
de Bary A (1884) Comparative anatomy of the vegetative organs of the phanerogams and ferns. Bower FO, Scott DH (translators). Clarendon Press, Oxford
DeMason DA (1983) The primary thickening meristem: definition and function in monocotyledons. Am J Bot 70:955–962
Dormer KJ (1972) Shoot organisation in vascular plants. Chapman & Hall, London
Eckardt T (1976) Classical morphological features of centrospermous families. Plant Syst Evol 126:5–25
Esau K (1965a) Plant anatomy, 2nd edn. Wiley, New York
Esau K (1965b) Vascular differentiation in plants. Holt, Rinehart and Winston, New York
Esau K (1969) The phloem. In: Zimmermann W, Ozenda P, Wulff HD (eds) Handbuch der Pflanzenanatomie, vol V, pt 2. Borntraeger, Berlin
Esau K (1977) Anatomy of seed plants, 2nd edn. Wiley, New York
Esau K, Cheadle VI (1969) Secondary growth in *Bougainvillea*. Ann Bot 33:807–819

Fahn A, Arzee T (1959) Vascularization of articulated Chenopodiaceae and the nature of their fleshy cortex. Am J Bot 46:330–338
Fahn A, Broido S (1963) The primary vascularization of the stems and leaves of the genera *Salsola* and *Suaeda* (Chenopodiaceae). Phytomorphology 13:156–165
Fahn A, Shchori Y (1968) The organization of the secondary conducting tissues in some species of the Chenopodiaceae. Phytomorphology 17:147–154
Fahn A, Zimmermann MH (1982) Development of the successive cambia in *Atriplex halimus* (Chenopodiaceae). Bot Gaz 143:353–357
Fron G (1899) Recherches anatomiques sur la racine et la tige des Chenopodiacées. Ann Sci Nat Bot Ser 8, 9:157–240
Georghieff S (1887) Beitrag zur vergleichenden Anatomie der Chenopodiaceen. Bot Centralbl 31:53–57
Gibson AC (1973) Comparative anatomy of secondary xylem in Cactoideae (Cactaceae). Biotropica 5:29–65
Gibson AC (1975) Another look at the cactus research of Irving Widmer Bailey. In: Glass C, Foster R (eds) Suppl 1975 Yearbook Cactus Succulent J, pp 76–85
Gibson AC (1976) Vascular organization of shoots of Cactaceae. I. Development and morphology of primary vasculature in Pereskioideae and Opuntioideae. Am J Bot 63:414–426
Gibson AC (1977a) Vegetative anatomy of *Maihuenia* (Cactaceae) with some theoretical discussions of ontogenetic changes in xylem cell type. Bull Torrey Bot Club 104:35–48
Gibson AC (1977b) Wood anatomy of opuntias with cylindrical to globular stems. Bot Gaz 138:334–351
Gibson AC (1978a) Rayless secondary xylem of *Halophytum*. Bull Torrey Bot Club 105:39–44
Gibson AC (1978b) Wood anatomy of platyopuntias. Aliso 9:279–303
Gibson AC (1978c) Architectural designs of wood skeletons in cacti. Cactus Succulent J (Great Britain) 40:73–80
Gibson AC (1982) The anatomy of succulence. In: Ting IP, Gibbs M (eds) Crassulacean acid metabolism. Am Soc Plant Phys, Rockville, Maryland, pp 1–17
Gibson AC, Nobel PS (1986) The cactus primer. Harvard University Press, Cambridge
Heimsch C (1942) Comparative anatomy of the secondary xylem in the "Gruinales" and "Terebinthales" of Wettstein with reference to taxonomic grouping. Lilloa 8:83–198
Hershkovitz MA (1991) Leaf morphology of *Cistanthe* Spach (Portulacaceae). Ann Mo Bot Gard 78:1022–1060
Hickey LJ, Wolfe JA (1975) The bases of angiosperm phylogeny: Vegetative morphology. Ann Mo Bot Gard 62:538–589
Hofmann U (1973) Morphologische Untersuchungen zur Umgrenzung und Gliederung der Aizoaceen. Bot Jahrb Syst 93:247–324
Horak KE (1981a) Anomalous secondary thickening in *Stegnosperma* (Phytolaccaceae). Bull Torrey Bot Club 108:189–197
Horak K (1981b) The three-dimensional structure of vascular tissues in *Stegnosperma* (Phytolaccaceae). Bot Gaz 142:545–549
Howard RA (1974) The stem-node-leaf continuum of the Dicotyledoneae. J Arnold Arbor 55:125–181
Inouye R (1956) Anatomical studies on the vascular system of *Mirabilis jalapa* L. Bot Mag Tokyo 69:554–559
Jacobs M, Gilbert SF (1983) Basal localization of the presumptive auxin transport carrier in pea stem cells. Science 220:1297–1300
Jacobs M, Rubery PH (1988) Naturally occurring auxin transport regulators. Science 241:346–349
Joshi AC (1931a) Contributions to the anatomy of the Chenopodiaceae and Amaranthaceae. I. Anatomy of *Alternanthera sessilis* R. Br. J Ind Bot Soc 10:213–231
Joshi AC (1931b) Contributions to the anatomy of Chenopodiaceae and Amaranthaceae. II. Primary vascular system of *Achyranthes aspera* L., *Cyathula prostrata* Blume and *Pupalia lappacca* Juss. J Ind Bot Soc 10:265–291

Joshi AC (1934) Variations in the medullary bundles of *Achyranthes aspera* L. and the original home of the species. New Phytol 33:53–57
Joshi AC (1937) Some salient points in the evolution of the secondary vascular cylinder in the Amaranthaceae and Chenopodiaceae. Am J Bot 24:3–9
Kirchoff BK, Fahn A (1984) The primary vascular system and medullary bundle structure of *Phytolacca dioica* (Phytolaccaceae). Can J Bot 62:2432–2440
Larson PR (1975) Development and organization of the primary vascular system in *Populus deltoides* according to phyllotaxy. Am J Bot 62:1084–1099
Mabry TJ (1977) The order Centrospermae. Ann Mo Bot Gard 64:210–220
Maheshwari P (1930) Contributions to the morphology of *Boerhaavia diffusa* (II). J Ind Bot Soc 9:42–61
Metcalfe CR (1987) Anatomy of the dicotyledons, 2nd edn, vol 3. Magnoliales, Illiciales, and Laurales. Clarendon Press, Oxford
Metcalfe CR, Chalk L (1950) Anatomy of the dicotyledons. Clarendon Press, Oxford
Metcalfe CR, Chalk L (1983) Anatomy of the dicotyledons, 2nd edn, vol 2. Wood structure and conclusion of the general information. Clarendon Press, Oxford
Mikesell JE (1979) Anomalous secondary thickening in *Phytolacca americana* L. (Phytolacceceae). Am J Bot 66:997–1005
Mikesell JE, Popham RA (1976) Ontogeny and correlative relationship of the primary thickening meristem in four-o'clock plants (Nyctaginaceae) maintained under long and short photoperiods. Am J Bot 63:427–437
Millner ME (1934) Anatomy of *Silene vulgaris* and *Silene maritima* as related to ecological and genetic problems. New Phytol 33:77–95
Morris DA, Thomas AG (1978) A microautoradiographic study of auxin transport in the stem of intact pea seelings (*Pisum sativum* L.). J Exp Bot 29:147–157
Pant DD, Mehra B (1961) Nodal anatomy of *Boerhaavia diffusa* L. Phytomorphology 11:385–405
Pate JS, Gunning BES (1969) Vascular transfer cells in angiosperm leaves. A taxonomic and morphological survey. Protoplasma 68:135–156
Payne MA (1933) Morphology and anatomy of *Mollugo verticillata* L. Univ Kans Sci Bull 21:399–419
Pfeiffer H (1926) Das abnorme Dickenwachstum. In: Linsbauer K (ed) Handbuch der Pflanzenanatomie, vol XV, pt 9. Borntraeger, Berlin
Philipson WR, Balfour EE (1963) Vascular patterns in dicotyledons. Biol Rev Camb Philos Soc 24:21–50
Philipson WR, Ward JM (1965) The ontogeny of the vascular cambium in the stem of seed plants. Biol Rev 40:534–579
Prabhakar M, Ramayya N (1979) Anatomical studies in the Indian Portulacaceae. I. Stem wood. Ind J Bot 2:96–106
Pulawska Z (1973) The parenchymo-vascular cambium and its derivative tissues in stems and roots of *Bougainvillea glabra* Choisy (Nyctaginaceae). Acta Soc Bot Pol 42:41–61
Raj DN, Nagar SP (1980) On medullary bundles of *Achyranthes aspera* L. Flora 169: 530–534
Raj DN, Nagar SP (1989) Primary vascular differentiation in *Achyranthes aspera* L. Flora 183:327–335
Rauh W (1961) Weitere Untersuchungen an Didiereaceen. I. Teil: Beitrag zur Kenntnis der Wuchsformen unter besonderer Berücksichtigung neuer Arten. Sitzungsb Heidelberger Akad Wiss Math-Naturwiss Kl 7. Abh, pp 1–118
Rodman JE, Oliver MK, Nakamura RR, McClammer JU, Bledsoe AH (1984) A taxonomic analysis and revised classification of Centrospermae. Syst Bot 9:297–323
Rogers GK (1985) The genera of Phytolaccaceae in the southeastern United States. J Arnold Arbor 66:1–39
Rubery PH (1981) Auxin receptors. Annu Rev Plant Phys 32:569–596
Sanio C (1863) Vergleichende Untersuchungen über die Zusammensetzung des Holzkörpers. Bot Zeit 21:377–387

Schleiden MJ (1845) Beiträge zur Anatomie der Cacteen. Mem Acad Imp Sci St Petersbourg Sér 6, 4:335–380
Shininger TL (1979) The control of vascular development. Annu Rev Plant Phys 30:313–337
Skipworth JP (1962) The primary vascular system and phyllotaxis in *Hectorella caepitosa* Hook. f. N Z J Sci 5:253–258
Solereder H (1908) Systematic anatomy of the dicotyledons. Clarendon Press, Oxford
Srivastava LM, Bailey IW (1962) The comparative anatomy of the leaf-bearing Cactaceae. V. The secondary phloem. J Arnold Arbor 43:234–278
Stevenson DW, Popham RA (1973) Ontogeny of the primary thickening meristem in seedlings of *Bougainvillea spectabilis*. Am J Bot 60:1–9
Studholme WP, Philipson WR (1966) A comparison of the cambium in two woods with included phloem: *Heimerliodendron brunonianum* and *Avicennia resinifera*. N Z J Bot 4:355–365
Thorne RF (1983) Proposed new realignments in the angiosperms. Nord J Bot 3:85–117
Webster JS, Wilson RC (1980) Anatomical diversity in roots of seven species of *Abronia* from California and its ecological implications. Aliso 9:567–579
Wheat D (1977) Successive cambia in the stem of *Phytolacca dioica*. Am J Bot 64: 1209–1217
Wilson CL (1924) Medullary bundle in relation to primary vascular system in Chenopodiaceae and Amaranthaceae. Bot Gaz 78:175–199
Yarrow GL, Popham RA (1981) The ontogeny of the primary thickening meristem of *Atriplex hortensis* L. (Chenopodiaceae). Am J Bot 68:1042–1049
Zamski E (1980) Vascular continuity in the primary and secondary tissues of *Bougainvillea* (Nyctaginaceae). Ann Bot 45:561–567
Zamski E, Azenkot A (1981) Sugarbeet vasculature. I. Cambial development and three-dimensional structure of the vascular system. Bot Gaz 142:334–343

4 Epicuticular Wax Ultrastructure and Systematics

WILHELM BARTHLOTT

4.1 Introduction

Based on scanning electron microscopy (SEM) techniques a vast amount of systematically relevant information on the structure and composition of cuticular surfaces has been accumulated (Behnke and Barthlott 1983; Juniper and Jeffree 1983; Barthlott 1990). At present no cuticular feature is known as an apomorphy to characterize the Caryophyllales as an order. However, there are certain structures to delimit taxa between subfamiliar and familiar level. Well-known examples are the "bladder cells" of certain higher taxa in the Aizoaceae (Hartmann 1991) or depressed cell-junctions, a synapomorphy in the seed coats of most Cactaceae (Barthlott and Voit 1979; Barthlott and Hunt 1993).

Epicuticular "waxes" are chemically-diverse solid lipophilic substances which occur throughout all major taxa of angiosperms. They often cause the "glaucous" appearance of many plants. Chemically these waxes consist of alkanes, long-chain alcohols, ketones, and esters of long-chain fatty acids; cyclic compounds like phytosterols, pentacyclic triterpenoids, and epicuticular flavonoids occur more frequently than was previously supposed (Kolattukudy 1980). Structurally, SEM methods have revealed a high diversity in the micromorphology of these substances, mostly occurring as microscopic crystalloids on the plant surface (Barthlott and Wollenweber 1981). They had never been thought to be of systematic significance. However, high-resolution SEM of the orientation and ultrastructure of the crystalloids allows the characterization and delimitation of taxa between the family and subclass levels in some groups and has proven to be a new character in angiosperm systematics (Barthlott 1990).

A first detailed survey of centrospermous epicuticular waxes was published in an earlier paper (Engel and Barthlott 1988), which still provides the main data-base for the following conclusions. Some of the results published in 1988 must be reinterpreted today on the basis of two aspects. First, more recent examinations of many taxa using the highest resolution reveals that many, or possibly all, crystalloid types thought to be massive and described earlier as "rodlets" are obviously hollow tubular structures (tubules). Second, possibly related orders of the Ranunculidae, Magnoliidae, and Dilleniidae have been examined systematically since 1988 and these results allow a better interpretation of the data found in the Caryophyllales with reference to possibly related subclasses.

The following conclusions are based on the examination by high-resolution SEM of some 550 species of 225 genera of all families in the Caryophyllales; they are interpreted in comparison with data of about 11 000 species of angiosperms analysed by the same method over the last 15 years.

4.2 Wax Ultrastructure of Caryophyllales

The circumscription of the families follows Cronquist (1988). The numbers following the family name in brackets indicate the number of genera and species examined within the family (gen./spp.). Altogether, 550 species from 225 genera, representing all families involved, were analysed by high-resolution SEM. The "high degree of seemingly patternless variation" within the Centrospermae (Hershkovitz 1989) applies to some extent also for the micromorphology of epicuticular waxes of several families, whereas others are very uniform and homogeneous (compare, e.g., Amaranthaceae with Chenopodiaceae). The main types of wax crystalloids are illustrated by SEM micrographs (Figs. 4.1 and 4.2) and additional schematic drawings are provided in Frölich and Barthlott (1988).

Achatocarpaceae (2/5). Dense irregular platelets; almost rosette-like orientation. Resembling Phytolaccaceae.

Agdestidaceae. See Phytolaccaceae.

Aizoaceae (28/55). Epicuticular waxes highly diverse (survey in Hartmann 1983). Platelets (sometimes regularly arranged) are less frequent than rodlets (tubules), which may be curled or ribbon-like flattened (Fig. 4.1c), a character occurring also in the Cactaceae and Didereaceae (Fig. 4.2e and f). Platelets are less frequent within this family. *Wooleya* (Ihlenfeldt and Hartmann 1982) shows a unique type of massive compound crystalloids similar to the Strelitzia type in monocotyledons (Fig. 4.2c). Seed coats of many Ruschioideae and Mesembryanthemoideae show a peculiar type of large massive rodlets not soluble in organic solvents (Ehler and Barthlott 1978), an epicuticular structure which is probably unique within angiosperms.

Amaranthaceae (12/22). No epicuticular wax crystalloids known: the absence is possibly a family character and delimitation from Chenopodiacae.

Basellaceae (4/9). Epicuticular wax crystalloids absent.

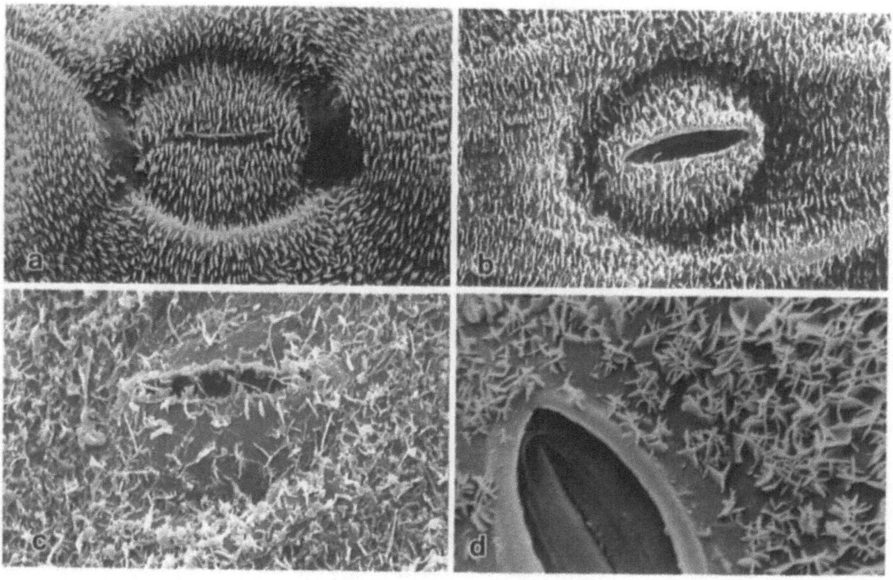

Fig. 4.1a–d. Epicuticular wax at low magnifications. **a** *Convallaria majalis* L. (Convallariaceae), oriented wax platelets (Convallaira type) around stoma; magn. ×1900. **b** *Dianthus carthusianorum* L. (Caryophyllaceae), oriented platelets around stoma; magn. ×1380 (Engel and Barthlott 1988). **c** *Oscularia deltoides* (L.) Schwant. (Aizoaceae), non-oriented wax ribbons and curled rodlets; magn. ×1290 (Engel and Barthlott 1988). **d** *Cercis siliquastrum* L. (Fabaceae), rosette-like clusters of wax platelets (Fabales type) characteristic for all waxy Fabales; magn. ×3200

Cactaceae (80/270). Epicuticular waxes highly diverse. Common are curled ribbons or rodlets (tubules ?) with all transitions (Fig. 4.2e) into irregular platelets (e.g. glaucous appearance of many cereoid genera). Massive incrustations of the cuticle with wax (continuous wax – cutin layers with lamellate ultrastructure) are common in highly xeromorphic taxa (e.g. *Ariocarpus* and *Copiapoa*).

Caryophyllaceae (25/51). Irregular-shaped platelets are common; occasionally they show a parallel orientation around the stomata (e.g. *Dianthus carthusianorum*, Fig. 4.1b) similar to the Convallaria waxes of many liliiflorous families of monocotyledons (Fig. 4.1a). Rodlets are less frequent, and high resolution always reveals them as tubes (possibly formed by densely spiralled band-like structures). In a few taxa (e.g. *Silene*, Fig. 4.2b) transversely ridged rodlets occur similar to the Aristolochia type (Fig. 4.2a) characteristic for most Magnoliidae. *Illecebrum* has a unique type of elongated platelets and *Herniaria* is unwaxy.

Chenopodiaceae (20/38). Almost uniformly well characterized by very small platelets, often with parallel orientation. In some species platelets in clusters

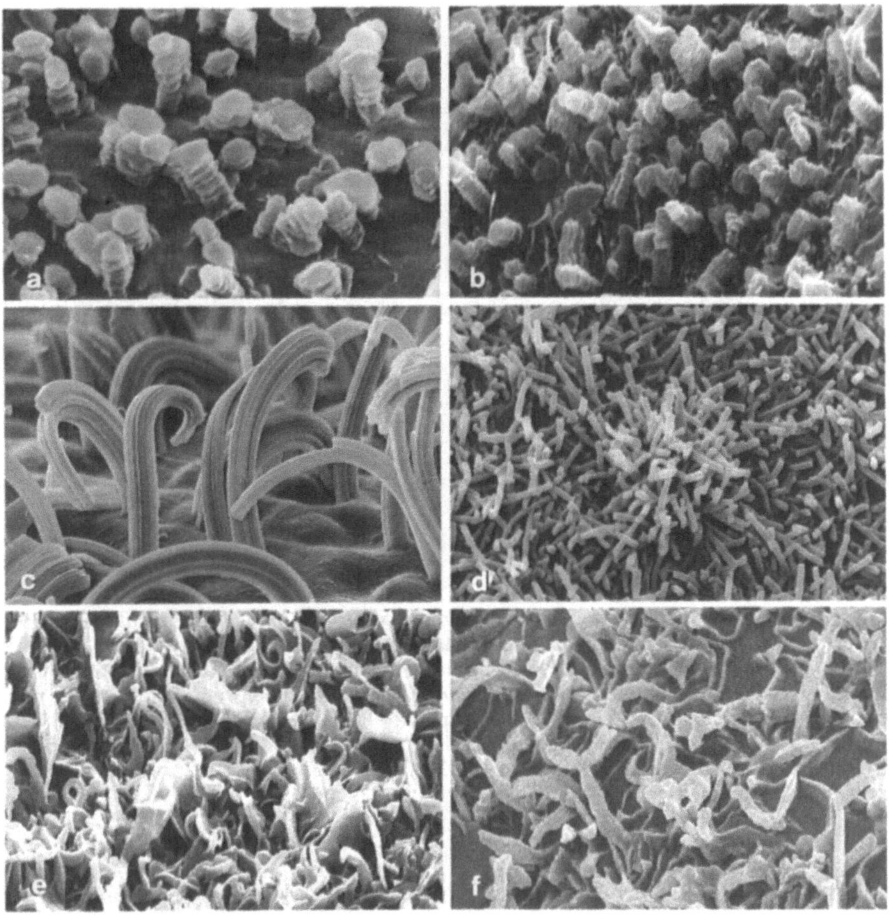

Fig. 4.2a–f. Details of epicuticular wax. **a** *Annona squamosa* L. (Annonaceae), transversally ridged wax rodlets (Aristolochia type), characteristic for the core groups of Magnoliidae; magn. ×13350. **b** *Silene bupleuroides* L. (Caryophyllaceae), transversally ridged wax rodlets; magn. ×6800. **c** *Phenakospermum guyanense* (L. Richter) Endl. ex Miq. (Strelitziaceae), compound massive wax rodlets with longitudinal striations (Strelitzia type); magn. ×1600. **d** *Thalictrum flavum* L. (Ranunculaceae), clustered wax tubules (Berberis type), characteristic for all waxy Ranunculidae; magn. ×7260. **e** *Parodia magnifica* (Ritter) F. Brandt, ribbons to curled rodlets and platelets; magn. ×4500 (Engel and Barthlott 1988). **f** *Didierea trollii* Cap. & Rauh (Didiereaceae), ribbons and curled rodlets; magn. ×6900 (Engel and Barthlott 1988)

occur, resembling Phytolaccaceae and Achatocarpaceae. *Dysphania* is without wax crystalloids.

Didiereaceae (4/10). Irregular platelets and curled ribbons (Fig. 4.2f) similar to Aizoaceae and Cactaceae.

Dysphaniaceae. See Chenopodiaceae.

Gisekiaceae. See Phytolaccaceae.

Hectorellaceae. See Portulacaceae.

Molluginaceae (10/18). Platelets and rodlets (tubules) are frequent and similar to those of the Caryophyllaceae.

Nyctaginaceae (12/20). Epicuticular crystalloids absent; only eroded platelets (herbarium artifact?) present in *Oxybaphus* and *Pisonia*.

Phytolaccaceae (16/25). Uniformly irregular platelets in all taxa examined (incl. *Agdestis*, *Gisekia*, *Lophiocarpus* and *Stegnosperma*).

Portulacaceae (12/28). Epicuticular crystalloids mostly absent. *Ceraria*, *Portulacaria*, and *Hectorella* with large platelets not found elsewhere in the Caryophyllales.

Stegnospermataceae. See Phytolaccaceae.

4.3 Relations Within the Order

The Caryophyllales show a wide structural range of epicuticular wax crystalloids. One family (Amaranthaceae) seems to be characterized by the absence of these microsculptures. The following discussion follows the classification of Takhtajan (1980) in three suborders.

Caryophyllineae. The two anthocyanin-bearing families Caryophyllaceae and Molluginaceae closely resemble each other by the frequent occurrence of irregular platelets and rodlets (tubules). The Caryophyllaceae show in a few taxa two features they share with possibly related subclasses (see Sect. 4.4) which could be considered as a plesiomorphic condition. This would emphasize a basal or "ancestral" position of the family within the order. *Illecebrum* differs from the Caryophyllaceae by the elongated platelets.

Chenopodiineae. The two families Amaranthaceae and Chenopodiaceae are easily distinguished and both are uniform. Amaranthaceae and *Dysphania* lack epicuticular crystalloids, while Chenopodiaceae are characterized by very small regular platelets. The clustering of platelets in some Chenopodiaceae distantly resembles certain Phytolaccaceae and Achatocarpaceae.

Phytolaccineae. Large irregular platelets characterize the Phytolaccaceae (incl. *Agdestis*, *Gisekia*, *Lophiocarpus*, and *Stegnosperma*); Achatocarpaceae do not differ from Phytolaccaceae. The Basellaceae and probably also Nyctaginaceae lack epicuticular crystalloids. Most Portulacaceae are also unwaxy; if crystalloids are present (*Ceraria*, *Portulacaria*, and *Hectorella*) they are large platelets unique in the Caryophyllales.

The Aizoaceae, Cactaceae, and Didiereaceae show a high diversity of epicuticular wax crystalloids; however, they share many features. The peculiar type of "curled ribbons" and irregular platelets (Figs. 4.1c and 4.2e and f) circumscribes the three families and could be considered a derived synapomorphic condition. On the other hand, it might represent only an adaptation as a part of the xeromorphic succulent syndrome; however, these features are lacking in the Portulacaceae.

4.4 Wax Ultrastructure and Postition of the Order

In some cases wax crystalloid ultrastructure enables a characterization of superorders and subclasses within the angiosperms (survey in Barthlott 1990). For an interpretation of the structures observed in the Caryophyllales a short characterization of the possibly related subclasses is provided.The data are based on our current studies for forthcoming treatments of the angiosperm wax crystalloid ultrastructure. Several of the smaller families formerly associated with the Caryophyllales (e.g. Plumbaginaceae and Polygonanceae) are described in more detail in Engel and Barthlott (1988).

The occurrence of systematically relevant wax crystalloid types is plotted against the classification of Dahlgren (1989) in Fig. 4.3 and against an alternative spatial diagram of angiosperm classification (Fig. 4.4) in Fig. 4.5. These diagrams provide for the first time general information regarding the distribution of different wax types in angiosperms.

Plumbaginaceae and Polygonaceae. Very unspecific types of platelets and rodlets occur throughout the two families. No elements which might be considered characteristic for the Caryophyllales occur; the data available at present provide no information concerning the relationship of these families.

Monocotyledoneae. Apart from the common wax types occurring throughout angiosperms (e.g. Arales, Dioscoroeales, and Orchidales), the Liliopsida are characterized by the occurrence of two specific types (Barthlott and Frölich 1983; Frölich and Barthlott 1988 and unpubl. data). Commelinidae, Zingiberidae, and Arecidae are characterized by the occurrence of compound rodlets with longitudinal striations (Strelitzia type, Fig. 4.2c); this wax type was never observed in the Caryophyllales (only *Wooleya* in the Aizoaceae bears a distant resemblance). In contrast, most liliiflorous families in the orders Liliales, Asparagales, Burmanniales, and also Triuridales (placed within the Liliidae in Fig. 4.4) are characterized by platelets strictly oriented in one direction (Convallaria type, Fig. 4.1a). Within the "Polycarpicae" and "Apetalae" oriented platelets of a slightly modified Convallaria type occur only in a few Caryophyllaceae (Fig. 4.1b) and Chenopodiaceae, and thus possibly indicate a distant relationship to the liliiflorous monocotyledons.

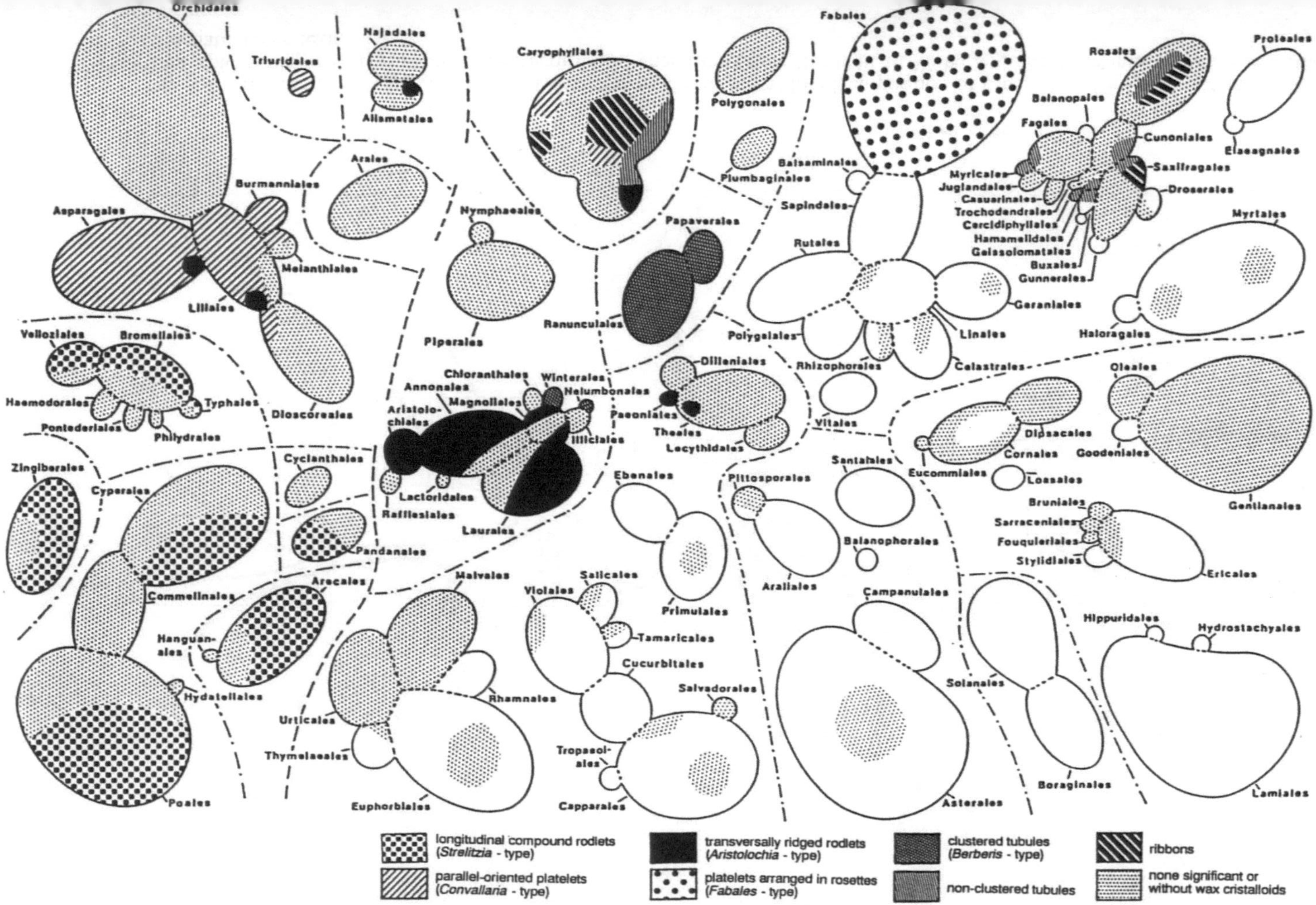

Fig. 4.3. Distribution of major types of epicuticular wax crystalloids (see Figs. 4.1 and 4.2) in the angiosperm classification diagram of Dahlgren (1989)

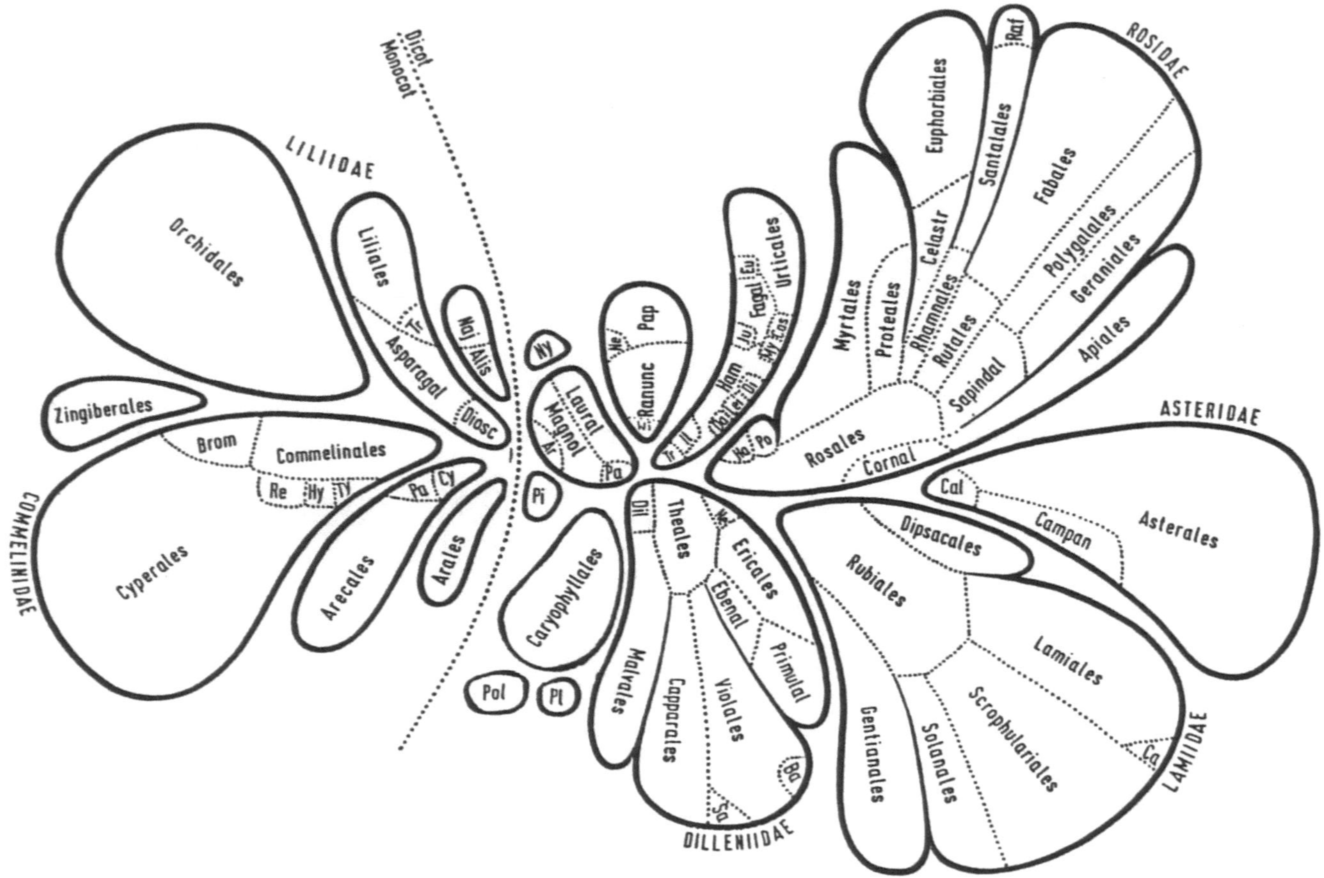

Fig. 4.4. An alternative spatial diagram of angiosperm classification. Modified position of orders based mainly on data available from epicuticular wax crystalloid ultrastructure and chemistry (after Barthlott in Weberling and Schwantes 1992). For further explanation, see text

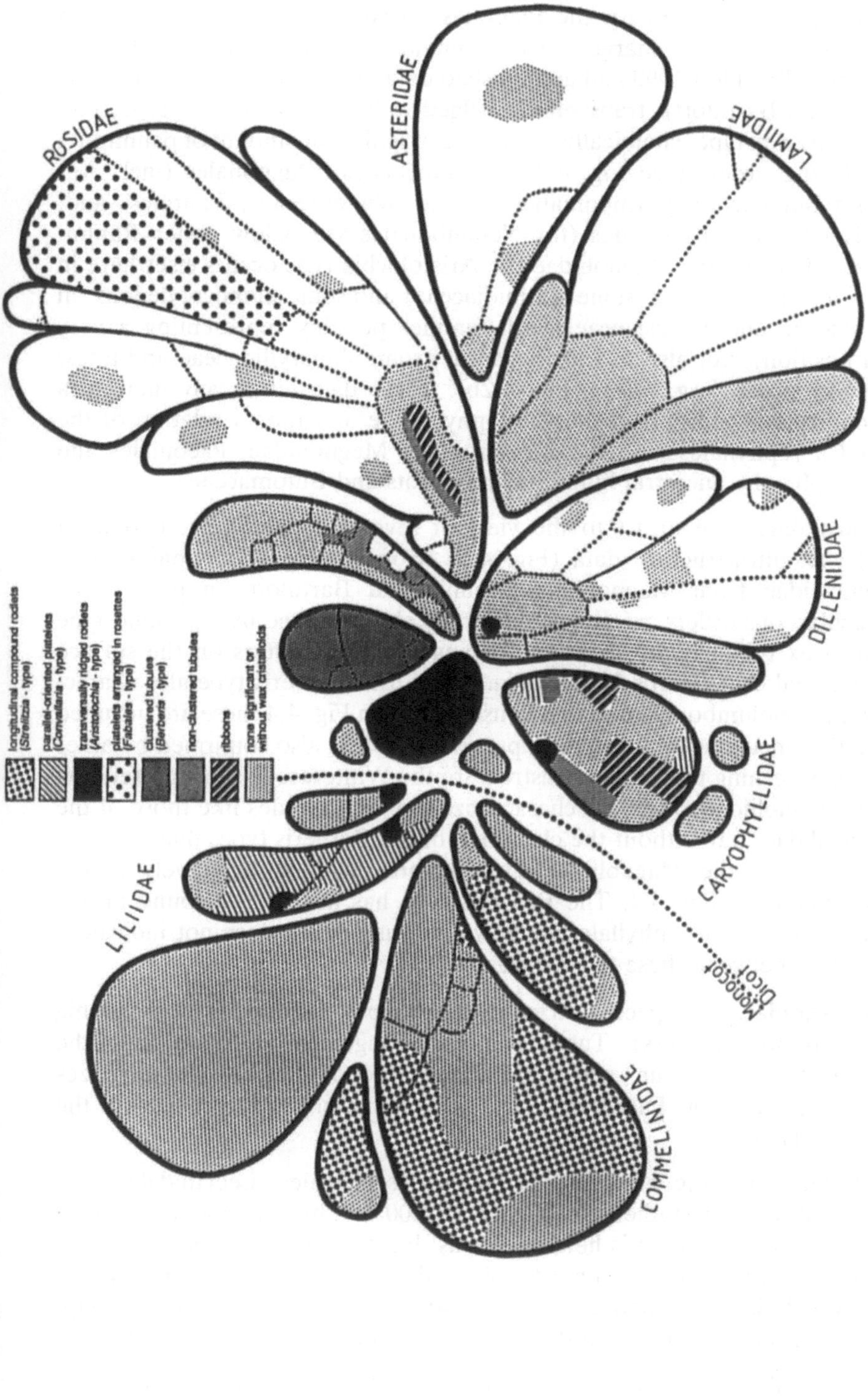

Fig. 4.5. Distribution of the major types of epicuticular wax crystalloids, based on Fig. 4.4

Magnoliidae s.s. Except for the Piperales and Nymphaeales the core group of Magnoliidae s.s. is characterized by unique epicuticular wax crystalloids (data from Barthlott 1990 and unpublished data by Ditsch, Hennig, Leistner, Meusel and Barthlott): transversally ridged rodlets (Aristolochia type, Fig. 4.2a). This wax type, chemically containing a significant amount of palmitone, delimits the orders (see Fig. 4.4) Aristolochiales, Magnoliales (incl. Annonales but excluding Eupomatiaceae and Winteraceae), Laurales, and, surprisingly, also Paeoniales (for Nelumbonales see below under Ranunculidae). Outside the Magnoliidae the Aristolochia type occurs elsewhere in only a few Theales (e.g. some Actinidiaceae) and some monocotyledons: in Butomaceae and Philesiaceae (both families possibly representing a very basal position), but also in Luzuriagaceae, some Amaryllidaceae and a few Caryophyllaceae (e.g. *Silene*, Fig. 4.2b). These data tentatively indicate a plesiomorphic condition in the Caryophyllaceae and thus a relation of the order Caryophyllales with the core group of Magnoliidae, Paeoniales and Theales plus the ancestral liliiflorous monocots and Butomaceae.

Ranunculidae. In contrast to the view of several authors (e.g. Cronquist 1988) wax ultrastructure data (Fig. 4.5) strongly support a separation of Ranunculidae from Magnoliidae (Hennig and Barthlott, in prep.) The ranunculaceous orders are almost uniformly characterized by a peculiar type of fine wax tubules with a specific arrangement in clusters on the surface of epidermal cells (Berberis type, Fig. 4.2d). The Berberis type also characterized the Nelumbonales, so that consequently in Fig. 4.4 these are included in the Ranunculidae adjacent to Papaverales (a view also confirmed by other data). Concerning their wax ultrastructure the Winteraceae show no relation to the Magnoliidae. They are characterized by wax tubules like those of the Ranunculidae, but without the clustering of the Berberis type; thus they are removed from the Magnoliidae s.s. and are included tentatively in the Ranunculidae in Fig. 4.4. The Berberis type has never been found in any member of the Caryophyllales; thus wax ultrastructure does not indicate a connection between these two subclasses.

Hamamelididae. As expected waxes in this group are heterogeneous (Hennig and Barthlott, in press). Tubules in an arrangement different from the Berberis type do occur, and a particular type of platelets characterizes several families (e.g. Fagaceae). No connection can be discerned with the Caryophyllales.

Dilleniidae. Only the basal orders Dilleniales, Theales, Lecythidales, and Malvales have been studied thoroughly (ca. 800 spp. by Ditsch and Barthlott, in press). The wax cover is heterogeneous, but the occasional occurrence of the Aristolochia type and almost ribbon-like crystalloids in a few Theales tentatively indicates a connection of the Caryophyllales with this basal group. This view is also expressed in the phylogenetic diagram of Thorne (1992).

Rosidae. Only a few orders have been studied in detail (e.g. Fehrenbach and Barthlott 1988). Some orders, like the Fabales, are extremely uniform

(Fabales type, Fig. 4.1d). Ribbon-like crystalloids in certain Rosales s.l. (e.g. Crassulaceae and Rosaceae) and slightly arranged platelets in the Pittosporaceae and Brunelliaceae are a feature shared with some Caryophyllales. If there should be a relation between the Rosidae and Caryophyllidae we would expect it here; however, it is questionable if these data reflect more than convergence.

Acknowledgements. I am indebted to Gertrud Dahlgren (Lund) for permission to reproduce the angiosperm classification system in Fig. 4.3 and in particular to Thomas Engel (Berlin) for providing many of the centrospermous wax data and several SEM micrographs. Heidrun EK Hartmann (Hamburg) has contributed important data concerning the waxes of Aizoaceae. Many of the unpublished results are part of forthcoming papers by students of my own SEM laboratory at the University of Bonn: Friedrich Ditsch (Dilleniidae), Sabine Hennig (Magnoliidae, Ranunuculidae, and Hamamelididae), Iris Meusel (chemistry of Aristolochia and Strelitzia waxes), Christoph Neinhuis and Hans-Jürgen Ensikat (high-resolution SEM); special thanks to Ingeborg Theisen (Lamiidae and Asteridae) who also has prepared data for the diagrams.

References

Barthlott W (1990) Scanning electron microscopy of the epidermal surface in plants. In: Claugher D (ed) Application of the scanning EM in taxonomy and functional morphology. Clarendon Press, Oxford, pp 69–94

Barthlott W, Frölich D (1983) Mikromorphologie und Orientierungsmuster epicuticularer Wachs-Kristalloide: Ein neues sytematisches Merkmal bei Monokotylen. Plant Syst Evol 142:171–185

Barthlott W, Hunt DR (1993) Cactaceae. In: Kubitzki K (ed) The families and genera of vascular plants, vol 2. Springer, Berlin Heidelberg New York, pp 161–197

Barthlott W, Voit G (1979) Mikromorphologie der Samenschalen und Taxonomie der Cactaceae: Ein raster-elektronenmikroskopischer Überblick. Plant Syst Evol 132: 205–229

Barthlott W, Wollenweber E (1981) Zur Feinstruktur, Chemie und taxonomischen Signifikanz epicuticularer Wachse und ähnlicher Sekrete. Trop Subtrop Pflanzenwelt 32. Steiner, Wiesbaden

Behnke H-D, Barthlott W (1983) New evidence from the ultrastructural and micromorphological fields in angiosperm classification. Nord J Bot 3:43–66

Cronquist A (1988) The evolution and classification of flowering plants, 2nd edn. New York Botanical Garden, New York

Dahlgren G (1989) An updated angiosperm classification. Bot J Linn Soc 100:197–203

Ditsch F, Barthlott W Mikromorphologie der Epicuticularwachse und die Systematik der Dilleniales, Lecythidales, Malvales und Theales. Trop Subtrop Pflanzenwelt, Steiner, Wiesbaden (in press)

Ehler N, Barthlott W (1978) Die epicuticulare Skulptur der Testa-Zellwände einiger Mesembryanthemaceae. Bot Jahrb Syst 99:329–340

Engel T, Barthlott W (1988) Micromorphology of epicuticular waxes in centrosperms. Plant Syst Evol 161:71–85

Fehrenbach S, Barthlott W (1988) Mikromorphologie der Epicuticularwachse der Rosales s.l. und deren systematische Gliederung. Bot Jahrb Syst 109:407–428

Frölich D, Barthlott W (1988) Die Mikromorphologie der epicuticularen Wachse und das System der Monocotylen. Trop Subtrop Pflanzenwelt 63. Steiner, Wiesbaden

Hartmann HEK (1983) Untersuchungen zum Merkmalsbestand und zur Taxonomie der Subtribus Leipoldtiinae (Mesembryanthemaceae). Bibl Bot 136:1–81

Hartmann HEK (1991) Mesembryanthema. Contrib Bolus Herb 13:75–157
Hennig S, Barthlott W Mikromorphologie der Epicuticularwachse und das System der Magnoliidae, Ranunculidae und Hamamelididae. Trop Subtrop Pflanzenwelt, Steiner, Wiesbaden (in press)
Hershkovitz MA (1989) Phylogenetic studies in Centrospermae: A brief appraisal. Taxon 38:602–610
Ihlenfeldt HD, Hartmann HEK (1982) Leaf surfaces in Mesembryanthemaceae. In: Cutler DF, Alvin FL, Price CE (eds) The plant cuticle. Academic Press, London, pp 397–423
Juniper BE, Jeffree CE (1983) Plant surfaces. Arnold, London
Kolattukudy PE (1980) Cutin, suberin, and waxes. In: Stumpf PK (ed) The biochemistry of plants, vol 4. Academic Press, London, pp 571–645
Takhtajan A (1980) Outline of the classification of flowering plants (Magnoliophyta). Bot Rev 46:225–359
Thorne RF (1992) An updated phylogenetic classification of the flowering plants. Aliso 13:365–389
Weberling F, Schwantes HO (1992) Pflanzensystematik, 6th edn. Ulmer, Stuttgart

5 Sieve-Element Plastids: Their Significance for the Evolution and Systematics of the Order

H.-Dietmar Behnke

5.1 Introduction

Systematic ultrastructural studies of sieve-element plastids were initiated in the monocotyledons (Behnke 1968) and have, over the last 25 years, been extended to the entire angiosperms (Behnke 1981, 1991), now covering some 2500 species from almost 500 families.

A detailed description and careful comparison of qualitative and quantitative data, obtained by ultrathin sectioning and transmission electron microscopy of sieve-element plastids (see Behnke 1991 for methods), furnish the systematist with a set of characters of potential use in the classification of higher taxa (cf. Behnke 1988a,b, 1989).

Early reports on peculiar plastids in *Tetragonia tetragonioides* O. Kuntze (*T. expansa* Murr.; Falk 1964) and *Beta vulgaris* L. (Esau 1965) prompted an investigation of 17 species in the Caryophyllales (Behnke 1969), making this order one of the first dicotyledon taxa to be systematically screened for their sieve-element plastids. The continued effort towards an enlargement of the number of investigated species (Behnke and Turner 1971; Behnke et al. 1975, 1983a,b; Behnke 1976a,b, 1993), an incorporation of critical genera as they became available (Behnke 1974a, 1975a, 1982a; Behnke et al. 1974; Hunziker et al. 1974; Mabry et al. 1976), and an extension of sieve-element plastid studies to putatively related taxa (Behnke and Turner 1971; Behnke 1975b, 1976b,c; Behnke and Mabry 1977) eventually resulted in our present knowledge of the sieve-element plastids of almost 300 species, representing all families of the Caryophyllales and most of their sometimes allied groups.

5.2 The Sieve-Element Plastid Characters

Sieve elements, while differentiating from meristematic initials into mature conduits for long-distance assimilate transport, undergo drastic structural changes, sometimes compared to a selective autophagy (Evert 1990). Their

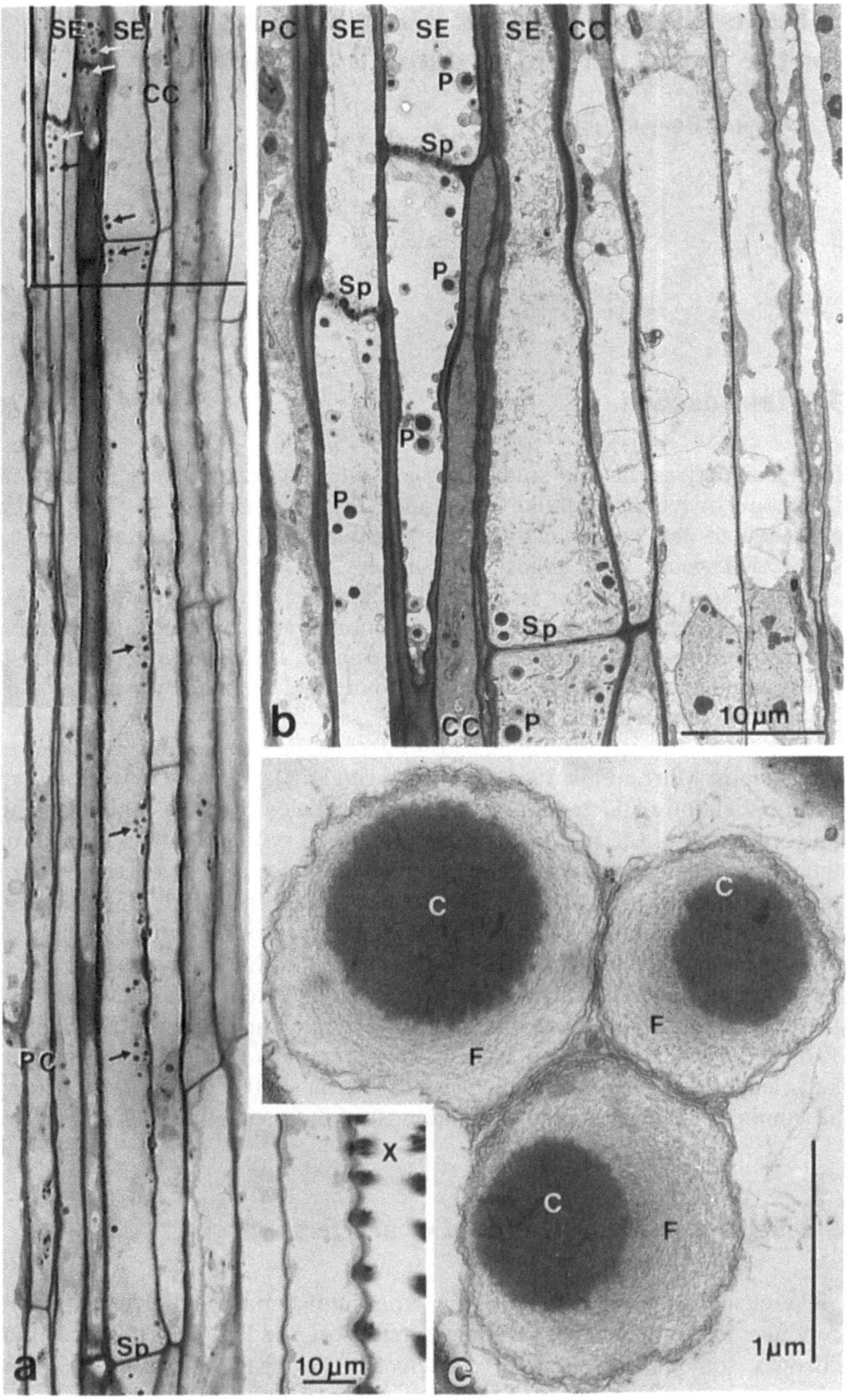
SE
SE
CC
PC
Sp
a
PC
SE
SE
SE
CC
P
Sp
P
Sp
P
P
Sp
b
CC
P
10 µm
C
C
F
F
C
F
1 µm
c
X
10 µm

plastids are among the few organelles that are present throughout the entire life span of the cell.

The (pro)plastids of young sieve elements have elongate amoeboid shapes and are characterized by a dense matrix and a few thylakoids. Often, the accumulation of ergastic material (protein or starch) indicates that the specific differentiation of sieve-element plastids has been initiated. This prominently applies to plastids of the Caryophyllales whose distinctive features, such as protein filaments or globular or polygonal protein crystals, are conspicuous very early on during plastid development (Figs. 5.1b, 5.2, 5.4 and 5.8). A gradual depletion of the matrix, an increase of starch and/or protein deposits, and the turning into spherical shapes are further steps associated with the differentiation of sieve-element plastids (cf. Behnke 1990).

Sieve-element plastids are stable organelles that only break down during final degeneration of the cell. Sampling different vegetative and generative parts of a few plants indicates that only one plastid form is present in the sieve elements of all organs of a given plant and in different individuals of a given species. Moreover, wounding and grafting experiments have demonstrated that the specificity of sieve-element plastids is genetically determined and part of a programme that is expressed wherever a sieve element is formed (see Behnke and Schulz 1983; Kollmann and Glockmann 1990; Schulz 1990). Therefore, taxonomic conclusions supported by sieve-element plastid characters imply a high degree of reliability.

5.2.1 Forms and Types of Sieve-Element Plastids

Sieve-element plastids of angiosperms are characterized by their inclusions, i.e. protein crystals, protein filaments and starch grains. Different combinations and quantities of these inclusions determine the specific sieve-element plastid possessed by a taxon. Any attempt to assign taxonomic significance to different sieve-element plastids has to define criteria and categories to explain their distinctness.

The form of a sieve-element plastid is the basic unit in the hierarchy of categories. Forms, as redefined by Behnke (1991), are distinct by the presence or absence of protein crystals (c), protein filaments (f) or starch

Fig. 5.1a–c. *Oxybaphus viscosus* L'Herit ex Choisy (Nyctaginaceae). **a** Semi-thin longitudinal section of phloem (stained with toluidin blue) viewed with the light microscope and showing sieve elements (*SE*), companion cells (*CC*) and phloem-parenchyma cells (*PC*). Sieve elements are connected to longitudinal conduits by sieve plates (*SP*) and contain plastids (*arrows*) as prominent markers. *X* xylem vessel. ×700. **b** Ultrathin section following semi-thin section of **a** and showing part of it (part **a**, *above*) enlarged and viewed with the electron microscope. Many chararcteristic plastids (*P*) are present in young (*right SE*) and mature (*left*) sieve elements. ×2000. **c** Three sieve-element plastids from mature cell, each with a peripheral ring of protein filaments (*F*) and a globular central protein crystal (*C*). ×30 000 (s), irrespective of morphological or quantitative differences. Eight qualitatively distinct forms are possible and have been recorded somewhere among the angiosperms (see Fig. 5.12): one form incorporates all three kinds of inclusion (cfs), three forms have two (cf, cs, fs), three have only one inclusion (c, f, s), and one form is devoid of any (o).

The six forms containing protein inclusions are classed as P-type plastids, while the remaining two (s, o) make up the S-type plastids.

Between type and form of sieve-element plastids the category *subtype* was introduced to allow simple reference to related higher taxa whose plastids belong to several forms, but are distinct from other taxa with the same forms by the morphology of their protein inclusions. Subtype-P3 of the Caryophyllales is a prominent example (see Sect. 5.3).

5.2.2 Sizes of Sieve-Element Plastids

Sizes of sieve-element plastids, expressed as averages of diameter measurements, and the amount of their protein and starch contents have been introduced as quantitative characters to help describe and classify sieve-element plastids among magnoliid families (Behnke 1988a). Since then, innumerable measurements taken from more than 2000 species have corroborated the usefulness of the size of sieve-element plastids as an additional character.

Within a species sizes of sieve-element plastids vary little, within the same plant part almost not at all (Behnke, unpubl. results). Therefore, at the family level size calculations should be based on measurements of plastids studied in sieve elements of comparable origin, e.g. from stem phloem. Subject to these precautions, an evaluation of average sizes of sieve-element plastids may be useful in: (1) providing additional information on the homogeneity of a family – uniform sizes in a great number of species (expressed by low standard deviations) are found in presumably monophyletic families; and (2) assigning families a relative position within higher taxa.

5.3 The Distinctive Characters of Sieve-Element Plastids in the Caryophyllales

All members of the Caryophyllales contain P-type sieve-element plastids. Their common distinctive character is a broad peripheral ring of protein filaments (F in Fig. 5.2). This character is order specific, i.e. in this configuration it is found nowhere else within the angiosperms (see Sect. 5.7), and therefore is used to define the subtype-P3 with its forms P3cf, P3cfs, P3f, P3fs (this description replaces the earlier and more complicated formula given in Behnke 1976b).

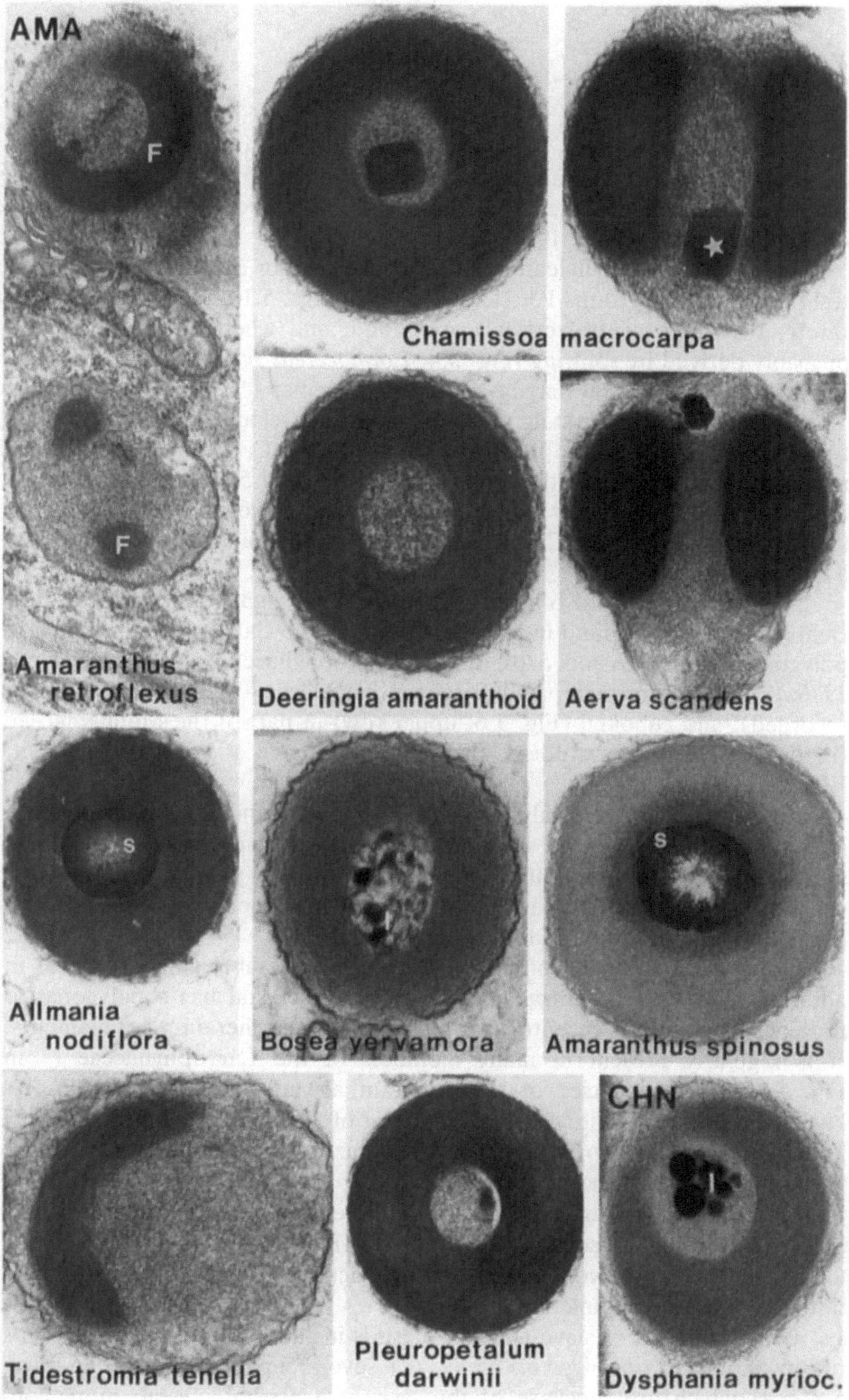

Fig. 5.2. Form-Pf(s) sieve-element plastids of Amaranthaceae and *Dysphania*. The peripheral ring of filaments (*F*) develops very early in young sieve elements (*upper left* micrograph). A small rodlike crystal (*) and lipid droplets (*l*) may be additional inclusions.

For two families, viz. the Amaranthaceae and Chenopodiaceae (incl. *Dysphania* and *Microtea*), the peripheral ring is the only characteristic inclusion found (forms P3f and P3fs; see Figs. 5.2 and 5.3) – a proteinaceous ring had already been detected with the light microscope in sieve elements of *Beta vulgaris* L. (Esau 1934, 1965).

All other caryophyllalean families incorporate in addition a single central protein crystal (forms P3cf and P3cfs; see Figs. 5.1 and 5.4–5.10). Very rarely, two large central crystals are found, and when present these are probably derived by splitting (Fig. 5.4; *Lophocereus schottii* Britt. et Rose). The shape of the crystal may be globular or angular.

Globular central crystals are another unique character, occurring virtually nowhere else in the flowering plants. In the Caryophyllales they are more frequent than the other shapes (Figs. 5.1 and 5.4–5.8). Their edges are not always smooth – they are often undate, serrate, dentate or of another form (e.g. in the Cactaceae; Fig. 5.4) – and hence their shape is not a real sphere.

Angular crystals, also very common among magnoliid families (see Behnke 1988a), are found in two shapes among the Caryophyllales: they are polygonal (Figs. 5.9 and 5.10) in the Caryophyllaceae (see also Behnke 1976a), Achatocarpaceae, and Stegnospermataceae (with no other shape recorded), and cuboidal (Fig. 5.6, upper row) in part of the Molluginaceae (in two of six genera studied, the other four have a globular shape; cf. Behnke et al. 1983b).

The size of the crystals ranges between 0.2 and 1.6 μm. These are maximum values recorded from numerous measurements taken within a species. Single views from ultrathin sections may easily yield incorrect values, especially if the section passed through the periphery of a plastid; this particularly applies to globular shapes.

Quite a few species contain a second, considerably smaller crystal in addition to the large central one. This smaller crystal has a polygonal to rodlike shape and sits eccentrically, closer to the peripheral ring of filaments. It is regularly present in many members of the Caryophyllaceae (* in Fig. 5.10), some species of the Amaranthaceae (* in Fig. 5.2), a few Nyctaginaceae (* in Fig. 5.8), in *Seguieria* of Petiveriaceae (* in Fig. 5.7) and in *Sarcobatus* (* in Fig. 5.3). The taxonomic significance of these findings is not yet clear.

Small starch grains were detected in sieve-element plastids of a few species belonging to several families (s in Figs. 5.2, 5.6–5.8 and 5.10). Obviously, in the Caryophyllales, the production of sieve-element starch is not of primary taxonomic importance.

Other contents of sieve-element plastids include: (1) vesicular remnants of thylakoids (T in Figs. 5.3–5.5); (2) several lipid droplets (l in Figs. 5.2–5.4) which are, because of the lack of a central crystal, more conspicuous in the Amaranthaceae and Chenopodiaceae than in other families; and (3), very rarely, aggregates of phytoferritin (pf in Fig. 5.4; see also Behnke 1971, 1977a, 1978).

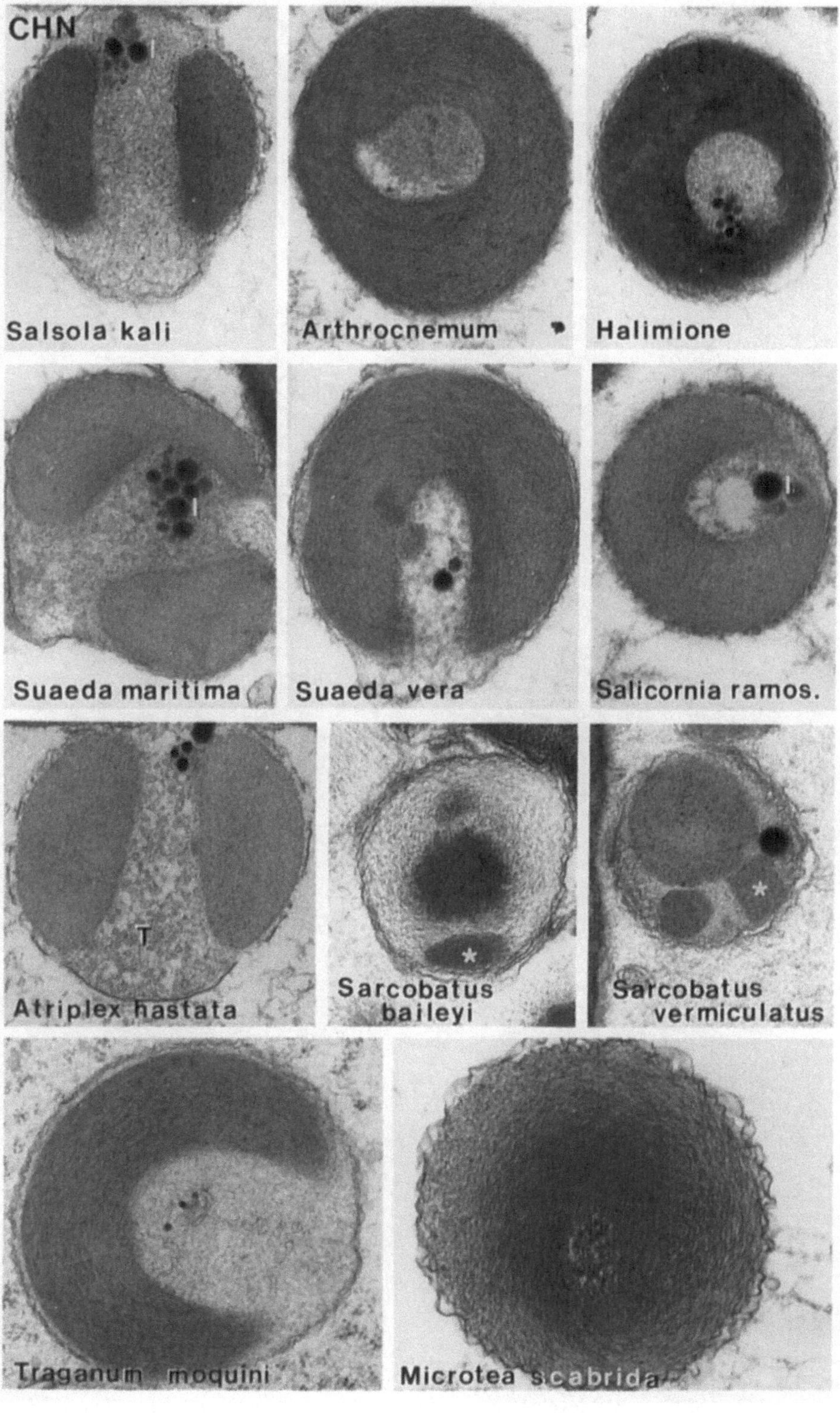

Fig. 5.3. Form-Pf(s) sieve-element plastids of Chenopodiaceae with peripheral ring of filaments sectioned parallel or at right angles to the ring. *Sarcobatus* deviates from all other genera by containing form-P3cf plastids with a globular central and an additional

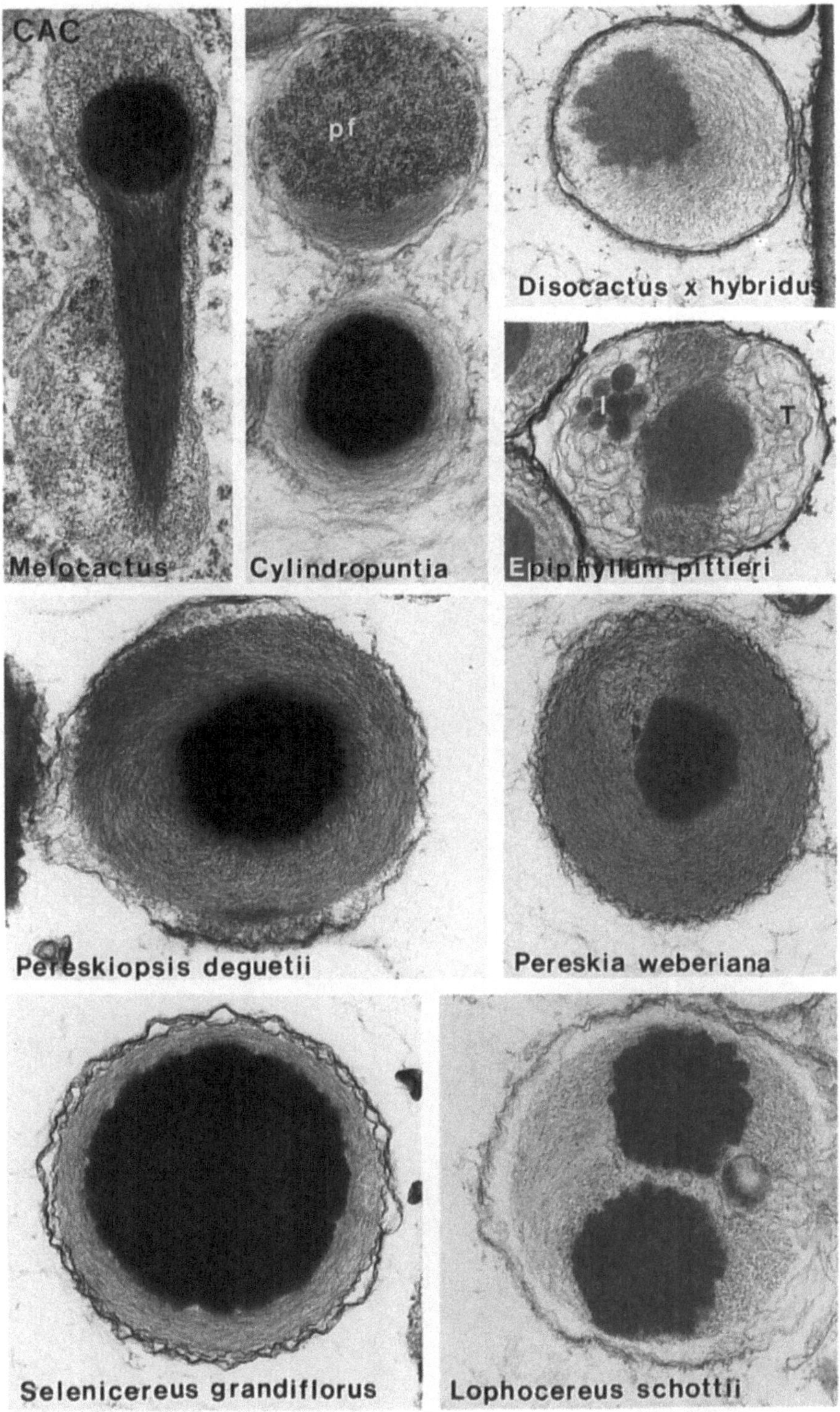

Fig. 5.4. Form-P3cf sieve-element plastids of Cactaceae. Globular central crystal and peripheral ring of filaments develop very early during sieve-element ontogeny (*upper left* micrograph). Two globular crystals are rarely found in *Lophocereus schotti* Britt. et Rose. One of the two plastids depicted of *Cylindropuntia acanthocarpa* (Engelm. et B.) F.M. Knuth contains massive phytoferritin depostis (*pf*). *l* Lipid droplets; T remnants of thylakoids. ×30 000

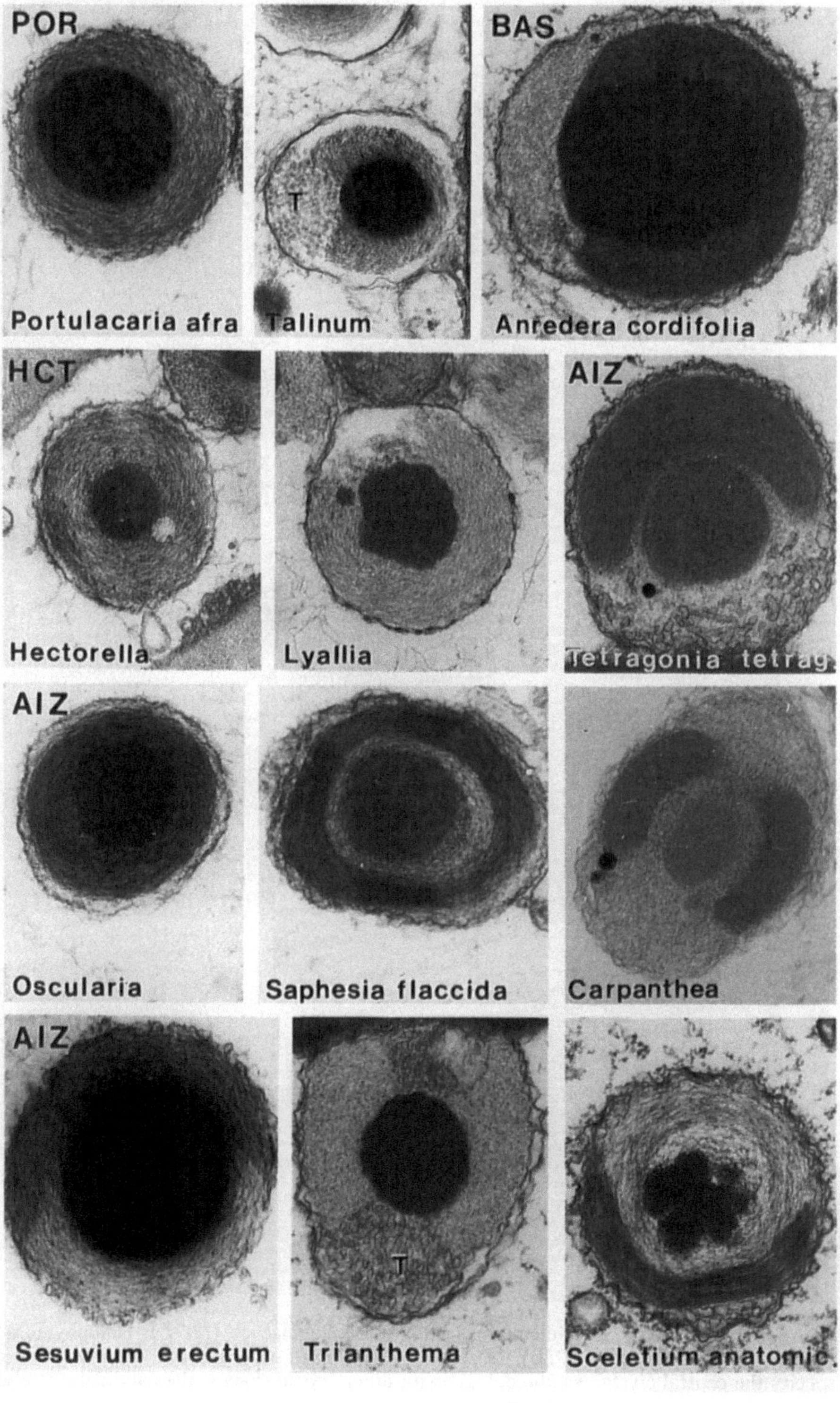

Fig. 5.5. Form-P3cf sieve-element plastids of Portulacaceae (*POR*), Basellaceae (*BAS*), Hectorellaceae (*HCT*), and Aizoaceae (*AIZ*). T Remnants of thylakoids. ×30 000

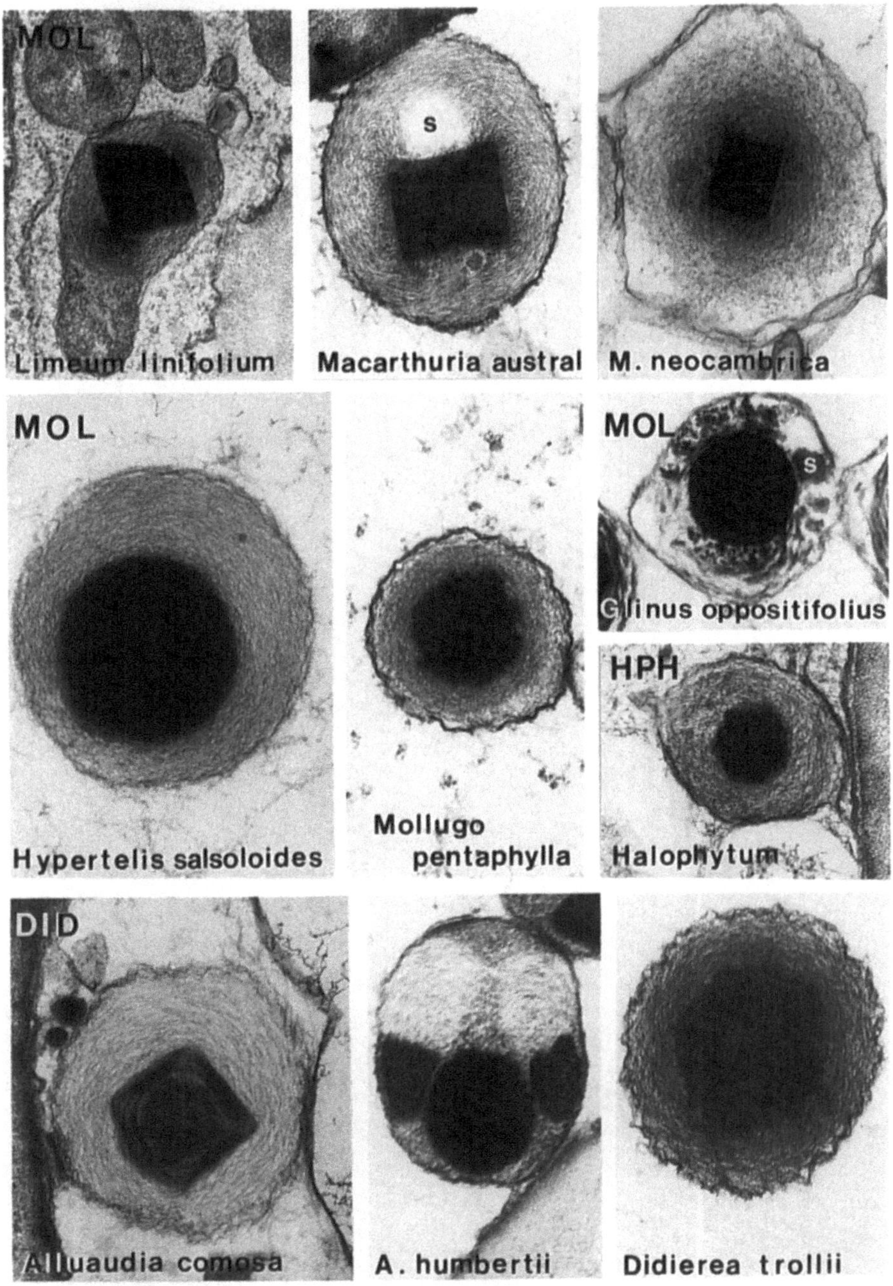

Fig. 5.6. Form-P3cf(s) sieve-element plastids of Molluginaceae (*MOL*), Halophytaceae (*HPH*), and Didiereaceae (*DID*). In *Limeum linifolium* Fenzl. and the two *Macarthuria* species the central crystal is cuboidal. As with other crystal shapes, the cuboidal one also appears early during sieve-element differentiation. Other genera of Molluginaceae and the studied Didiereaceae and *Halophytum ameghinoi* Speg. contain globular crystals. *s* Starch grains. ×30 000

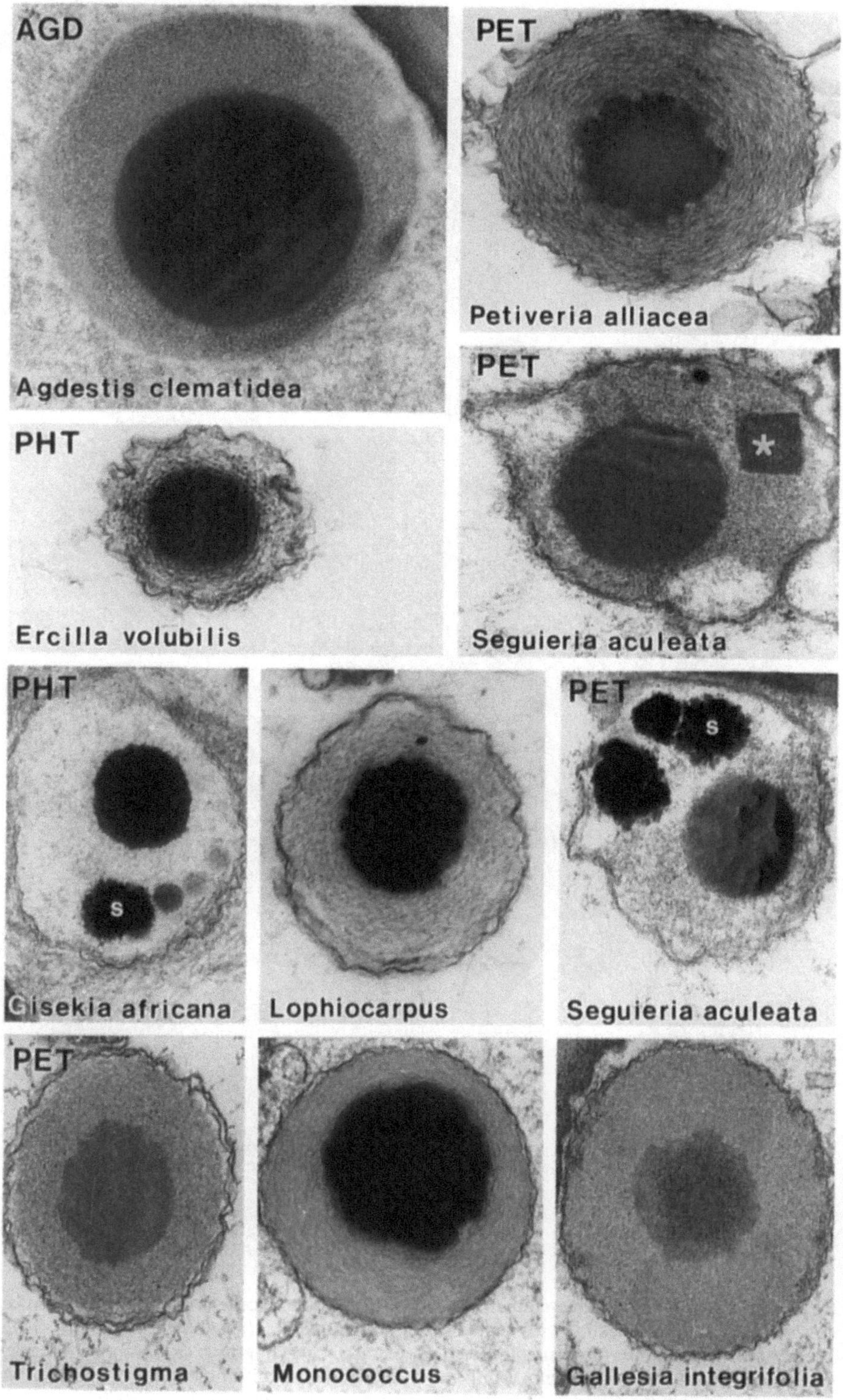

Fig. 5.7. Form-P3cf(s) sieve-element plastids of Agdestidaceae (*AGD*), Petiveriaceae (*PET*), and Phytolaccaceae (*PHT*), all with a globular central crystal. An additional (rodlike) crystal (*) is found in *Seguieria aculeata* Jacq. *s* Starch granis. ×30 000

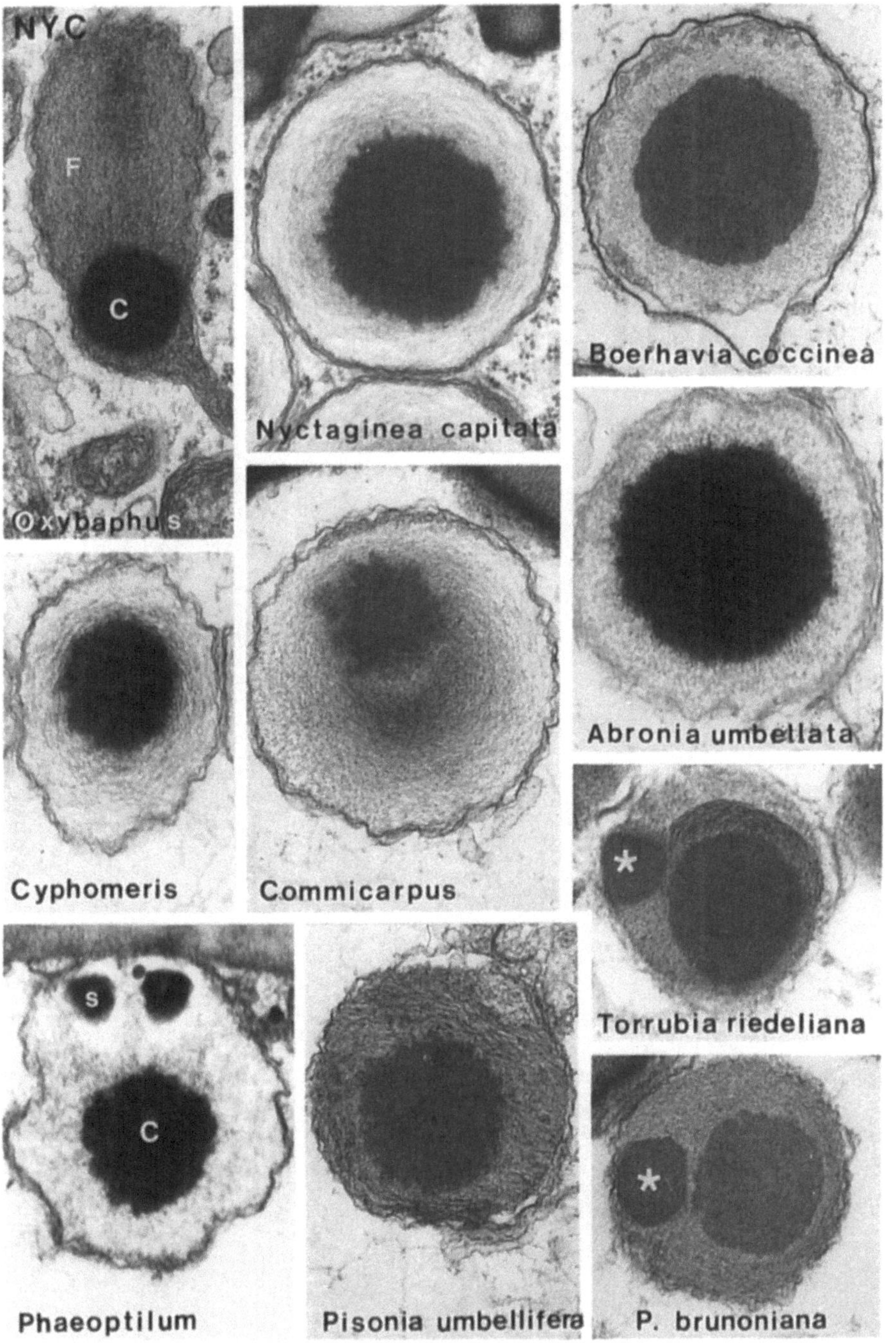

Fig. 5.8. Form-P3cf(s) sieve-element plastids of Nyctaginaceae with globular central crystals. Amoeboid plastid with crystal (*C*) and filaments (*F*) shown in young sieve element of *Oxybaphus viscosus* L'Herit ex Choisy. Starch grains (*s*) are prominently present in *Phaeoptilum spinosum* Radlk. and additional (rodlike) crystals (*) are present in *Torrubia riedeliana* (Fisch.) Standley and *Pisonia brunoniana* Endl. ×30 000

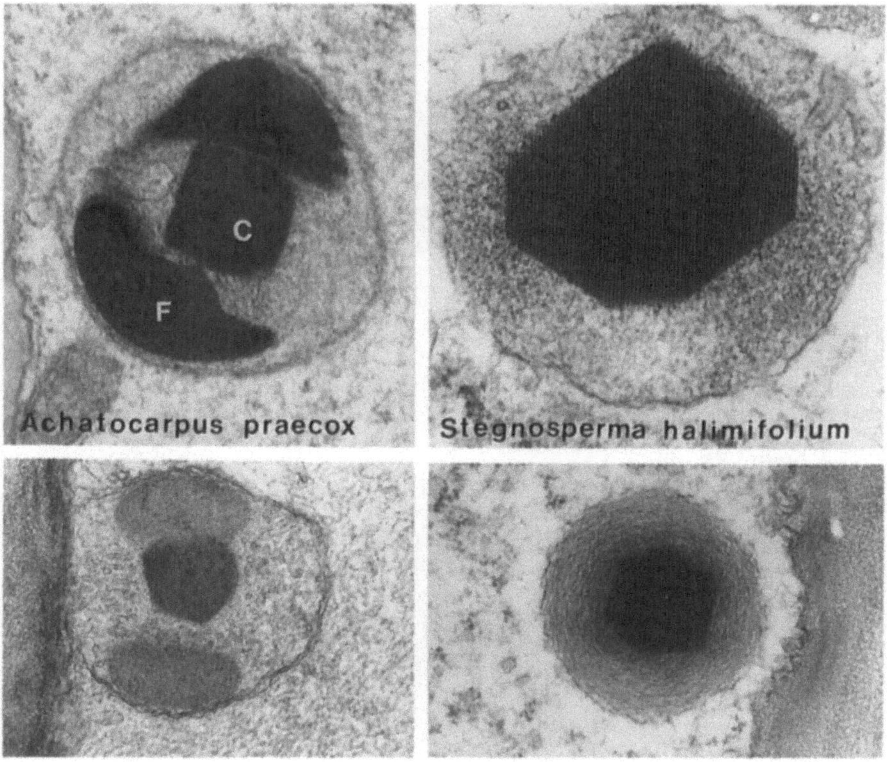

Fig. 5.9. Form-P3cf sieve-element plastids with polygonal central crystal (*c*) of Achatocarpaceae (*left* two micrographs) and Stegnospermataceae (*right* two micrographs). The protein crystal is composed of regularly arranged subunits. *F* Peripheral ring of protein filaments. *Upper right* micrograph ×55 000; all others ×30 000

The size of the sieve-element plastids in the stems of the Caryophyllales compared here ranges between 0.63 and 1.74 μm with an order average (1.06 μm) that is much less than that of other higher taxa in the angiosperms (Behnke, unpubl. results). However, their size is still large enough to make the sieve-element plastids very prominent in any monitoring light microscopic section (see semi-thin section in Fig. 5.1a and compare with ultrathin section in Fig 5.1b).

Among the eight qualitatively distinct and logically interconnected forms of sieve-element plastids, the four caryophyllalean forms reside in the F-plane of the "plastid cube" (see Fig. 5.12; Behnke 1991). The great majority of investigated taxa contain form-P3cf (160 species belonging to all but two of the families) and a few (16 species) contain form-P3cfs sieve-element plastids. The size of the latter does not differ from the former, another indication that in the Caryophyllales the difference between the two forms (i.e. the presence of starch grains) is really not taxonomically relevant. From the Amaranthaceae and Chenopodiaceae 44 investigated species were found

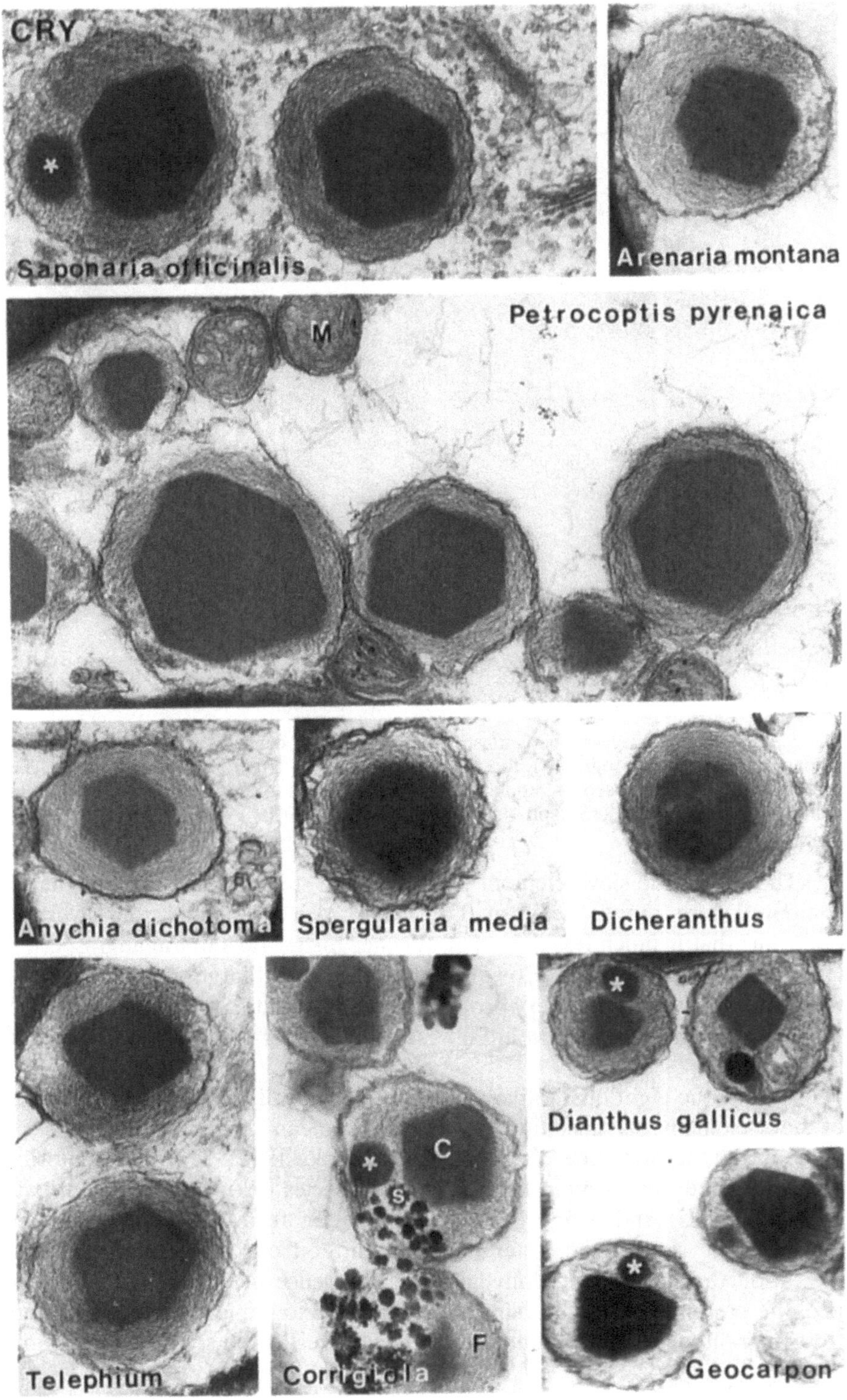
CRY
Saponaria officinalis
Arenaria montana
Petrocoptis pyrenaica
M
Anychia dichotoma
Spergularia media
Dicheranthus
Dianthus gallicus
C
S
F
Telephium
Corrigiola
Geocarpon

to contain form-P3f plastids, while five additional species had form-P3fs. Again, there is no size difference between the two forms. The average size of form-P3f(s) plastids is considerably larger than that of the form-P3cf(s): 1.25 vs 1.01 μm.

In addition to these subtype-P3 and form characteristics, some species have such distinct features that their sieve-element plastids might be easily distinguished from those of other species (e.g. *Sceletium anatomicum* L. Bolus with particulate crystal and partly condensed filaments; see Fig. 5.5).

5.4 The Distribution of Forms and Sizes of Sieve-Element Plastids in the Higher Taxa of the Caryophyllales

The results of ultrastructural studies, measurements and calculations of the sieve-element plastids of the 225 investigated species are shown in Table 5.1 and discussed in this section. While the grouping of families, subfamilies and genera presented in Table 5.1 is as suggested by the sieve-element plastid characters and (with respect to the taxa containing globular crystals) by traditional data complemented with calculations of the averages of sieve-element plastids, the sequence in the following text is alphabetical.

Achatocarpaceae (AHT; Fig. 5.9). The two species investigated, *Achatocarpus praecox* Griseb. and *Phaulothamnus spinescens* A. Gray, contain form-P3cf plastids with polygonal crystals that sometimes show transitions to globular ones. Their size is above the average of the order with almost no difference between the two species. Ultrathin sections through the crystal often yield rectangular outlines (Fig. 5.9), but a cuboidal shape cannot be proven so far (see Sect. 5.6). Inadequate preservation of field-collected material resulted in the report of globular crystals (Behnke 1976b, 1977b). The present report is based on additional field collections and garden-grown material.

Agdestidaceae (AGD; Fig. 5.7). This unigeneric family has the largest sieve-element plastids (average diameter 1.74 μm, form-P3cf) in the Caryophyllales. Its globular crystals are likewise among the largest in the order (see also Fig. 6 in Behnke 1976a). These plastid characters certainly add to those separating *Agdestis* from Phytolaccaceae s.l.

Aizoaceae (AIZ; Fig. 5.5). Twenty species have been investigated, belonging to all the five subfamilies recognized by Bittrich and Hartmann (1988), all with form-P3cf plastids containing a globular crystal. The average plastid

←

Fig. 5.10. Form-P3cf(s) sieve-element plastids with polygonal central crystal (*c*) of Caryophyllaceae. Many species contain an additional (rodlike) crystal (*); *Corrigiola litoralis* L. is one of the only two species found so far in the Caryophyllaceae to contain small starch grains (*s*). *F* Peripheral ring of protein filaments; *M* mitochondrium ×30 000

Table 5.1. Sieve-element plastid data of the Caryophyllales

Taxon	No. species recorded for taxon	Sieve-element plastid form	Crystal shape[a]	No. species investigated[b]	Average plastid diameter in stem phloem (μm)	Reference[c]
Caryophyllineae		P3cf, P3cfs	g/a	48*	0.87	
Molluginaceae	90	P3cf, P3cfs	g/c	11*	1.03	
Limeeae		P3cf, P3cfs	c	3	1.03	
Limeum		P3cf	c	1	0.91	Behnke (1976a)
Macarthuria		P3cfs	c	2	1.09	Behnke et al. (1983b)
Mollugineae		P3cf	g	8*	1.03	
Glinus		P3cfs	g	1		Behnke et al. (1983a)
Hypertelis		P3cf	g	2	1.17	Behnke (1976a)
Mollugo		P3cf, P3cfs	g	4*	0.88	Behnke and Turner (1971)
Pharnaceum		P3cf	g	1		Behnke et al. (1983a)
Corbichonieae				–		
Caryophyllaceae	1750	P3cf, P3cfs	p	37*	0.84	
Paronychioideae		P3cf, P3cfs	p	16*	0.81	
Anychia		P3cf	p	1	0.80	Behnke (1993)
Corrigiola		P3cfs	p	1	0.77	Behnke (1993)
Dicheranthus		P3cf	p	1	0.80	Behnke (1993)
Drymaria		P3cf	p	1	0.84	Behnke et al. (1983a)
Herniaria		P3cf	p	2	0.76	Behnke (1976a)
Illecebrum		P3cf	p	1	0.77	Behnke (1976a)
Krauseola		P3cfs	p	1	0.89	Behnke (1993)
Loeflingia		P3cf	p	1	0.79	Behnke (1976a)
Paronychia		P3cf	p	1	0.95	Behnke (1976a)
Polycarpaea		P3cf	p	1	0.76	Behnke et al. (1983a)
Polycarpon		P3cf	p	2*	0.83	Behnke (1976a)
Spergula		P3cf	p	1		Behnke and Turner (1971)
Spergularia		P3cf	p	1	0.91	Behnke (1993)
Telephium		P3cf	p	1	0.77	Behnke (1993)
Alsinoideae		P3cf	p	5	0.81	
Arenaria		P3cf	p	1	0.93	Behnke (1993)
Cerastium		P3cf	p	1		Behnke and Turner (1971)
Geocarpon		P3cf	p	1	0.63	Behnke (1982a)
Honkenya		P3cf	p	1	0.89	Behnke (1976a)
Schiedea		P3cf	p	1	0.79	Behnke (1976a)
Caryophylloideae		P3cf	p	16*	0.87	
Agrostemma		P3cf	p	1	0.98	Behnke (1976a)
Cucubalus		P3cf	p	1	0.97	Behnke (1969)
Dianthus		P3cf	p	3	0.72	Behnke (1969)
Drypis		P3cf	p	1	0.95	Behnke (1993)
Gypsophila		P3cf	p	1	0.80	Behnke (1976a)
Lychnis		P3cf	p	2	0.85	Behnke (1969)
Petrocoptis		P3cf	p	1	0.84	Behnke (1993)
Saponaria		P3cf	p	1	0.96	Behnke (1975a)
Silene		P3cf	p	4*	0.92	Behnke and Turner (1971)
Vaccaria		P3cf	p	1		Behnke and Turner (1971)
Phytolaccineae		P3cf, P3cfs	g/p	42*	1.12	
Phytolaccaceae	45	P3cf, P3cfs	g	9*	0.99	
Phytolaccoideae		P3cf, P3cfs	g	6*	0.95	
Ercilla		P3cf	g	1	0.89	Behnke (1976b)
Phytolacca		P3cf, P3cfs	p	5*	0.96	Behnke (1969)

Table 5.1. *Continued*

Taxon	No. species recorded for taxon	Sieve-element plastid form	Crystal shape[a]	No. species invest-igated[b]	Average plastid diameter in stem phloem (μm)	Reference[c]
Gisekioideae		P3cf, P3cfs	g	2	1.06	
Gisekia		P3cf, P3cfs	g	2	1.06	Mabry et al. (1976)
Lophiocarpoideae		P3cf	g	1	1.04	
Lophiocarpus		P3cf	g	1	1.04	Behnke (1974a)
Petiveriaceae	40	P3cf, P3cfs	g	7	1.17	
Gallesia		P3cf	g	1	1.30	Behnke (1993)
Hilleria		P3cf	g	1	1.23	Behnke (1976b)
Monococcus		P3cf	g	1	1.28	Behnke (1991)
Petiveria		P3cf	g	1	1.07	Behnke et al. (1974)
Rivina		P3cfs	g	1	1.06	Behnke (1974a)
Seguieria		P3cfs	g	1	1.17	Behnke et al. (1983a)
Trichostigma		P3cf	g	1	1.09	Behnke (1969)
Barbeuiaceae	1	P3cfs	p	1	1.09	
Barbeuia		P3cfs	p	1	1.09	Behnke (1993)
Nyctaginaceae	290	P3cf, P3cfs	g	21*	1.13	
Abronia		P3cf	g	2	1.18	Behnke (1993)
Acleisanthes		P3cf	g	1		Behnke (1993)
Allionia		P3cf	g	1		Behnke (1993)
Boerhavia		P3cf	g	1	1.04	Behnke (1993)
Bougainvillea		P3cfs	g	1	1.08	Behnke (1976b)
Commicarpus		P3cf	g	1	1.23	Behnke (1993)
Cyphomeris		P3cf	g	1	0.92	Behnke (1993)
Leucaster		P3cf	g	1		Behnke (1993)
Mirabilis		P3cf	g	2	1.02	Behnke (1969)
Neea		P3cf	g	1		Behnke (1993)
Nyctaginea		P3cf	g	1	1.50	Behnke (1993)
Oxybaphus		P3cf	g	1	1.22	Behnke (1976b)
Phaeoptilum		P3cfs	g	1	1.04	Behnke (1993)
Pisonia		P3cf	g	3*	1.07	Behnke (1969)
Selinocarpus		P3cf	g	1	1.07	Behnke (1993)
Torrubia		P3cf	g	2	1.19	Behnke et al. (1983a)
Agdestidaceae	1	P3cf	g	1	1.74	
Agdestis		P3cf	g	1	1.74	Behnke et al. (1974)
Stegnospermataceae	3	P3cf	p	1	1.10	
Stegnosperma		P3cf	p	1	1.10	Behnke (1976a)
Achatocarpaceae	10	P3cf	p	2	1.11	
Achatocarpus		P3cf	p	1	1.11	Behnke (1976b)
Phaulothamnus		P3cf	p	1	1.10	Behnke (1976b)
Portulacineae		P3cf	g	84*	1.03	
Aizoaceae	1900	P3cf	g	20	1.02	
Aizoideae		P3cf	g	3	1.11	
Aizoon		P3cf	g	1	0.99	Behnke (1993)
Galenia		P3cf	g	1	0.85	Behnke (1993)
Gunniopsis		P3cf	g	1	1.50	Behnke (1993)
Aptenioideae		P3cf	g	1	1.05	
Aptenia		P3cf	g	1	1.05	Behnke (1976b)
Ruschioideae		P3cf	g	12	0.95	
Carpanthea		P3cf	g	1	1.01	Behnke (1993)
Carpobrotus		P3cf	g	1	0.97	Behnke (1976b)
Conophytum		P3cf	g	1	0.76	Behnke (1976b)
Delosperma		P3cf	g	1	0.86	Behnke (1976b)
Gibbaeum		P3cf	g	1	1.08	Behnke (1976b)
Lampranthus		P3cf	g	1	1.00	Behnke (1976b)
Lithops		P3cf	g	1	0.81	Behnke (1976b)

Table 5.1. *Continued*

Taxon	No. species recorded for taxon	Sieve-element plastid form	Crystal shape[a]	No. species invest-igated[b]	Average plastid diameter in stem phloem (μm)	Reference[c]
Oscularia		P3cf	g	1	0.89	Behnke et al. (1983a)
Ruschia		P3cf	g	1	0.90	Behnke (1976b)
Saphesia		P3cf	g	1	1.05	Behnke et al. (1983a)
Sceletium		P3cf	g	2	1.04	Behnke (1969)
Sesuvioideae		P3cf	g	2	1.14	
Sesuvium		P3cf	g	1	1.15	Behnke and Turner (1971)
Trianthema		P3cf	g	1	1.13	Behnke (1976b)
Tetragonioideae		P3cf	g	2	1.14	
Tetragonia		P3cf	g	2	1.14	Falk (1964); Behnke (1976b)
Halophytaceae	1	P3cf	g	1	0.92	
Halophytum		P3cf	g	1	0.92	Hunziker et al. (1974)
Cactaceae	1550	P3cf	g	26*	1.07	
Pereskioideae		P3cf	g	3*	1.00	
Pereskia		P3cf	g	3*	1.00	Behnke (1969)
Opuntioideae		P3cf	g	4	1.10	
Cylindropuntia		P3cf	g	1	0.88	Behnke (1993)
Opuntia		P3cf	g	1		Behnke (1976b)
Pereskiopsis		P3cf	g	1	1.37	Behnke (1993)
Tephrocactus		P3cf	g	1	1.06	Behnke (1976b)
Cactoideae		P3cf	g	19*	1.08	
Armatocereus		P3cf	g	1	1.21	Behnke (1993)
Cereus		P3cf	g	1		Behnke (1976b)
Discocactus		P3cf	g	1	1.02	Behnke (1993)
Disocactus		P3cf	g	1	1.07	Behnke (1993)
Echinocereus		P3cf	g	1	1.05	Behnke (1993)
Epiphyllum		P3cf	g	1	1.10	Behnke (1976b)
Ferocactus		P3cf	g	1	0.94	Behnke (1993)
Lophocereus		P3cf	g	1	1.23	Behnke (1993)
Mammillaria		P3cf	g	1	1.08	Behnke (1993)
Matucana		P3cf	g	1	0.85	Behnke (1993)
Melocactus		P3cf	g	1	1.05	Behnke et al. (1983a)
Myrtillocactus		P3cf	g	1	1.22	Behnke (1993)
Parodia		P3cf	g	1	0.81	Behnke (1993)
Phyllocactus		P3cf	g	1		Behnke (1976b)
Rebutia		P3cf	g	1	0.92	Behnke (1993)
Rhipsalis		P3cf	g	1	1.02	Behnke (1976b)
Selenicereus		P3cf	g	1	1.38	Behnke (1993)
Thrixanthocereus		P3cf	g	1	1.28	Behnke (1993)
Wittia		P3cf	g	1		Behnke et al. (1983a)
Basellaceae	20	P3cf	g	4*	1.17	
Anredera		P3cf	g	2	1.20	Behnke (1993)
Basella		P3cf	g	1	1.11	Behnke (1976b)
Boussingaultia		P3cf	g	1		Behnke (1969)
Portulacaceae	575	P3cf	g	20*	0.95	
Anacampseros		P3cf	g	1	0.92	Behnke et al. (1975)
Calandrinia		P3cf	g	3	0.90	Behnke et al. (1975)
Ceraria		P3cf	g	2	0.92	Behnke et al. (1975)
Claytonia		P3cf	g	2*	0.96	Behnke et al. (1975)

Table 5.1. *Continued*

Taxon	No. species recorded for taxon	Sieve-element plastid form	Crystal shape[a]	No. species invest-igated[b]	Average plastid diameter in stem phloem (μm)	Reference[c]
Lewisia		P3cf	g	3		Behnke et al. (1975)
Montia		P3cf	g	1	0.99	Behnke et al. (1975)
Portulaca		P3cf	g	3	0.94	Behnke et al. (1975)
Portulacaria		P3cf	g	2	0.91	Behnke (1969)
Talinella		P3cf	g	1	0.91	Behnke et al. (1975)
Talinum		P3cf	g	2	1.10	Behnke et al. (1975)
Hectorellaceae	2	P3cf, P3cfs	g	2*	1.00	
Hectorella		P3cf	g	1		Behnke (1975a)
Lyallia		P3cfs	g	1	1.00	Behnke (1993)
Didiereaceae	11	P3cf	g	11	1.04	
Alluaudia		P3cf	g	6	1.14	Behnke (1969)
Alluaudiopsis		P3cf	g	2	0.87	Behnke (1978)
Decaryia		P3cf	g	1	0.91	Behnke (1978)
Didierea		P3cf	g	2	0.98	Behnke (1978)
Chenopodiineae		P3f, P3fs		51*	1.24	
Amaranthaceae	850	P3f, P3fs		23*	1.21	
Achyranthes		P3f		1	1.39	Behnke (1976b)
Aerva		P3f		1	1.13	Behnke (1969)
Allmania		P3fs		1	1.06	Behnke et al. (1983a)
Alternanthera		P3f		3	1.22	Behnke (1976b)
Amaranthus		P3f, P3fs		3	1.36	Behnke (1976b)
Bosea		P3f		1	1.26	Behnke (1976b)
Celosia		P3fs		1	1.19	Behnke (1974a)
Chamissoa		P3f		1	1.26	Behnke (1993)
Cyathula		P3f		1		Behnke et al. (1983a)
Deeringia		P3f		1	1.34	Behnke (1976b)
Froelichia		P3f		1	1.15	Behnke (1976b)
Gomphrena		P3f		1	1.23	Behnke (1976b)
Iresine		P3fs		4	1.10	Behnke (1976b)
Pleuropetalum		P3f		1	1.04	Behnke (1991)
Tidestromia		P3f		2	1.21	Behnke (1976b)
Chenopodiaceae (excl. Sarcobatoideae)	1500	P3f, P3fs		26*	1.28	
Chenopodioideae		P3f, P3fs		16*	1.26	
Atriplex		P3f		5	1.23	Behnke (1971)
Bassia		P3f		1	1.23	Behnke (1993)
Beta		P3f		1	1.41	Esau (1965); Behnke (1993)
Chenopodium		P3f		1	1.33	Behnke (1976b)
Dysphania		P3f		1	1.04	Behnke and Turner (1971)
Grayia		P3f		1	1.47	Behnke (1993)
Hablitzia		P3f		1	1.48	Behnke (1969)
Halimione		P3f		1	1.03	Behnke (1976b)
Kochia		P3f		1	1.20	Behnke (1976b)
Microtea		P3fs		1	1.34	Behnke (1976b)
Rhagodia		P3f		1	1.05	Behnke (1993)
Spinacia		P3f		1		Behnke (1976b)
Salicornioideae		P3f		5	1.27	
Allenrolfea		P3f		1	1.09	Behnke (1976b)
Arthrocnemum		P3f		1	1.34	Behnke (1993)
Salicornia		P3f		3	1.32	Behnke and Turner (1971)

Table 5.1. *Continued*

Taxon	No. species recorded for taxon	Sieve-element plastid form	Crystal shape[a]	No. species investigated[b]	Average plastid diameter in stem phloem (μm)	Reference[c]
Salsoloideae		P3f		5	1.38	
Haloxylon		P3f		1	1.62	Behnke (1993)
Salsola		P3f		1	1.07	Behnke (1976b)
Suaeda		P3f		2	1.36	Behnke (1976b)
Traganum		P3f		1	1.51	Behnke (1993)
(Sarcobatoideae)						
Sarcobatus		P3cf	g	2	0.92	Behnke (1993)
Caryophyllales	8700	P3f, P3fs, P3cf, P3cfs		225*	1.06	

[a] Crystal shape: a, angular (cuboidal or polygonal); c, cuboidal; g, globular; p, polygonal.
[b] Total number may also include species in which leaves or other parts (not stem phloem) were investigated, denoted by *.
[c] First documentation of sieve-element plastids at generic level.

diameter in the family is 1.02 μm. Species from the subfamilies Sesuvioideae, according to Bittrich (1990) "an earlier branch within the Aizoaceae", and Tetragonioideae have larger plastids than do the other groups. The first electron micrographs of a caryophyllalean plastid published were from *Tetragonia expansa* Murr. (*tetragonioides* O. Kuntze) and did clearly depict the peripheral ring, but no globular crystal (Falk 1964). Our own micrographs of sieve-element plastids from stems of both *Tetragonia echinata* Ait and *T. tetragonioides* (Fig. 5.5) showed a central crystal.

The globular crystals of some Aizoaceae are very small (e.g. in *Conophytum hians* N.E.Br.) and obviously very sensitive to fixation media: *Carpanthea pomeridiana* N.E.Br. (Fig. 5.5), routinely fixed with glutaraldehyde and osmium tetroxide, contains only a small, faintly stained central crystal; plastids of *Lampranthus conspicuus* (Haw.) N.E.Br., fixed with osmium tetroxide only, did not contain a central crystal. Further investigations with different fixation media acting on different species are necessary to clarify these aberrant results.

Amaranthaceae (AMA; Fig. 5.2). Studies have been made of 23 species distributed over the two subfamilies; 19 with form-P3f and four with form-P3fs plastids. The sizes of the sieve-element plastids in all categories (family, subfamilies, form-P3f, form-P3fs) are very homogeneous and much larger than the average in the order (for *Cyathula* the number of measurements is insufficient). The absence of a central crystal applies to all investigated taxa, without exception. A few species, notably *Chamissoa macrocarpa* H.B.K. (* in Fig. 5.2), contain a small rodlike crystal of the type found in some other families, in addition to the characteristic large crystal (see Sect. 5.3). Fisher and Evert (1982) depict sieve-element plastids in *Amaranthus retroflexus*.

Barbeuiaceae (BBU). The monotypic, endemic Madagascan *Barbeuia madagascariensis* Steud. has form-P3cfs sieve-element plastids much like those of the Phytolaccaceae (cf. Behnke 1993).

Basellaceae (BAS; Fig. 5.5). The four species investigated contain form-P3cf plastids with globular crystals and an average diameter of 1.17 µm in the stem phloem, which is considerably larger than that of its sister family Portulacaceae.

Cactaceae (CAC; Fig. 5.4). The sieve-element plastids of all the 26 species studied belong to form P3cf with globular crystals. As is typical for many Cactaceae, the crystal surface is often irregular. The average size of the plastids is 1.07 µm in the family, 1.0 µm in the Pereskioideae, 1.10 µm in the Opuntioideae, and 1.08 µm in the Cactoideae. Due to insufficient material within a few species only 21 species are represented in the size calculations. Delay and Darmanaden (1973) depict sieve-element plastids of *Opuntia subulata*, Gibson and Nobel (1986) those of *Ferocactus wislizenii*.

Caryophyllaceae (CRY; Fig. 5.10). Studies have been made of 37 species from the three subfamilies, of which 32 species represent studies from stem phloem and are included in the size average. Almost all contain form-P3cf plastids with a polygonal crystal; plastids of *Corrigiola* and *Krauseola* also have several small starch grains (form-P3cfs). The average plastid diameter (0.84 µm) is the smallest in the order. The family and subfamily diameters (see Table 5.1) do not differ much and are considerably smaller than the order average. *Geocarpon minimum* Mackenzie (Fig. 5.10), previously doubtfully referred to either Aizoaceae or Caryophyllaceae (cf. Pax and Hoffmann 1934), is included here in the Alsinoideae, as proposed by Steyermarck et al. (1959) and corroborated by the presence of anthocyanin pigments (Bogle et al. 1971). Its sieve-element plastids (form-P3cf with polygonal crystals; average diameter 0.63 µm) conform to the family characters (see also Behnke 1982a). *Corrigiola litoralis* L. and *Telephium imperati* L. are discussed by Gilbert (1987) as possible members of the Molluginaceae. The size of their sieve-element plastids (average diameter 0.77 µm) and their polygonal crystals fit well into the Caryophyllaceae.

Chenopodiaceae (CHN; Figs. 5.2 and 5.3). Twenty-six species investigated (including *Dysphania* and *Microtea*, but excluding *Sarcobatus*), all with form-P3f, except for *Microtea* (P3fs). The average diameter of 25 critically measured species is 1.29 µm, but differs between the subfamilies (see Table 5.1). *Microtea* is included in the Chenopodioideae; neither form-P3fs nor size (1.34 µm) find a counterpart in the Phytolaccaceae (see Table 5.1). *Sarcobatus* is excluded on the basis of its form-P3cf plastids (see below), found nowhere else in the family, although a wide coverage of different tribes has been reached (see Behnke 1993). Esau (1965, 1976) depicted sieve-element plastids of *Beta vulgaris* and *Spinacia oleracea*, Danilova and Kozubov (1980) those of *Aellenia subaphylla*.

The so-defined Chenopodiaceae are very close to the Amaranthaceae, both in form-P3f(s) and the comparatively large sizes of its plastids.

Didiereaceae (DID; Fig. 5.6). All 11 species of this endemic Madagascan family have form-P3cf plastids with globular crystals (cf. Behnke 1978). Often, condensed filaments surround the globoid crystal. The average size of the plastids in the different species falls between 0.8 and 1.2 μm, the family average is 1.04 μm.

Halophytaceae (HPH; Fig. 5.6). The monotypic family has been found to contain form-P3cf plastids with globular crystals (Hunziker et al. 1974). Their size is 0.92 μm. These characters do not contradict a position near Aizoaceae, as proposed by Gibson (1978), but negate its inclusion into Chenopodiaceae, as proposed by Ulbrich (1934) and followed by Cronquist (1981, 1988).

Hectorellaceae (HCT; Fig. 5.5). The two species of the digeneric family contain form-P3cf plastids (average diameter in stem phloem 1.0 μm). Globular central crystals found in *Hectorella* (Behnke 1975a) and *Lyallia* (Behnke 1993) oppose their inclusion into the Caryophyllaceae as treated by Pax and Hoffmann (1934), but neither form nor size of the plastids would conflict with their treatment as a family of the suborder Portulacineae as proposed by Thorne (1992).

Molluginaceae (MOL; Fig. 5.6). Of 11 species investigated all have form-P3cf/P3cfs plastids, but the central crystal is either globular or cuboidal. The average family diameter of the plastids is 1.03 μm (critical measurements in stem phloem of seven species incorporated), but ranges from 0.8 to 1.2 μm for the species averages. The great diversity of the sieve-element plastid characters underlines the suggestion that the Molluginaceae are not monophyletic (Gilbert 1987). The tribe Limeeae sensu Hofmann (1973) is so far the only one to show cuboidal crystals (cf. Behnke et al. 1983b). However, its separation as a distinct higher taxon would not make the rest of the family more homogeneous with respect to the plastid sizes. The transfer of *Limeum* (with cuboidal crystals) to the Phytolaccaceae, as proposed by Dahlgren (1980), could be supported by plastid data only if Achatocarpaceae and Stegnospermataceae (with polygonal crystals) are retained within Phytolaccaceae s.l.

Nyctaginaceae (NYC; Figs. 5.1 and 5.8). Of the 21 species investigated 20 belong to the two tribes Mirabileae and Pisonieae, the majority of which contain form-P3cf plastids with globular central crystals. Additional starch grains were found in *Bougainvillea* and *Phaeoptilum* (form-P3cfs). Sieve-element plastids of *Pisonia* and *Torrubia* (Fig. 5.8) contain an additional small, roughly polygonal crystal. *Leucaster caniflorus* Choisy (tribe Leucastereae) was found to contain form-P3cf plastids with globular crystals, but the preservation of the material was not good enough to be able to detect other specific inclusions or to measure average sizes. Quantitative measurements were made within stem phloem of 16 species and the family average is 1.13 μm.

Petiveriaceae (Rivinaceae) (PET; Fig. 5.7). The seven species investigated contain comparatively large sieve-element plastids (average size 1.17 μm) of form-P3cf with globular crystals (form-P3cfs in *Seguieria*). *Monococcus* is incorporated here because of its large plastids (see also treatments of Nowicke 1968; Brown and Varadarajan 1985); the placement of *Lophiocarpus* into Phytolaccaceae follows Takhtajan (1987) and Thorne (1992). A separation of the Petiveriaceae from the Phytolaccaceae would be favoured by the considerably larger sizes of their form-P3cf plastids.

Phytolaccaceae (PHT; Fig. 5.7). The family, as treated here, includes the subfamilies Phytolaccoideae, Gisekioideae, and Lophiocarpoideae, of which the nine species studied contain form-P3cf/P3cfs plastids with globular crystals and comparatively uniform sizes (0.99 μm). *Microtea* is placed into the Chenopodiaceae (treatment proposed by Cronquist 1981) on the basis of its form-P3fs and large plastids (see Table 5.1). *Lophiocarpus* (form-P3cf, 1.04 μm; cf. also Behnke 1974a; Eckardt 1974) is retained in the Phytolaccaceae following Takhtajan (1987) and Thorne (1992), but see Cronquist (1981). The transfer of *Gisekia* from Molluginaceae to the Phytolaccaceae was proposed by Takhtajan (1980) and Cronquist (1981) referring to the detection of betalains in this genus (Mabry et al. 1976). Sieve-element plastid data do not contradict this.

Portulacaceae (POR; Fig. 5.5). The sieve-element plastid characters of the 20 species studied (16 from stem phloem included in size calculations) are rather uniform: form-P3cf, small diameter of 0.95 μm (see also Behnke et al. 1975). *Talinella* has been very doubtfully referred to the Portulacaceae by Pax and Hoffmann (1934). Neither Carolin (1987) nor recent angiosperm systems (e.g. Cronquist 1981; Takhtajan 1987) mention this genus in their texts. Its small-sized form-P3cf plastids fit into the family (see also Behnke et al. 1975). Crookston and Ozhun (1975) depicted sieve-element plastids of *Portulaca oleracea*.

Sarcobatoideae (SRB; Fig. 5.3). The two species of *Sarcobatus* constituting Ulbrich's (1934) Sarcobatoideae, subfamily of Chenopodiaceae, contain form-P3cf plastids with a central globular crystal and identical sizes of 0.92 μm (Behnke 1993). Since form-P3cf plastids are otherwise absent from the Chenopodiaceae, *Sarcobatus* is placed here as *incertae sedis*, but close to Chenopodiaceae. On account of additional small crystals found in *Sarcobatus* a connection with *Pisonia*/*Torrubia* in the Nyctaginaceae may be considered (cf. Behnke 1993).

Sarcobatus also differs in its trichome types from most, but not all, other Chenopodiaceae (Carolin 1983). Pollen morphology supports, although not strongly, its distinction from other Chenopodiaceae (see Chap. 7).

Stegnospermataceae (STG; Fig. 5.9). The unigeneric family contains form-P3cf plastids with polygonal crystals; the plastid size is 1.10 µm. Gibson and Nobel (1986) depicted sieve-element plastids of *Stegnosperma cubense*.

5.5 The Sieve-Element Plastids of the Families Sometimes Included in or Most Often Allied to the Caryophyllales

Since Eichler's (1876) first mention of the order "Centrospermae" quite a number of families have, at one time or another, been incorporated into or associated with the order but now, for several reasons, are excluded in recent systems (cf. Eckardt 1976). Table 5.2 lists the sieve-element plastid data and some taxonomic treatments of these families. Almost all contain S-type plastids (Fig. 5.11).

Among the families that were most often treated as members of the Caryophyllales, but excluded after the combined action of phytochemical, ultrastructural and other investigations, are the Bataceae (Behnke and Turner 1971), Gyrostemonaceae (Goldblatt et al. 1976), Theligonaceae (Mabry et al. 1975), and Vivianiaceae (Behnke and Mabry 1977). Others, like the Polygonaceae and Plumbaginaceae, have most often been associated as the closest orders within the subclass Caryophyllidae (e.g. Takhtajan 1987).

The only family that contains P-type plastids and was previously sometimes doubtfully placed within the order (or even within Phytolaccaceae; cf. Takhtajan 1973) is the Rhabdodendraceae. Prance (1968) raised *Rhabdodendron* to the family rank and, after discussing different affinities, placed the new family into the Caryophyllales. Previously, Metcalfe and Chalk (1950), with a strong emphasis on wood anatomical characters, had incorporated *Rhabdodendron* into Phytolaccaceae. Its position in recent systems is within the Rosales (Cronquist 1981, 1988; Dahlgren 1989) or Rutales (Takhtajan 1987; Thorne 1983, 1992). The form-Pcs sieve-element plastids with a single small angular crystal and large starch grains (Fig. 5.11; cf. Behnke 1976c) do not fit into the S-type Rutales, find distant similarities in a few P-type families of the Rosales (e.g. Neuradaceae; cf. Behnke 1988b) and would better be at home in the Magnolianae, if other characters confirm.

5.6 The Putative Evolution of the Sieve-Element Plastids in the Caryophyllales

Any suggestions of the evolutionary importance of sieve-element plastids in the Caryophyllales have to consider primarily their specific forms and the morphological distinction of their protein crystals into globular or angular

Table 5.2. Sieve-element plastids of families sometimes allied with the Caryophyllales

Family	Taxonomic treatment	Sieve-element plastid form	No. species investigated	Reference[a]
Bataceae	Centrospermae (Bessey 1915; Takhtajan 1959, 1969, 1973; Airy Shaw 1973)	Ss	1	Behnke and Turner (1971)
Begoniaceae	Centrospermae? (Eichler 1876)	Ss	2	Behnke (1981)
Crassulaceae	Caryophyllinae (Hallier 1912)	Ss, So	6	Behnke (1988b)
Droseraceae	See Chapter 8	Ss	2	Fig. 5.11
Elatinaceae	Caryophyllales (Bessey 1915)	Ss	1	Fig. 5.11
Fouquieriaceae	Caryophyllinae (Hallier 1903); marginal to Centropermae (Airy Shaw 1973)	Ss	3	Behnke (1976c)
Frankeniaceae	Caryophyllales (Hallier 1903; Bessey 1915)	Ss	3	Behnke (1976c)
Gyrostemonaceae	Centrospermae (Engler 1934; Takhtajan 1959, 1969, 1973; Airy Shaw 1973) Phytolaccaceae (Cronquist 1968)	Ss	6	Behnke (1977b)
Hydrostachyaceae	Caryophyllales (Bessey 1915)	–		
Limoniaceae	(Included in Plumbaginaceae)	Ss	5	Behnke (1976b)
Nepenthaceae	See Chapter 8	Ss	2	Behnke (1981)
Plumbaginaceae	Caryophyllinae (Hallier 1912); in most systems peripheral to Caryophyllales	Ss	3	Behnke (1976b)
Podostemonaceae	Caryophyllales (Bessey 1915)	Ss	1	Behnke (1988b)
Polygonaceae	Centrospermae (Eichler 1876, 1878; Hallier 1912; Bessey 1915; Cronquist 1957); in most systems peripheral to Caryophyllales	Ss	15	Behnke (1976b)
Primulaceae	Marginal to Centrospermae (Airy Shaw 1973)	Ss	7	Behnke (1981)
Rhabdodendraceae	Caryophyllales (Prance 1968)	Pcs	2	Behnke (1976c)
Salicaceae	Caryophyllales (Bessey 1915)	Ss	2	Behnke (1981)
Simmondsiaceae	Caryophyllales (Takhtajan 1959)	Ss	1	Behnke (1982b)
Sphenocleaceae	Doubtful or marginal to Centrospermae (Emberger 1960; Airy Shaw 1973)	Ss	1	Behnke (unpubl.)
Tamaricaceae	Caryophyllales (Hallier 1903; Bessey 1915)	Ss	3	Behnke (1976c)
Theligonaceae	Centrospermae (Eichler 1878; Bessey 1915; Cronquist 1957; Takhtajan 1959)	Ss	1	Behnke (1975b)
Vivianiaceae	Caryophyllales (Takhtajan 1973)	Ss	1	Behnke and Mabry (1977)

[a] First documentation of sieve-element plastids.

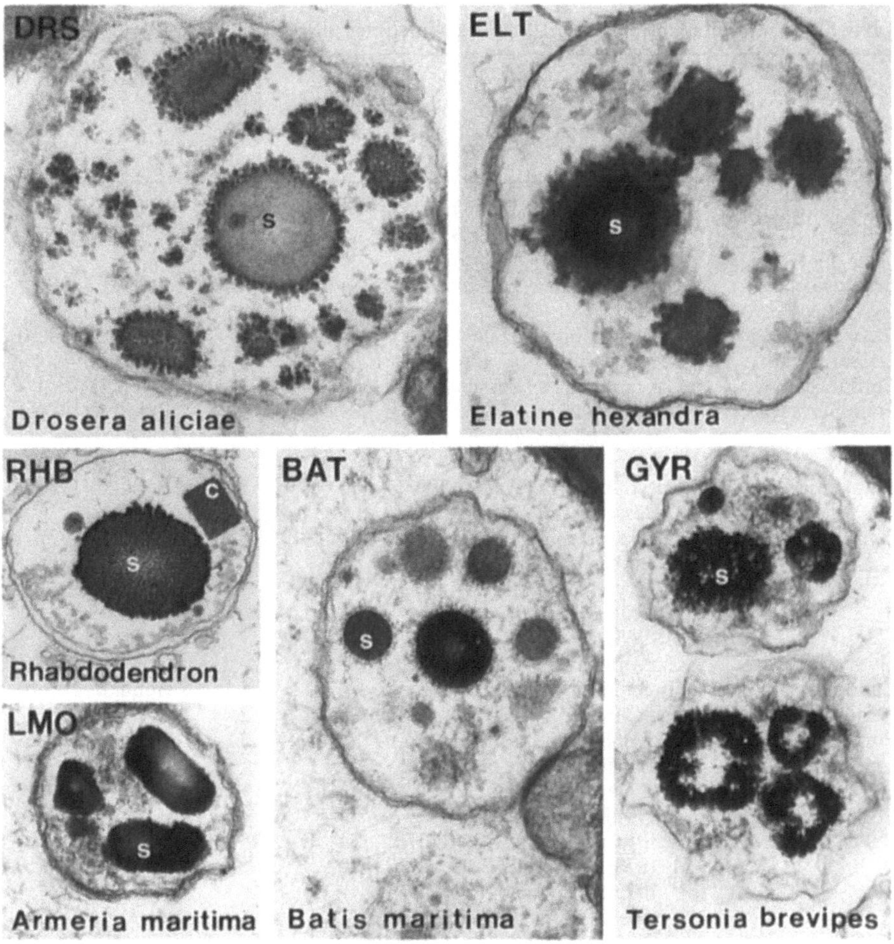

Fig. 5.11. Sieve-element plastids of families sometimes allied with the Caryophyllales. *Rhabdodendron amazonicum* (Spruce ex Benth.) Hub. (Rhabdodendraceae) is the only one to contain P-type plastids with small protein crystals (form-Pcs); *Drosera aliciae* Hamet (Droseraceae), *Elatine hexandra* DC. (Elatinaceae), *Batis maritima* L. (Batidaceae), *Tersonia brevipes* Moq. (Gyrostemonaceae), and *Armeria maritima* (Mill.) Willd. (Limoniaceae) have S-type plastids. *c* Protein crystal; *s* Starch grains. ×30 000

(polygonal/cuboidal) shapes (see crystal shape "g", "p" or "c" in Table 5.1), and consider only secondarily their sizes. The sporadic occurrence of starch grains (forms-P3cfs or -P3fs) in some species of only a few caryophyllalean families, otherwise characterized by their starchless "sister forms" (P3cf, P3f), justifies their neglect and a combined treatment as the forms P3cf/P3cfs and P3f/P3fs, respectively. Similarly, the shape of the crystals may be defined as either globular or non-globular (angular; i.e. polygonal or cuboidal). The distinction between polygonal and cuboidal shapes can only

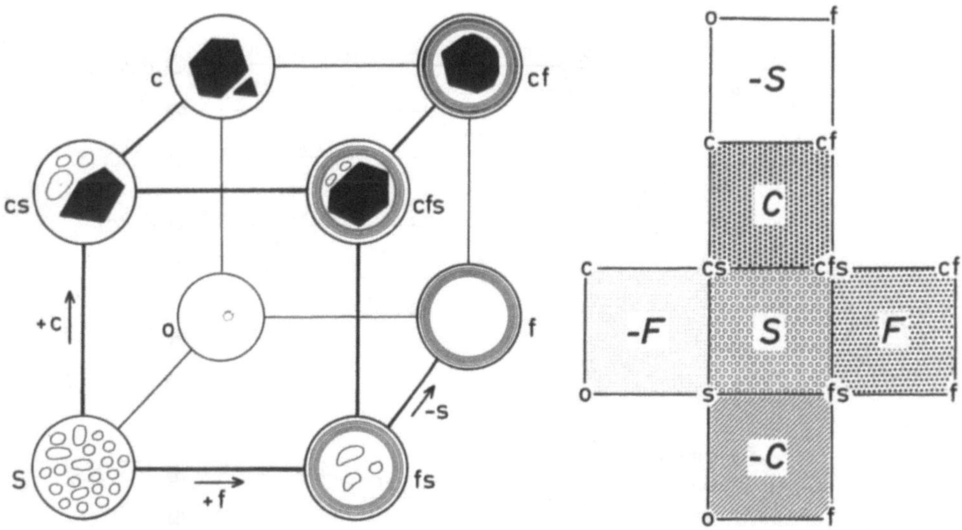

Fig. 5.12. Cubic model of the interconnections between the eight forms of sieve-element plastids. Forms of plastids are located at the corners of the cube. A transition of any form into any other along the edges is accompanied by a gain or loss of only one character (*c* protein crystals; *f* protein filaments; *s* starch). Unfolding of the six faces of the cube enables the demonstration of their genuine features, i.e. presence or absence of one character (e.g. the *F* face is bordered by all forms of sieve-element plastids containing protein filaments, as in Caryophyllales) (Behnke 1991)

be made by serial sectioning (see Behnke et al. 1983b), although single electron micrographs may rarely show quadrangular outlines as a result of the plane of sectioning through a polygonal crystal or polygonal sectional planes from a cuboidal crystal.

From the data presented in Sect. 5.4 and Table 5.1 it follows quite clearly that the peripheral ring of protein filaments has to be treated as a synapomorphy of the order. It is equally logical that the different sieve-element plastids of the order evolved on this basis.

Using the "plastid cube" as a model for the interconnections of the different forms of sieve-element plastids (see Fig. 5.12; Behnke 1991) the putative course of evolution is discussed in two hypotheses:

A. The unknown ancestors of the Caryophyllales had form-Pfs plastids that evolved from S-type plastids by addition of protein filaments (alongside the "*+f*" edge of the cube; cf. Fig. 5.12), and the ancestors of the recent families, characterized by form-P3f(s) plastids, separated comparatively early, i.e. before central crystals were developed in the main line. Globular central crystals developed next and were characteristic for the ancestors of all other families. A further step in plastid evolution included the change from globular to angular crystals, a split that resulted in the evolution of the Achatocarpaceae, Stegnospermataceae, Molluginaceae-Limeeae, and

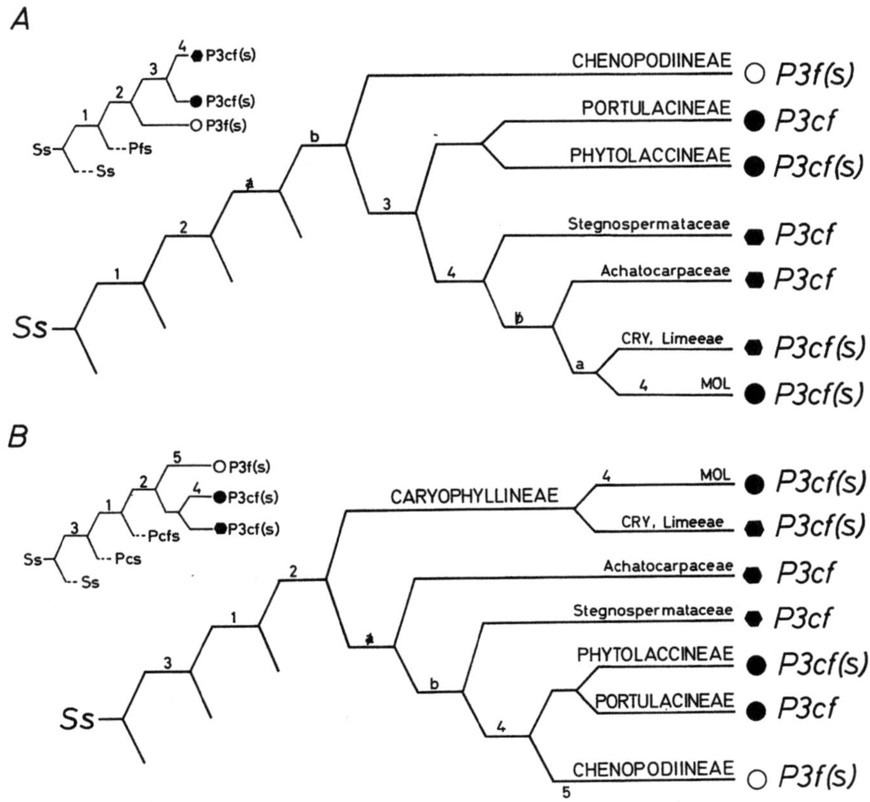

Fig. 5.13A and B. Two cladograms of the putative evolution of sieve-element plastids in the Caryophyllales. Both start with S-type plastids that are common to all angiosperms. While the *insets* only explain the two possibilities of the evolution of the different caryophyllalean plastid forms as suggested in Fig. 5.12, the *main diagrams* also consider the distribution of betalains and anthocyanins in the different taxa of the order.

The following steps are marked: *1* acquisition of protein filaments; *2* formation of Caryophyllales-specific peripheral ring of filaments; *3* acquisition of protein crystal; *4* change of crystal shape (either angular into globular or vice versa); *5* loss of central crystal; *a* acquisition of anthocyanins; *a̸* loss of anthocyanins; *b* acquisition of betalains; *b̸* loss of betalains

Caryophyllaceae in the one branch and the remaining P3cf(s) families in the other branch (see Fig. 5.13A, inset).

B. The hypothetical ancestors of the Caryophyllales contained form-Pcfs plastids with angular central crystals that had evolved by at least two steps, with form-Pcs as an intermediate (alongside the "+*c*" and "+*f*" edges of the cube; cf. Fig. 5.12). The ancestors of the four taxa containing form-P3cf(s) plastids with non-globular crystals separated early on from the main line which subsequently evolved their non-globular crystals into globular crystals. Another early branch gave rise to form-P3f(s), here characterized by the loss of the central crystal (see Fig. 5.13B, inset).

Comparing the two hypotheses, the first is distinguished as the more parsimonious to explain the evolution of all forms of sieve-element plastids in the Caryophyllales (cf. Fig. 5.13A, inset). The four steps necessary include: (1) the change from form-Ss into form-Pfs; (2) the formation of the Caryophyllales-specific ring of filaments; (3) the acquisition of a central crystal (either globular or angular) and thus transfer into form-P3cfs; and (4) a change from one crystal shape to the other. This cladogram is drawn to show the globular crystals being developed first, but if angular crystals were first, it would need the same number of steps.

In B one more step is required to evolve all sieve-element plastids found, since two steps (3 and 1) are used to change the ancestral angiosperm form-Ss into the form-Pcfs of the hypothetical ancestor of the Caryophyllales (see Behnke 1991). The formation of the peripheral ring of filaments (2) precedes two changes from angular crystals into globular ones (4) and the eventual loss of the central crystal (5).

However, if both sieve-element plastids and pigments are considered, the two cladograms are much more branched with the result that in the large versions A would require an additional step while B would become the more parsimonious (cf. Fig. 5.13A and B).

The early formation of the characteristic ring of filaments (step 2) and globular crystals (step 3) are in favour of cladogram A. Unsatisfactory, however, is the late branching of the Caryophyllineae and the required loss of betalains (step b) before anthocyanins (step a) are expressed.

Although requiring one step more, cladogram B would be supported by the following: (i) form-Pcs plastids (with various crystal shapes and number) are by far the most frequent among the P-type plastids of angiosperms; (ii) form-Pcfs plastids with angular (most often polygonal) crystals (as formed by step 3) are the third most abundant and widespread in the Magnolianae (see Sect. 5.7 and Behnke 1988a); (iii) the Caryophyllineae are separated early on, i.e. anthocyanins in the Caryophyllales could even be traced to the ancestral complex and were lost not much prior to the formation of betalains; and (iv) no pigment reversal is necessary.

Figure 5.14 is a balloon diagram showing some interrelationships between the sieve-element plastids of the order. It specifically demonstrates possible links between: (1) Chenopodiineae and Phytolaccineae via *Sarcobatus*, *Lophiocarpus* and *Microtea*; (2) Phytolaccineae and Caryophyllineae via *Gisekia* and *Stegnosperma* (cf. Hofmann 1977); (3) Caryophyllineae and Portulacineae via Aizoaceae-Molluginaceae; and (4) Portulacineae and Chenopodiineae via Halophytaceae (see Gibson 1978).

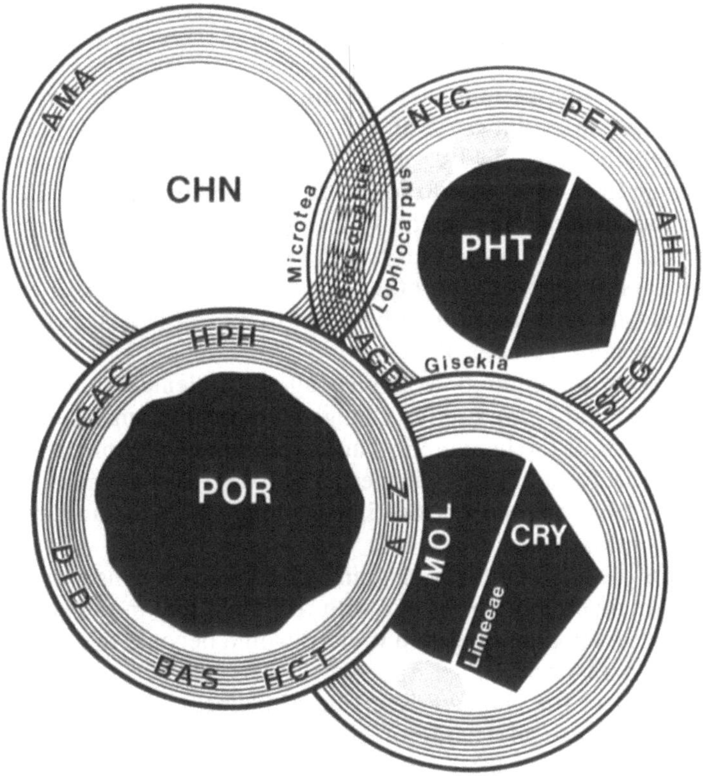

Fig. 5.14. Interrelationships of the major taxa of the Caryophyllales and their forms of sieve-element plastids (*shaded areas* represent protein crystals; *dotted areas* starch grains)

5.7 Relationships of the Order Caryophyllales

The homogeneity of the sieve-element plastids – and the presumption that form-Ss plastids are ancestral (see Behnke 1991, revising an earlier, 1981, contrary view) – supports the opinion that the specific subtype-P3 represents a synapomorphy of the Caryophyllales.

Although P-type plastids are widespread among the dicotyledons (cf. Behnke 1991), neither form-P3cf nor form-P3f plastids were recorded outside the Caryophyllales. Furthermore, among the comparatively many species with form-Pcfs only a few (belonging to the Annonaceae and the Aristolochiaceae) are characterized by a single angular crystal. Outside the Caryophyllales form-Pfs plastids are even restricted to a single genus (*Canella* in the Canellaceae, but other genera of this family contain form-Pcfs plastids with many small crystals; cf. Behnke 1988a). None of these species, however, conform to the subtype-P3 characteristics, i.e. show a specific broad ring of filaments.

This situation makes the search for outgroups of the Caryophyllales almost impossible.

Among the dicotyledonous taxa above the order level, the Magnolianae contain the greatest number of families with P-type plastids. This might suggest that the ancestors of the two superorders (Magnolianae and Caryophyllanae) have had closer connections than any other two. However, the report of form-Pcfs plastids in the Pinaceae, but not in other families of the gymnosperms (Behnke 1974b), with characteristics of both the Magnolianae and Caryophyllanae transfers this problem to yet another level.

Thus, outgroup comparison for the plastid types of angiosperms favours the S-type being the plesiomorphous state and the P-type being apomorphous, but it does not offer a solution for the origin of the Caryophyllales-specific subtype-P3 plastids.

5.8 Addendum: On Phytoferritin in Plastids of Phloem Cells

Phytoferritin, an iron-containing protein, is not uncommon in plastids, its presence often being recorded in photosynthetically inactive plastids [e.g. sieve-element plastids, cf. Fig. 5.4 (*Cylindropuntia*) and Behnke 1971, 1977a, 1978] or chloroplasts of senescing plant parts (cf. Wildman and Hunt 1976). Heavy deposits of phytoferritin were reported to occur in the phloem-parenchyma plastids of many succulent species of the Caryophyllales and were discussed as possibly characterizing the succulent taxa of the order (Behnke 1977a, 1978).

Subsequently, the phytoferritin data were used, among others, as being indicative of a synapomorphy of an advanced group of succulents, the "cohort Portulacares" (Rodman et al. 1984). This interpretation has been criticized by Bittrich (1993). In between, the detection of phytoferritin accumulations in the phloem-parenchyma plastids of the succulent *Dicheranthus plucamioides* Webb. (Caryophyllaceae) corroborates what has previously been said, that (1) the presence of phytoferritin accumulations "adds to the peculiar features of the Centrospermae" (Behnke 1977a) and (2) "like other characters (e.g., anomalous secondary thickening, C4-photosynthesis, presence of betalains) is widespread, but not always present" (translated from Behnke 1978, p. 349).

Acknowledgements. The author wishes to thank again all those who were engaged in collecting and mailing fresh plant parts and who have been mentioned in the original publications of the author reviewed herein. Mrs. B. Moraw carried out most of the additional ultrathin sectioning and Mrs. D. Laupp the photographic processing required for the quantitative analyses of the many species.

The study of sieve-element plastids was continuously supported by grants from the Deutsche Forschungsgemeinschaft.

Refereneces

Airy Shaw HG (ed) (1973) JC Willis' A dictionary of the flowering plants and ferns, 8th edn. University Press, Cambridge

Behnke H-D (1968) Zum Feinbau der Siebröhren-Plastiden bei Monokotylen. Naturwissenschaften 55:140–141

Behnke H-D (1969) Über Siebröhren-Plastiden und Plastidenfilamente der Caryophyllales. Untersuchugen zum Feinbau und zur Verbreitung eines weiteren spezifischen Plastidentyps. Planta 89:275–283

Behnke H-D (1971) Phytoferritin in sieve-tube plastids. Naturwissenschaften 58:151–152

Behnke H-D (1974a) Elektronenmikroskopische Untersuchungen an Siebröhren-Plastiden und ihre Aussage über die systematische Stellung von *Lophiocarpus*. Bot Jahrb Syst 94:114–119

Behnke H-D (1974b) Sieve-element plastids of Gymnospermae: Their ultrastructure in relation to systematics. Plant Syst Evol 123:1–12

Behnke H-D (1975a) *Hectorella caespitosa*: Ultrastructural evidence against its inclusion into Caryophyllaceae. Plant Syst Evol 124:31–34

Behnke H-D (1975b) Elektronenmikroskopische Untersuchungen zur Frage der verwandtschaftlichen Beziehungen zwischen *Theligonum* und Rubiaceae: Feinbau der Siebelement-Plastiden und Anmerkungen zur Struktur der Pollenexine. Plant Syst Evol 123:317–326

Behnke H-D (1976a) Die Siebelement-Plastiden der Caryophyllaceae, eine weitere spezifische Form der P-Typ Plastiden bei Centrospermen. Bot Jahrb Syst 95:327–333

Behnke H-D (1976b) Ultrastructure of sieve-element plastids in Caryophyllales (Centrospermae), evidence for the delimitation and classification of the order. Plant Syst Evol 126:31–54

Behnke H-D (1976c) Sieve-element plastids of *Fouquieria*, *Frankenia* (Tamaricales), and *Rhabdodendron* (Rutaceae), taxa sometimes allied with Centrospermae (Caryophyllales). Taxon 25:265–268

Behnke H-D (1977a) Regular occurring massive deposits of phytoferritin in the phloem of succulent Centrospermae. Z Pflanzenphysiol 85:89–92

Behnke H-D (1977b) Phloem ultrastructure and the systematic position of Gyrostemonaceae. Bot Not 130:255–260

Behnke H-D (1978) Elektronenoptische Untersuchungen am Phloem sukkulenter Centrospermen (incl. Didiereaceen). Bot Jahrb Syst 99:341–352

Behnke H-D (1981) Sieve-element characters. Nord J Bot 1:381–400

Behnke H-D (1982a) *Geocarpon minimum*: sieve-element plastids as additional evidence for its inclusion in the Caryophyllaceae. Taxon 31:45–47

Behnke H-D (1982b) Sieve-element plastids, exine sculpturing and the systematic affinities of the Buxaceae. Plant Syst Evol 139:257–266

Behnke H-D (1988a) Sieve-element plastids, phloem protein and evolution of flowering plants. III. Magnoliidae. Taxon 37:699–732

Behnke H-D (1988b) Sieve-element plastids and systematic relationships of Rhizophoraceae, Anisophylleaceae, and allied groups. Ann Mo Bot Gard 75:1387–1409

Behnke H-D (1989) Sieve-element plastids, phloem proteins, and the evolution of flowering plants. IV. Hamamelidae. In: Crane PR, Blackmore S (eds) Evolution, Systematics, and fossil history of the Hamamelidae. Systematics Association, Spec Vol 40A, Clarendon Press, Oxford, pp 105–128

Behnke H-D (1990) Siebelemente. Kernlose Spezialisten für den Stofftransport in Pflanzen. Naturwissenschaften 77:1–11

Behnke H-D (1991) Distribution and evolution of forms and types of sieve-element plastids in the dicotyledons. Aliso 13:167–182

Behnke H-D (1993) Further studies of the sieve-element plastids of the Caryophyllales including *Barbeuia*, *Corrigiola*, *Lyallia*, *Microtea*, *Sarcobatus* and *Telephium*. Plant Syst Evol 186:231–243

Behnke H-D, Mabry TJ (1977) S-type sieve-element plastids and anthocyanins in Vivianiaceae: Evidence against its inclusion into Centrospermae. Plant Syst Evol 126:371–375
Behnke H-D, Schulz A (1983) The development of specific sieve-element plastids in wound phloem of *Coleus blumei* (S-type) and *Pisum sativum* (P-type), regenerated from amyloplast-containing parenchyma cells. Protoplasma 144:125–132
Behnke H-D, Turner BL (1971) On specific sieve-tube plastids in Caryophyllales. Further investigations with special reference to the Bataceae. Taxon 20:731–737
Behnke H-D, Chang C, Eifert IJ, Mabry TJ (1974) Betalains and P-type sieve-tube plastids in *Petiveria* and *Agdestis* (Phytolaccaceae). Taxon 23:541–542
Behnke H-D, Mabry TJ, Eifert IJ, Pop L (1975) P-type sieve-element plastids and betalains in Portulacaceae (including *Ceraria*, *Portulacaria*, *Talinella*). Can J Bot 53:2103–2109
Behnke H-D, Pop L, Sivarajan VV (1983a) Sieve-element plastids of Caryophyllales: Additional investigations with special reference to the Caryophyllaceae and Molluginaceae. Plant Syst Evol 142:109–115
Behnke H-D, Mabry TJ, Neuman P, Barthlott W (1983b) Ultrastructural, micromorphological and phytochemical evidence for a "central positon" of *Macarthuria* (Molluginaceae) within the Caryophyllales. Plant Syst Evol 143:151–161
Bessey C (1915) The phylogenetic taxonomy of flowering plants. Ann Mo Bot Gard 2:109–164
Bittrich V (1990) Systematic studies in Aizoaceae. Mitt Inst Allg Bot Hamb 23b:491–507
Bittrich V (1993) Introduction to Centrospermae. In: Kubitzki K, Rohwer J, Bittrich V (eds) Families and genera of flowering plants, vol 2. Springer, Berlin Heidelberg New York, pp 13–19
Bittrich V, Hartmann H (1988) The Aizoaceae – a new approach. Bot J Linn Soc 97:239–254
Bogle AL, Swain T, Thomas RD, Kohn ED (1971) *Geocarpon*: Aizoaceae or Caryophyllaceae? Taxon 20:473–477
Brown GK, Varadarajan GS (1985) Studies in Caryophyllales. I. Re-evaluation of classification of Phytolaccaceae s.l. Syst Bot 10:49–63
Carolin RC (1983) The trichomes of the Chenopodiaceae and Amaranthaceae. Bot Jahrb Syst 103:451–466
Carolin R (1987) A review of the family Portulacaceae. Aust J Bot 35:383–412
Cronquist A (1957) Outline of a new system of families and orders of dicotyledons. Bull Jard Bot 27:13–40
Cronquist A (1968) The evolution and classification of flowering plants. Nelson, London
Cronquist A (1981) An integrated system of classification of flowering plants. Columbia University Press, New York
Cronquist A (1988) The evolution and classification of flowering plants, 2nd edn. New York Botanical Garden, Bronx
Crookston RK, Ozbun JL (1975) The occurrence and ultrastructure of chloroplasts in the phloem parenchyma of leaves of C4 dicotyledons. Planta 123:247–255
Dahlgren G (1989) The last Dahlgrenogram – system of classification of the dicotyledons. In: Tan K (ed) The Davis and Hedge Festschrift. Edinburgh University Press, pp 249–260
Dahlgren RMT (1980) A revised system of classification of the angiosperms. Bot J Linn Soc 80:91–124
Danilova MP, Kozubov GM (eds) (1980) Atlas of plant tissue ultrastructure. Publ House Karelia, Petrozavordsk (in Russian)
Delay C, Darmanaden J (1973) Association entre les caractères morphologiques de la varieté horticole "minor" d'*Opuntia subulata* et la présence de microorganismes de type mycoplasme dans le phloème. Etude experimentale et infrastructurale. Ann Sci Nat Bot (Paris) 14:407–464
Eckardt T (1974) Vom Blütenbau der Centrospermen-Gattung *Lophiocarpus* Turcz. Phyton (Austria) 16:13–27

Eckardt T (1976) Classical morphological features of centrospermous families. Plant Syst Evol 126:5–25
Eichler AW (1876) Syllabus der Vorlesungen über Phanerogamenkunde. Schwer'sche Buchhandlung, Kiel
Eichler AW (1878) Blüthendiagramme. Zweiter Teil. Engelmann, Leipzig
Emberger L (1960) Les végétaux vasculaires. Vol 2 of Chadefaud M, Emberger L. Traité de botanique. Systématique. Masson & Cie, Paris
Engler A (ed) (1934) Die natürlichen Pflanzenfamilien, 2nd edn, vol 16c. Centrospermen. Engelmann, Leipzig
Esau K (1934) Ontogeny of phloem in sugar beet (*Beta vulgaris* L.). Am J Bot 21:632–644
Esau K (1965) Fixation images of sieve element plastids in *Beta*. Proc Natl Acad Sci (Wash) 54:429–437
Esau K (1976) Hyperplastic phloem and its plastids in spinach infected with the curly top virus. Ann Bot 40:637–644
Evert RF (1990) Dicotyledons. In: Behnke H-D, Sjolund RD (eds) Sieve elements. Comparative structure, induction and development. Springer, Berlin Heidelberg New York, pp 103–137
Falk H (1964) Zur Herkunft des Siebröhrenschleimes bei *Tetragonia expansa* Murr. Planta 60:558–567
Fisher DG, Evert RF (1982) Studies on the leaf of *Amaranthus retroflexus* (Amaranthaceae): Chloroplast polymorphism. Bot Gaz 143:146–155
Gibson AC (1978) Rayless secondary xylem of *Halophytum*. Bull Torrey Bot Club 105:39–44
Gibson AC, Nobel PS (1986) The cactus primer. Harvard University Press, Cambridge
Gilbert MG (1987) The taxonomic position of the genera *Telephium* and *Corrigiola*. Taxon 36:47–49
Goldblatt P, Nowicke JW, Mabry TJ, Behnke H-D (1976) Gyrostemonaceae: Status and affinity. Bot Not 129:201–206
Hallier H (1903) Über Verwandschaftsverhältnisse bei Englers Rosalen, Parietalen, Myrtifloren und and. Ordnungen d. Dicotylen. Abh Naturwiss Ver Hamb 18:1–98
Hallier H (1912) L'origine et le système phylétique des Angiospermes exposés à l'aide de leur arbre généalogique. Arch Néerl Sci Exact Nat Sér 3B:146–234
Hofmann U (1973) Morphologische Untersuchungen zur Umgrenzung und Gliederung der Aizoaceen. Bot Jahrb Syst 93:247–324
Hofmann U (1977) Die Stellung von *Stegnosperma* innerhalb der Centrospermen. Centrospermenstudien 9. Ber Dtsch Bot Ges 90:39–52
Hunziker JH, Behnke H-D, Eifert IJ, Mabry TJ (1974) *Halophytum ameghinoi*: A betalain-containing and P-type sieve-tube plastid species. Taxon 23:537–539
Kollmann R, Glockmann C (1990) Sieve elements of graft unions. In: Behnke H-D, Sjolund RD (eds) Sieve elements. Comparative structure, induction and development. Springer, Berlin Heidelberg New York, pp 219–237
Mabry TJ, Behnke H-D (1976) Betalains and P-type sieve-element plastids: The systematic position of *Dysphania* R. BR. (Centrospermae). Taxon 25:109–111
Mabry TJ, Eifert IJ, Chang C, Mabry H, Kidd C, Behnke H-D (1975) Theligonaceae: pigment and ultrastructural evidence which excludes it from the order Centrospermae. Biochem Syst 3:53–55
Mabry TJ, Behnke, H-D, Eifert, IJ (1976) Betalains and P-type sieve-element plastids in *Gisekia* L. (Centrospermae). Taxon 25:112–114
Metcalfe CR, Chalk L (1950) Anatomy of the dicotyledons. Clarendon Press, Oxford
Nowicke JW (1968) Palynotaxonomic study of the Phytolaccaceae. Ann Mo Bot Gard 55:294–363
Pax F, Hoffmann K (1934) Caryophyllaceae. In: Engler A (ed) Die natürlichen Pflanzenfamilien, 2nd edn, vol 16c. Engelmann, Leipzig, pp 275–367
Prance GT (1968) The systematic position of *Rhabdodendron* Gilg a. Pilg. Bull J Bot Nat Belg 38:127–146

Rodman JE, Oliver MK, Nakamura RR, McClammer JU, Bledsoe AH (1984) A taxonomic analysis and revised classification of Centrospermae. Syst Bot 9:297–323

Schulz A (1990) Wound-sieve elements. In: Behnke H-D, Sjolund RD (eds) Sieve elements. Comparative structure, induction and development. Springer, Berlin Heidelberg New York, pp 199–217

Steyermarck JA, Voigt JW, Mohlenbrock H (1959) Present biological status of *Geocarpon minimum* Mackenzie. Bull Torrey Bot Club 85:124–127

Takhtajan A (1959) Die Evolution der Angiospermen. Fischer, Jena

Takhtajan A (1969) Flowering plants. Origin and dispersal. Oliver & Boyd, Edinburgh

Takhtajan A (1973) Evolution und Ausbreitung der Blütenpflanzen. Fischer, Stuttgart

Takhtajan A (1980) Outline of the classification of flowering plants (Magnoliophyta). Bot Rev 46:225–359

Takhtajan A (1987) Systema Magnoliophytorum. Nauka, Leningrad (in Russian)

Thorne RF (1983) Proposed new realignments in the angiosperms. Nord J Bot 3:85–117

Thorne RF (1992) An updated phylogenetic classification of the flowering plants. Aliso 13:365–389

Ulbrich E (1934) Chenopodiaceae. In: Engler A (ed) Die natürlichen Pflanzenfamilien, 2nd edn, vol 16c. Engelmann, Leipzig, pp 379–584

Wildman RB, Hunt P (1976) Phytoferritin associated with yellowing in leaves of *Cocos nucifera* (Arecaceae). Protoplasma 87:121–134

6 Flower Morphology and Ontogeny

Ursula Hofmann

6.1 Introduction

Delimitation of higher taxa in the Caryophyllales is made difficult by the great diversity in character expressions and by their high degree of reticulation. In flower morphology a very rich radiation can be observed, resulting in a wealth of forms. Furthermore, flowers (and the derived fruits) evolve functionally in reaction to surrounding conditions, e.g. pollinators and dispersers, thus representing highly adaptive, possibly partly analogous syndromes.

As a consequence, the use of data from flower morphology poses a great problem when circumscriptions of higher taxa with a sufficiently homogeneous set of character expressions are to be achieved. In the following I try to demonstrate the great variety within flower character, rather than propose a systematic rearrangement.

6.1.1 Materials and Methods

In addition to material used in earlier papers, flower buds collected in the wild and from the old botanical garden of Göttingen, as well as additional samples provided by several colleagues (Table 6.1), were fixed in a mixture of formalin, acetic acid, and alcohol (FAA). Some flower buds were prepared according to standard methods for serial sections, while others were dehydrated with formaldehyde-dimethyl-acetal for scanning electron microscopy (SEM) investigations with a Phillips 515 SEM. Furthermore, O. Rohweder (Zürich) passed on to me all his serial sections of Caryophyllales.

6.1.2 What Groups Can Be Used as Homogeneous Units?

In view of the difficulties in circumscription of higher taxa in the order, a pragmatic approach is used here. Family names are applied to those groups of genera that share a common flower morphological type. This allows some degree of variation (see, e.g., Table 6.2), but restricts the name to a defined

Table 6.1. Taxa recognized as sufficiently distinct in flower morphology (FM) and taxa of uncertain classification which are preliminarily supposed to be individual operational units (O)

Distinguished taxa	FM/O	Species critically investigated for FM	Method	Source	Reference
Phytolaccaceae	FM				
Phytolacceae	FM	*Phytolacca acinosa* Roxb.	S	BG Z	Rohweder (1965b)
		P. americana L.	S	BG MB	
		P. clavigera W.W. Smith	S	BG Z	
		Phytolacca sp.	S, SEM	BG GOET	
		Ercilla spicata (Bert.) Moq.	S	H M, B	Hofmann (1977)
		Anisomeria coriacea Moq.	S	H B, M	Hofmann (1977)
		A. litoralis Moq.	S	H B, Z	Hofmann (1977)
Rivineae	FM	*Seguieria* sp.	S	H GOET	
		Gallesia integrifolia (Spreng.) Harms	S	H M	
		Monococcus echinophorus F. Müll.	S	H H-D Behnke	
		Petiveria alliacea L.	S	BG Z	
		Rivina humilis L.	S	BG Z	
		Trichostigma peruvianum (Moq.) H. Walt.	S, SEM	BG Z, GOET	
		Ledenbergia seguierioides Klotzsch	S	H GOET	
		Hilleria latifolia (Lam.) H. Walt.	S	BG Z	
Agdestis	FM O	*A. clematidea* Mocino & Sesse	S	H M	Hofmann (1977)
Barbeuia	FM O	*B. madagascariensis* Steud.	S	H M	Hofmann (1977)
Stegnosperma	FM O	*S. halimifolium* Benth.	S	R Moran, Baja California	Hofmann (1977)
Microtea	O				
Lophiocarpus	O				Eckardt (1974)
Achatocarpaceae	O				
Nyctaginaceae	FM				
Mirabileae	FM	*Mirabilis jalapa* L.	S	BG GOET	Rohweder and Huber (1974)
		M. longiflora L.	S	BG GOET	
		Oxybaphus viscosus L'Herr.	S, SEM	V Bittrich	
		O. nyctagineus (Michx.) Sweet	S, SEM	BG GOET	
		Abronia latifolia Eschsch.	S, SEM	BG GOET	

Pisonieae		O	*Pisonia umbellifera* (Forster & G. Forster) Seemann	S		P Endress 9121 N Queensland	
			Neea sp.		SEM	V Bittrich	
Aizoaceae s.s.	FM						
Aizooideae	FM		*Acrosanthes teretifolia* Ecklon & Zeyher	S		See Hofmann (1973)	Hofmann (1973)
			Aizoon canariensis L.	S		See Hofmann (1973)	Hofmann (1973)
			A. hispanicum L.				Hofmann (1973)
			Aizoanthemum (*Aizoon*) *galenioides* (Fenzl ex Sond.) Friedrich	S		See Hofmann (1973)	Hofmann (1973)
			Aizoanthemum (*Aizoon*) *dinteri* (Schinz) Friedrich		SEM	V Bittrich	Bittrich (unpubl.)
			Gunniopsis (*Aizoon*) *kochii* (Wagner) Chinnock	S,	SEM	See Hofmann (1973)	Hofmann (1973)
			G. (*Aizoon*) *glabra* (Luehm. ex Ewart) Gardner		SEM	V. Brittrich	Bittrich (unpubl.)
			G. zygophylloides (F.v.M.) Diels		SEM	V. Brittrich	Bittrich (unpubl.)
			G. (*Neogunnia) septifraga* (F.v.M.) Chinnock	S		See Hofmann (1973)	Hofmann (1973)
			Galenia sarcophylla Fenzl	S		See Hofmann (1973)	Hofmann (1973)
			Plinthus sericeus Pax	S		See Hofmann (1973)	Hofmann (1973)
Tetragonioideae	FM		*Tetragonia calycina* Fenzl	S		See Hofmann (1973)	Hofmann (1973)
			T. echinata Ait.	S		See Hofmann (1973)	Hofmann (1973)
			T. eremaea Ostenfeld	S		See Hofmann (1973)	Hofmann (1973)
			T. teragonoides (Pallas) Kuntze	S		See Hofmann (1973)	Hofmann (1973)
Mesembryanthemoideae	FM						Ihlenfeldt (1960)
Ruschioideae	FM						Ihlenfeldt (1960)
Sesuvioideae	FM		*Sesuvium portulacastrum* (L.) L.	S		See Hofmann (1973)	Hofmann (1973)
			Zaleya galericulata (Melville) HJ. Eichler	S		See Hofmann (1973)	Hofmann (1973)
			Trianthema portulacastrum L.	S		See Hofmann (1973)	Hofmann (1973)
Didiereaceae	FM						Rauh and Schölch (1965)
Cactaceae	FM						Leins and Schwitalla (1986)
Chenopodiaceae		O	*Beta trigyna* Waldst. & Kit.		SEM	BG GOET	
			B. vulgaris L.		SEM	BG GOET	
Halophytum		O					

Table 6.1. *Continued*

Distinguished taxa	FM/O	Species critically investigated for FM	Method	Source	Reference
Sarcobatus	O	*S. vermiculatus* (Hook.) Torr.	S	H-D Behnke	
Dysphania	O				Eckardt (1967a,b, 1968, 1969)
Amaranthaceae	O				
Portulacaceae	FM				
Portulacoideae	O	*Portulaca grandiflora* Hooker	S	BG GOET, Z	
		P. oleracea L.	S, SEM	G GOET	
		Portulaca sp.	S	G GOET	
		Calandrinia grandiflora Lindl.	S, SEM	BG GOET	
		Anacampseros arachnoides (Haw.) Sims	S	BG Z	
		Talinum patens (Jacq.) Willd.	S	BG Z, GOET	
		Lewisia cotyledon (S. Wab.) Rob.	S	BG GOET	
		L. heckheri (Morton) Howell	S	BG GOET	
		L. pygmaea (Gray) Rob.	S	BG GOET	
		L. tweedii (Gray) Rob.	S	BG GOET	
		L. columbiana (Howell) Robinson var. *walowiensis* C. Hitchc.	S	BG GOET O Rohweder	
		Portulacaria afra Jacq.	S		Rauh and Schölch (1965)
		Ceraria			
Montioideae	O	*Montia perfoliata* (Donn ex Willd.) Howell	S, SEM	BG GOET, Z	
		M. sibirica (L.) Howell	S, SEM	BG GOET, Z	
		M. fontana L.	S	O Rohweder	
Basellaceae	FM	*Basella alba* L. (*B. rubra*)	S	BG GOET, Z	
Molluginaceae	FM				
Corbichonieae	FM	*Corbichonia decumbens* (Forsk.) Exell	S	See Hofmann (1973)	Hofmann (1973)
Limeeae	FM	*Macarthuria necocambrica* F.v.Müll.	S	See Hofmann (1973)	Hofmann (1973)
		M. australis Huegel ex Endl.	SEM	Bittrich & Jenssen 18585, W. Australia	Bittrich (unpubl.)
		Limeum linifolium Fenzl.	S	H-D Behnke	Hofmann (1973)
		L. sulcatum (Klotzsch) Hutch.	S	See Hofmann (1973)	Hofmann (1973)
		L. viscosum (J. Gay) Fenzl	S	See Hofmann (1973)	Hofmann (1973)

Mollugineae	FM		*Glinus lotoides* L.	S, SEM	Bittrich & Jenssen 18156, W. Australia	Hofmann (1973)
			Hypertelis bowkeriana Sond.	S	See Hofmann (1973)	Hofmann (1973)
			H. salsoloides (Burch.) Adamson	S, SEM	V Bittrich	Bittrich (unpubl.)
			Pharnaceum subtile E. Mey. ex Fenzl	S	See Hofmann (1973)	Hofmann (1973)
			Coelanthum semiquinquefidum (Hook. F.) Druce	S	See Hofmann (1973)	Hofmann (1973)
			Mollugo cerviana (L.) Ser.	S	See Hofmann (1973)	Hofmann (1973)
			M. verticillata L.	S	See Hofmann (1973)	Hofmann (1973)
			Glischrothamnus ulei Pilger	S	See Hofmann (1973)	Hofmann (1973)
			Psammotropha myriantha Sond.	S	See Hofmann (1973)	Hofmann (1973)
			Adenogramma glomerata (L. f.) Druce	S	See Hofmann (1973)	Hofmann (1973)
Gisekia	FM	O	*G. africana* (Lour.) O. Kuntze	S	See Hofmann (1973)	Hofmann (1973)
			G. pharnacioides L.	S	See Hofmann (1973)	Hofmann (1973)
Polpoda	FM	O	*P. capensis* Presl	S	See Hofmann (1973)	Hofmann (1973)
Caryophyllaceae	FM					
Paronychioideae		O	*Spergula arvensis* L.	SEM	BG GOET	Rohweder and König (1971) Rohweder and Urmi-König (1975)
Alsinoideae	FM		*Cerastium brachypetalum* Pers.	SEM	BG GOET	Rohweder (1970)
			C. vulgatum L.	SEM	Ricken (CH)	
			Myosoton aquaticum (L.) Moench	SEM	G GOET	
			Arenaria procera Sprengel	SEM	BG GOET	
			Stellaria graminea L.	SEM	Ricken (CH)	
			S. media (L.) Vill.	SEM	G GOET	
			Lepyrodiclis holosteoides (C. A. Mey) Fenzl ex Fisch.	SEM	BG GOET	
			Sagina procumbens L.	SEM	G GOET	
			Scleranthus annuus L.	SEM	BG GOET	

Table 6.1. *Continued*

Distinguished taxa	FM/O	Species critically investigated for FM	Method	Source	Reference
Silenoideae	FM	*Drypis spinosa* L.	SEM	BG GOET	Rohweder and Urmi (1978)
		Lychnis coronaria (L.) Desr.	SEM	G GOET	Rohweder (1965a, 1967)
		Lychnis flos-cuculi L.	SEM	Ricken (CH)	
		Silene dioica (L.) Clairv.	SEM	Ricken (CH)	
Telephium	O	*T. imperati* L.	SEM	BG GOET	
Corrigiola	O	*C. litoralis* L.	SEM	BG GOET	
Hectorella	O				
Lyallia	O				
Geocarpon	O	*G. minimum* Mackenzie			Bogle et al. (1971)

S, serial sections; SEM, scanning electron microscopy; BG, botanical garden; G, private garden; H, herbarium; City abbreviations according to *Index Herbariorum*. Family arrangement according to Cronquist (1981). Family division according to Eckardt (1964); Aizoaceae according to Bittrich and Hartmann (1988), Hartmann and Bittrich (1991); Molluginaceae according to Hofmann (1973), partly modified.

basic pattern. In some cases, due to a lack of material, the results refer only to those genera or species that have been studied (Table 6.1). These family delineations are certainly often more restricted than those of other authors (e.g. Cronquist 1981).

In general, it must not be forgotten that many major families of Caryophyllales require revision based on the new interpretation of known features and the discovery of new characters. At present, volume 16c (Reihe Centrospermae) of Engler's (1934) *Die natürlichen Pflanzenfamilien* is the latest comprehensive treatise of the Caryophyllales and is unfortunately still used as a reference work.

In all cases of genera of uncertain position, the genera are referred to as separate units at that rank, although all of them have been raised to familial rank at one point or another (e.g. Nakai 1942). In this way, relatively homogeneous groups can be compared.

6.1.2.1 Commentary on Families

Phytolaccaceae. As uncontested members I include only the tribes Phytolacceae and Rivineae sensu Eckardt (1964). The genera *Agdestis*, *Barbeuia* and *Stegnosperma* (often included in the family) as well as *Microtea* and *Lophiocarpus* (rarely included) appear individually. Today, the glucosinolate-containing Gyrostemonaceae are mostly assigned to totally different orders (see, e.g., Cronquist 1981; Dahlgren 1983); they are treated in section 6.2.2.3 here.

Nyctaginaceae. Only material of the Mirabileae (*Mirabilis*, *Oxybaphus* and *Allionia*) and Pisonieae (*Neea* and *Pisonia*) was available for examination; therefore I cannot comment on the remaining three tribes, Colignonieae, Boldoeae and Leucastereae.

Aizoaceae s.s. Bittrich and Hartmann (1988) give a comprehensive survey of the different treatments of this family, dividing it into five subfamilies. I have only worked on representatives of the *Aizoon*, *Tetragonia* and *Sesuvium* groups.

Portulacaceae. Delimitation of genera is still under dispute (e.g. Carolin 1987, or Hershkovitz 1991 combining in his genus *Cistanthe* species formerly treated as members of four different genera), as is the grouping of genera. Seven genera were investigated: *Portulaca*, *Anacampseros*, *Talinum*, *Calandrinia* (probably forming one group), *Lewisia*, *Portulacaria* and *Montia*.

Molluginaceae. Three groups from Hofmann (1973) are maintained, namely the *Corbichonia*, *Limeum*, and *Mollugo* groups; the *Sesuvium* group is treated under Aizoaceae; and *Gisekia* is discussed as an individually treated genus in relation to the Phytolaccaceae. Whether the genus *Polpoda* belongs to the Molluginaceae is still an open question, because the growth pattern does not agree and the flower character expressions are too general to permit conclusions to be drawn.

Bittrich and Hartmann (1988) state the lack of a synapomorphy for the Molluginaceae. However, I wish to emphasize that the peculiar sympodial ramification (Figs. 1–12 in Hofmann 1973) is found in all three groups included above. With the exception of the main stem, the plant consists of identical modules with a definite number of leaves. Only *Limeum* and *Corbichonia* deviate in having a variable numbers of leaves, and *Macarthuria* has thyrsical inflorescences.

Outside the Molluginaceae this pattern is found in *Gisekia*; similar patterns are found in some members of the Aizoaceae, in *Stellaria alsine* and *Corrigiola litoralis* (Rutishauser 1981) of the Caryophyllaceae, and in the tribe Mirabileae of the Nyctaginaceae.

Caryophyllaceae. Three subfamilies are traditionally accepted (Pax and Hoffmann 1934): Silenoideae and Alsinoideae seem well defined (excluding *Scleranthus*), whereas the Paronychioideae appear problematical (Bittrich 1993), in particular *Spergula*, *Spergularia* and *Drymaria* (with rather large flowers). The position of *Telephium* and *Corrigiola* with scattered leaves also remains controversial (Gilbert 1987). *Hectorella*, producing betalains (Behnke 1975; Mabry et al. 1978), and *Lyallia*, sometimes included here, must be treated separately, perhaps in relation to the Portulacaceae.

6.2 Results

6.2.1 Observations on some Individually Treated Genera

Agdestis: K2 + 2, A4 × 3 – 4as + 1es, G4es; 1 apotropous ovule/carpel (K denotes calyx/sepals; P, corolla/petals; A, androecium/stamens; G, gynoecium; es, episepalous position; ep, epipetalous position; as, alternisepalous position). Based on studies of three buds there are four alternisepalous groups of three to four stamens. In addition, a single stamen occurred in a strict episepalous position in two flowers. The "disc" represents the tissue at the top of the ovary and bears closely packed stomata (probably "Saftspalten"). The filaments insert at the marginal zone of the top of the ovary, surrounded by the free lobes of the calyx. There are no connate parts above the point of fusion of ovary and calyx. The styles are free and recurved toward the ovary (Fig. 73 in Heimerl 1934). They are covered by fascicles of laterally connected hairs; single papillae have never been found. As a rule, the dorsal bundles divide into two strands in their course over the top of the ovary from where they run laterally, close to one another, down the locules, to join again below them. Sometimes, a bundle loses connection with its counterpart. Long apical septa occur internally (not externally visible, as in *Phytolacca acinosa*); the apical porus is filled by small-celled pollen tube conducting tissue reaching into the long central channel. The pollen tube conducting papillae, which cover the margins of the septa and the funicles (forming a frill around each of them), block the ventral clefts of the short symplicate zone. In the synascidiate zone the ventral wall of each carpel is occupied by the point of insertion of the funicle. The septa are present at anthesis, although the mesophyll exhibits large intercellular spaces. No oxalate crystals were found in the serial sections.

Barbeuia: K5, A∞, G2; 1 ovule/carpel; oxalate in druses. Only one flower bud could be studied. Although the calyx forms a quincunx, the carpels are placed almost exactly in front of sepals 1 and 2, instead of in the usual position between sepals 1 and 3 and in front of sepal 2. Sepals 1 and 2 insert on a slightly, but distinct, lower level than the others. The flowers are hypogynous and a broad ring, from which the filaments arise, can be seen but there is no separate disc. In my specimen there were 52 stamens, in two rows in front of sepals 4 and 5, and in three to four rows in front of sepals 1–3. The styles are slightly flattened dorsiventrally and covered with large stigmatic papillae on the interior and along the margins. The styles join basally, from where they pass gradually into the thick conical top of the ovary. The dorsal bundles lie just below the outer surface. On top of the ovary, additional strands, sometimes anastomosing, occur next to the stylar channel lined with small papillae. Further down, the additional strands may be found passing approximately through the middle of the thick outer ovary wall. Even further down, next to the stylar channel, appear the locules, apically separated from it by the apical septa. Only a few sections below, the stylar channel is united with both locules. It was not possible to determine whether the apical septa were ruptured or the ventral clefts genuine. In the entire adjacent symplicate zone the thick septa do not meet in the centre. The margins are covered by rather fine papillae. The plane of symmetry of the campylotropous ovules is parallel to the plane of the septa (as in *Acrosanthes* and *Pollichia*). The points of insertion of the funicles occupy the whole ventral wall in the short synascidiate zone.

The herbarium specimen was totally black, a feature also observed in species of *Acrosanthes*.

Stegnosperma: K5, P5, A5es + 5ep, G5ep, not episepalous as often stated; 1 epitropous ovule/carpel; oxalate in druses (Plate 6.6,61). The ten stamens, arranged in one whorl, form a short, basal filament tube evenly lined inside with nectary tissue bearing "Saftspalten". The five episepalous stamens are the longer ones in buds, corresponding rather to the pattern of Caryophyllaceae than to that of Mollugineae (e.g. *Hypertelis*). The five carpels are epipetalous. *Stegnosperma* shares this rare arrangement of carpels only with some members of the Silenoideae and Alsinoideae; however, *Stegnosperma* differs from those groups in having an ovary without a thickened top. The ovules, soon covered by an aril, completely occupy the locules and seem to squash the septa very early (Fig. 6 in Hofmann 1977).

Gisekia: K5, A5 × 3as or A5as, G5es; 1 anatropous ovule/carpel; oxalate in raphides. This genus shares with the Phytolacceae the form of the ovary and the raphides, and with most Molluginaceae the modular growth form. The stamens occur in alternisepalous groups of three, wedged between carpels and sepals, with the longest one of each group innermost and the two shorter ones outermost. The five inner filaments link together at the base, forming a short tube containing nectary tissue (Hofmann 1973). The outer ones connect basally with this tube. The traces do not join, but enter the same leaf gap. The arrangement of the stamens in alternisepalous groups of three elements agrees neither with the Phytolaccaceae nor with the Molluginaceae. Some *Gisekia* species show only one stamen in each group.

Polpoda: K4, A4as, G2; 1 epitropous ovule/carpel. The monopodial shoots of this dwarf shrub are closely covered with alternate leaves. The scarious adnate stipules are larger than the reduced green blades, resulting in a scaly appearance of the branches (resembling species of *Anacampseros*). Covered by the leaf, a jolted short shoot grows from the axil. On this shoot two green leaves (the largest blades on the plant), in the position of bracteoles, and further smaller leaves of uncertain placement are found. All leaves have scarious stipules. The supposedly terminal inflorescence possesses at most three flowers. The perianth is composed of two outer and two inner sepals, four alternisepalous stamens with long filaments (pollination by wind?) and two connate carpels in front of the outer sepals. The styles are joined at the base, a condition not found in Molluginaceae. The two ovules are each placed a little below the "Querzone" on the ventral wall in the synascidiate zone.

Telephium and *Corrigiola*: K5, P5, A5es, G3. Both genera possess alternate leaves and cordate stipules. Their flowers contain five petals and only five episepalous stamens, like *Drypis*, but they differ in the number of ovules. In *Corrigiola* the single ovule appears very early (Plate 6.5,41) in the ontogeny. The fruit is a thick-walled nut. In the ovary of *Telephium*, the edges of the septa detach later than in other Caryophyllaceae and the basally connate styles continue into the ovary, not forming a dip on the top of the ovary. *Telephium* lacks the thickened and lignified apical ovary walls present in most Caryophyllaceae.

Sarcobatus: The plant is monoecious. The female flowers seem to grow solitary in the axils of the uppermost foliage leaves below the terminal male stand. This catkin-like part of the plant, apparently similar to the male inflorescence of *Batis*, is composed of the main stem covered with peltate bracts, joining at the apex of the male stand into a rather solid tip. In addition, many solitary stamens can be found. No pattern in the positions of bracts and stamens can be detected, and therefore no male flowers with several stamens can be distinguished. Ulbrich (1934) states that one stamen represents one male flower. In *Batis* the male flower is covered by a peltate bract and consists of an involucre of two connate median bracts, four small petaloid phyllomes, and four stamens with long filaments. In *Sarcobatus* the filaments are very short and the anthers very long, thus resembling those of Gyrostemonaceae. The radially symmetrical anthers have four locules with thickened walls at the top. The stamens of *Sarcobatus* differ greatly from those of *Chenopodium*, which have long filaments and short dithecic anthers.

The female flower consists of two carpels only, as can be seen from the two styles, lined inside with stigmatic papillae. At one-third of the length from the base of the unilocular ovary, there is a ring-fold developing into a wing around the fruit. This fold is obscured by a filt of squarrosely, irregularly branched hairs. The single ovule inserts basally. One ovary is accompanied by a very young stamen (see also Ulbrich 1934), the other by a female accessory flower. Indeed, each gynoecium is preceded by two small bract-like phyllomes in the positions of bracteoles. After Ulbrich, bracteoles are considered to be absent.

6.2.2 Ontogeny of the Flowers, Especially of the Androecium and Petals

This study is focused on two aspects of floral development in the Caryophyllales.

1. Are there examples of centrifugal differentiation of stamens outside the Aizoaceae (Payer 1857) and Cactaceae (Leins and Schwitalla 1986)? This might support a closer relationship between Caryophyllales and Dilleniidae, which are often characterized by centrifugal development in the androecium (Leins 1975).
2. In general, the outer green elements of a double perianth are called sepals, the inner coloured ones petals. In Caryophyllales this simple classification does not work because organs of equal function can be of different origin. Sepals and petals cannot be distinguished on the basis of presence or absence of pigmentation, e.g. the bracts of *Bougainvillea* are more showy than the perianth, and the calyx of *Sesuvium* is coloured inside. Rohweder (1967, 1970) attempted to determine anatomical differences between sepals and petals (e.g. formation of scarious or white margins), but found exceptions in many cases.

In this chapter, an ontogenetic definition is suggested: The term sepals is used for elements of the perianth differentiated earlier than the stamens. The term petals (stapetals) is used for those elements that appear simultaneously with the first primordia of individual stamens or later than these. This implies that petals (stapetals) meeting this definition are not homologous to petals of several other systematic groups. In *Hypericum*, for example, with a centrifugal androecium the petal primordia appear before the emergence of the primary stamen primordia (Leins 1964).

6.2.2.1 Families with Centrifugally Originating Stamens

Phytolaccaceae. In Phytolacceae there is often a leap in the number of organs after the 2/5 formation of the pentamerous perianth. The number 8, next in the Fibonacci series, is frequently, though inconsistently, observed in the androecium and in the alternating gynoecium (Plate 6.1,2). Rarely, 13 stamens can be seen. In some species of *Phytolacca* (and presumably in *Anisomeria* with about 20 stamens) there is not only one whorl with more elements in comparison with the perianth, but outside this whorl another one later arises, consisting of only some more elements than the inner one. There is no connection between the position of the elements of both whorls. In some specimens a distinct edge or rim of the floral apex, resembling a primary primordium, can be observed before the emergence of the stamens of the subsequent whorls (Plates 6.1,2;6.6,42).

Trichostigma peruviana was the only species studied from the Rivineae to exhibit a great number of stamens. Like all representatives of Rivineae, with the exception of *Seguieria*, *T. peruviana* has a single tetramerous perianth, which is composed of two outer median sepals (subsequent to the transversal bracteoles) and two inner transversal ones (Plate 6.1,3). Four alternisepalous individual stamen primordia appear simultaneously on the rather high floral apex near the top (Plate 6.1,4). On the flanks below, four episepalous and, finally, four alternisepalous primordia follow in sequence (Plates 6.1,5 and 6; Plate 6.6,43), if the whole pattern is performed regularly. Perhaps the other Rivineae (e.g. *Ledenbergia*) with a moderate number of stamens follow the same pattern of development. Studies of the ontogeny of *Trichostigma polyandrum*, *Gallesia* or *Seguieria* (for which Heimerl 1934 stated 25 or more stamens) would be of value. *Rivina* and two species of *Hilleria* develop only the first four alternisepalous stamens (Plate 6.6,44).

Aizoaceae. The following is based on unpublished SEM photographs of Bittrich. In the tetramerous *Gunniopsis zygophylloides* and *G. glabra* four distinct alternisepalous primary primordia arise on which individual primordia develop centrifugally and regularly (Plate 6.6,49), at least in

Plate. 6.1. Development of the androecium in Phytolaccaceae and Nyctaginaceae. *1* and *2*, *Phytoplacca* spec.; *3–6*, *Trichostigma peruviana*; *7* and *8*, *Neea* spec. *I*, *II*, Decussate or otherwise paired sepals; *1–5*, sepals in successive appearance; •*1* –•*8*, stamina in successive appearance; ●, innermost stamen (centrifugal appearance); •, outermost stamen (centrifugal appearance); *, alternisepalous stamen; *white arrowhead*, top of carpel; *B*, bract; *o*, ovule; *t*, (transversal) bracteole; *bar* denotes 0.1 mm

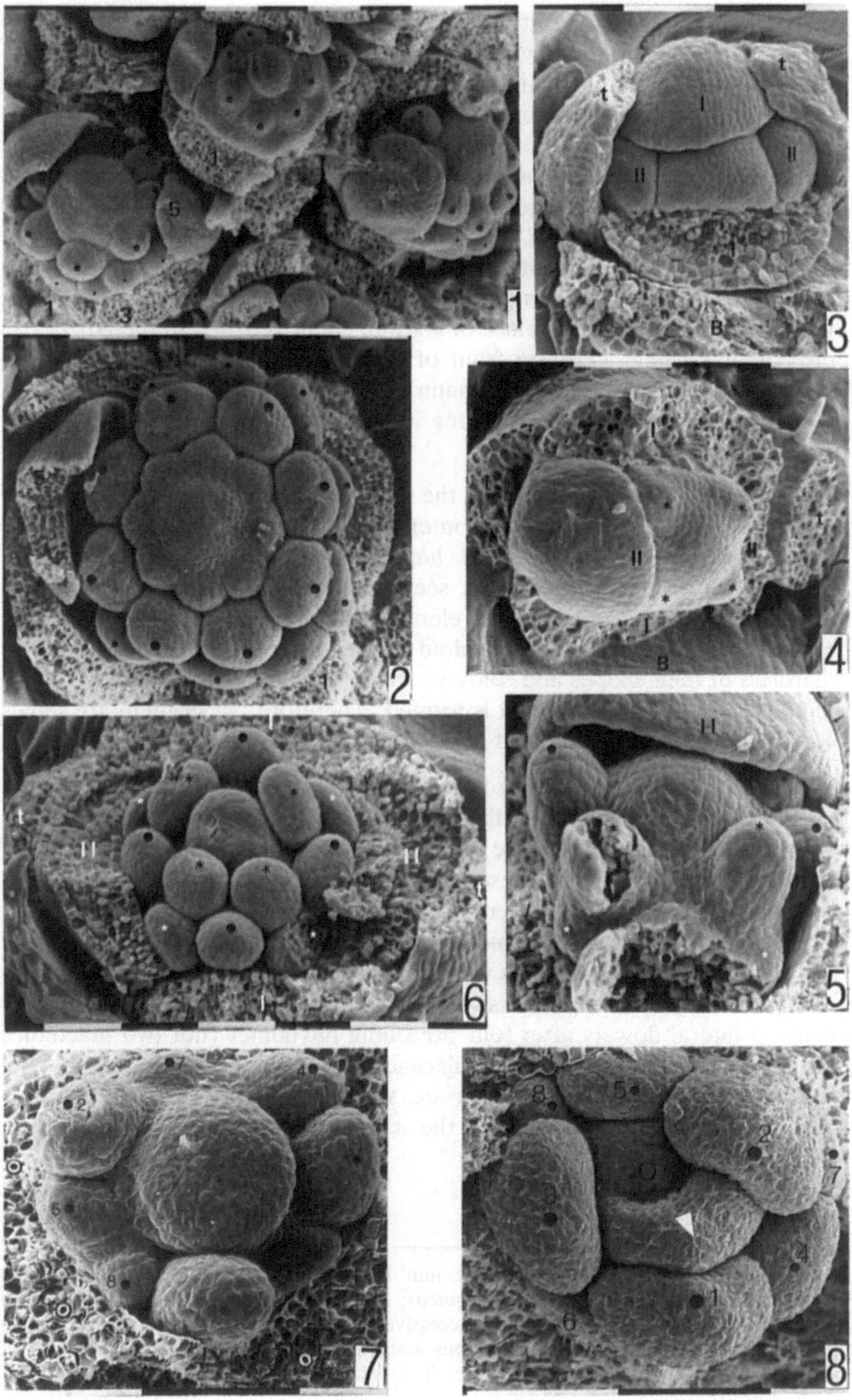

the beginning (Fig. 7 in Bittrich 1990). In contrast, *Aizoanthemum* shows centrifugally originating individual stamens not preceded by distinct primary primordia and without clear arrangement in the succeeding rows (Plate 6.6,48). Only in the Mesembryanthemoideae (Hartmann and Bittrich 1991) and Ruschioideae are the outermost elements of the centrifugal series able to differentiate into petals instead of stamens. In several genera transitional forms occur between stamens and petals (Ihlenfeldt 1960; filamentous staminodes after Hartmann 1991). Not all members of the Aizooideae have alternisepalous elements in large groups. In *Galenia sarcophylla*, for example, each group consists of only two stamens, while in *Plinthus sericeus* individual stamens take the place of a group (Plate 6.6,50). In the Aizoaceae episepalous stamens exist (in front of sepals 1–3) only in *Trianthema* belonging to the Sesuvioideae (Hofmann 1973).

In *Tetragonia* the species differ in numbers of stamens in the alternisepalous groups.

Cactaceae. In *Pereskia grandifolia* the stamens originate centrifugally without a recognizable pattern on the outer side of a common ring primordium. In *Pereskia diaz-romeroana* and *P. bleo* isolated primary primordia can be found (Leins and Schwitalla 1986; see also Chap. 13) alternating more or less obviously with the innermost elements of the perianth. The individual primordia do not develop into petaloid organs. All elements of the perianth, regardless of their shapes and colours, originate in spiral arrangement before the androecial ring primordium is formed. Therefore, all of them are sepals according to the definition used here.

Portulacaceae. Great variety in floral morphology characterizes this family. It is typical for all members of the Portulacaceae that the coloured elements of the perianth in all flowers are preceded by two, mostly green, broad leafy organs. In lateral flowers these are placed decussate to the two lower bracteoles (if present) and therefore meet the median plane. I call these leafy organs involucral phyllomes. These are followed apically by five or, rarely, more perianth members which in *Portulaca oleracea* and *Calandrinia grandiflora* are initiated successively in a 2/5 arrangement. This quincunx begins in lateral flowers after four preceding phyllomes (not two bracteoles only as, for example, in Caryophyllaceae) at a slightly different place. The elements 1 and 2 of the quincunx are situated laterally in front, 3 and 5 laterally at the back and 4 meets the median plane in the front. Because

⟶

Plate. 6.2. Development of the androecium in Nyctaginaceae, Chenopodiaceae and Molluginaceae. *9–12*, *Oxybaphus nyctagineus*; *13* and *14*, *Beta vulgaris*; *15–17*, *Glinus lotoides*. *I–V*, (Involucral) bracts in successive appearance; *1–5*, sepals in successive appearance; ●, stamen; *, alternisepalous stamen; ⧫, petal; *white arrowhead*, top of carpel; *gear*, flower primordium; *o*, ovule; *B*, bract; *bar* denotes 0.1 mm

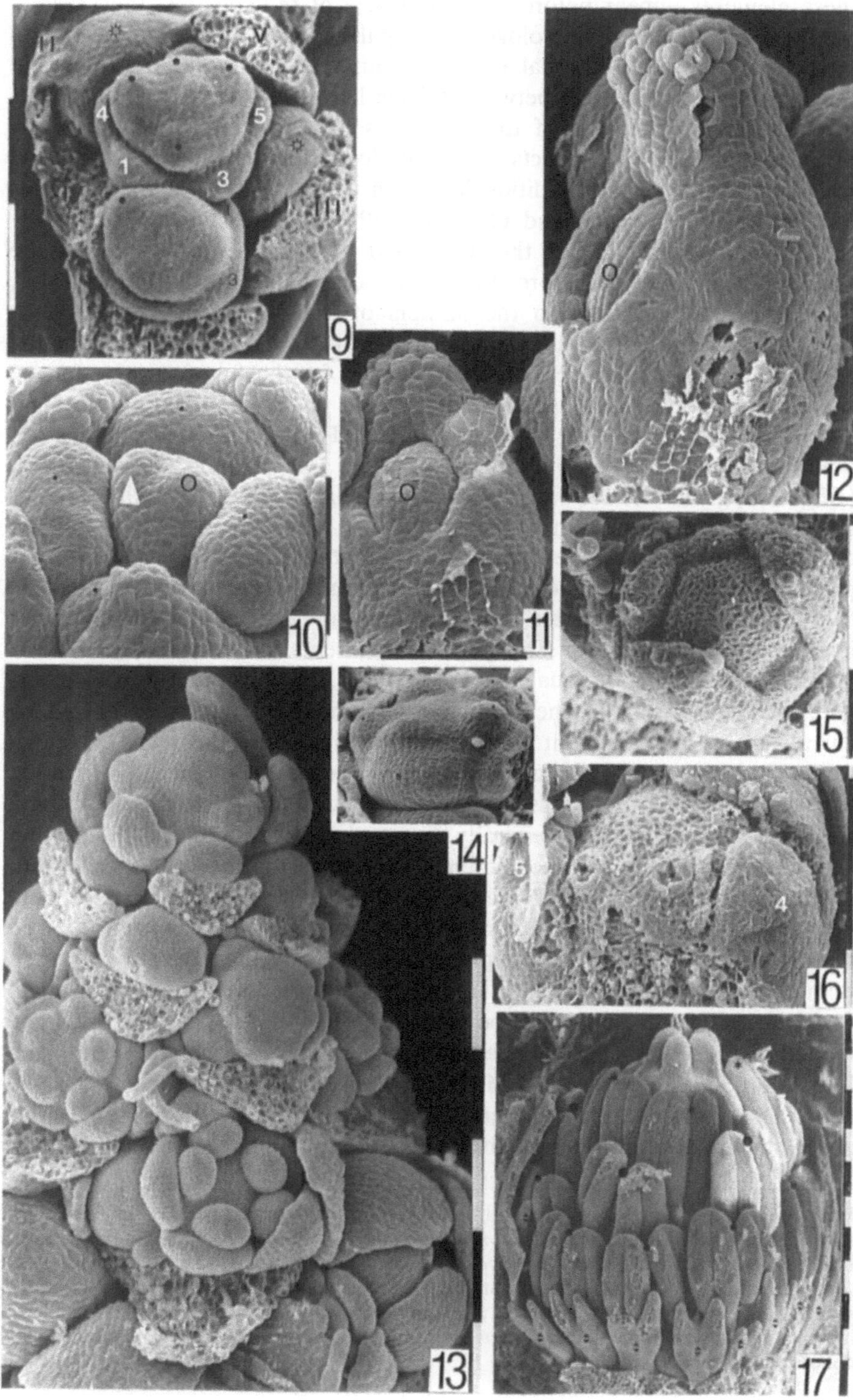
9
10
11
12
13
14
15
16
17

these elements appear before the emergence of the stamens, they are recognized here as being homologous to sepals (Plate 6.6,45). The involucral phyllomes may be additional phyllomes inserted between bracteoles and sepals (in terminal flowers between foliage leaves and sepals) and can thus be regarded as elements of the flower itself. This view is supported by the fact that axillary products are never formed. A similar case of floral extension by intercalary addition has been demonstrated for the calyculus of Dipsacales (Hofmann and Göttmann 1990). In the Portulacaceae, the involucral phyllomes cover the inner bud very early and take over the function of a calyx. Therefore, the sepals behave like petals. In particular, their growth, in relation to the anthers of the innermost (first formed) stamens, is retarded. In the androecium, a rather inexact leap from the sepals to a higher Fibonacci number is found for the first-formed stamens. In *Portulaca oleracea* no pattern exists in the positions of the five to eight carpels to the adjacent first-formed stamens. The later-formed stamens follow centrifugally but without a distinct pattern. This is valid for *Calandrinia (Cistanthe) grandiflora* as well (Plate 6.3,18; Plate 6.6,45). In both examples no primary primordia can be detected. For *Anacampseros* and *Talinum* the same sequence of development is suggested. The genus *Lewisia* is particularly variable in the number of perianth elements and stamens in the different species. In this genus the naming of the different phyllomes is especialy controversial. The perianth elements following the involucral phyllomes are finally of equal size and petaloid. The two outermost elements are placed decussate to the glandular phyllomes of the involucre. At an early stage of development they are larger than the others and cover all following elements. If five further elements occur, a quincuncial aestivation can often be found. In cases where six additional, pigmented perianth elements occur, three are generally arranged in an abaxial and three in an adaxial position. The two median ones are found innermost. In species with a still higher number of elements these are not arranged spirally. The elements situated between transversal and median planes possess covered margins towards the transverse plane; consequently their covering margins point towards the median plane. Therefore, the initiation appears to progress from the outermost transversal points anteriorly and posteriorly towards the median plane (Plate 6.3,21). The number of stamens varies in *Lewisia* from 5 to about 30. Unpublished SEM photographs by Rutishauser show that the two transversal perianth members are initiated after the involucre. At this point, the initiation of the stamens begins rather high up, at the bulged floral

Plate. 6.3. Development of the androecium in Portulacaceae and Basellaceae. *18*, *Calandrinia grandiflora*; *19* and *20*, *Montia perfoliata*; *21*, *Lewisia columbiana* (courtesy of R Rutishauser); *22* and *23*, *Basella alba*. *I*, *II* Decussate or otherwise paired sepals; *1–5*, sepals in successive appearance; ●, stamens formed first; •, stamens formed later; ♦, petal; *white arrowhead*, top of carpel; *B*, bract; *E*, terminal flower; *L*, leaf; *m*, (median) involucral bracts; *M*, involucral bracts of terminal flower; *t*, transversal bracteole; *bar* denotes 0.1 mm

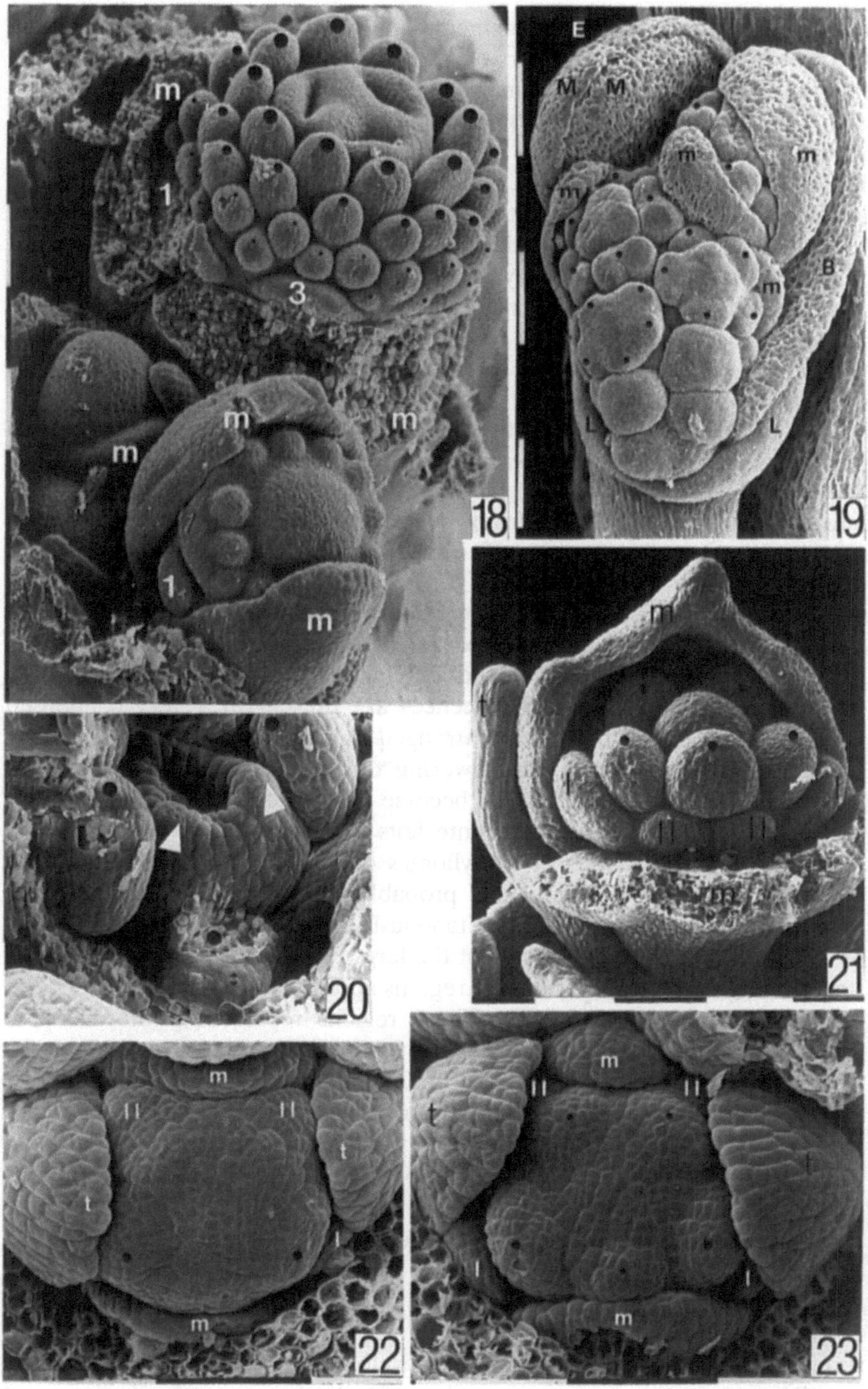
m
1
3
m
m
m
1
m
18
E
M
M
m
m
m
B
m
L
L
19
20
m
t
l
l
ll
ll
m
21
m
ll
ll
t
t
t
l
m
22
m
ll
ll
t
t
l
l
m
23

apex, before further perianth elements appear. The examined buds belong to taxa with rather low numbers of stamens, but the centrifugal origination in species possessing many stamens can be expected. The strict application of my definition only allows homology of the first two transversal elements with sepals (because only these are formed before the first stamens appear). It may be questioned, however, whether this leads to a sensible result, since in the mature flower all perianth elements appear equal. *Lewisia* obviously represents a good case for the shortcomings of artificially created categories. In resemblance to similar delays in the origination of leaves in Rubiaceae (Rutishauser 1985), it can be suggested that the perianth elements 3-n are already internally determined when the stamens appear on the surface, while their outer shaping lags behind. Therefore, I classify the perianth elements as sepals.

Serial microtome sections through buds of *Portulacaria afra* show two involucral phyllomes, five sepals in 2/5 arrangement and five alternisepalous stamens (Plate 6.6,47). The genus *Montia*, with the same number of elements as *Portulacaria*, differs from the latter by the arrangement of the five stamens in episepalous position. The superposed organs are attached to each other at the base.

In *Montia sibirica* only a single bracteole is formed on the pedicels of lateral flowers. The monochasial partial inflorescence is continued from its axil. In *Montia perfoliata* even this bracteole is absent. At the beginning of their development the inflorescences are present in a very condensed form. The still sessile flowers are arranged in a cincinnate, rolled-up zigzag sequence (Plate 6.3,19). Until flowering time the petioles and also some hypopodia elongate, the flowers become scattered, either solitary or in groups, and the inflorescence counterfeits a monopodium. The involucral phyllomes are initiated as first phyllomes of the flowers, in respect to the entire inflorescence; the outer and probably abaxial one is the larger. The involucre is followed by five simultaneously appearing staminal primordia, of which one is placed at the side of the larger involucral phyllome. Later in stamen development, meristematic regions appear on the outer basal surface of the filaments (Plate 6.3,20). These regions represent primordial protuberances of the perianth elements initiated rather tardily: after the gynoecium is initiated and never before the anthers begin to differentiate the archesporium. The resulting organs clearly fall into my definition of petals (stapetals) (Plate 6.6,46). This approach unfortunately does not offer an economical explanation since, on the one hand, the loss of a category of organs, the sepals, is assumed, while on the other hand, in the same place, petals, never mentioned before for Portulacaceae, are newly added.

In the mature flowers of *Montia*, these coloured perianth elements are attached to the superposed filaments, but otherwise they do not look essentially different from hypostaminal sepals of other Portulacaceae. In spite of their late and synchronous development, the petals cover often in a 2/5 arrangement, a situation characteristic for successive initiation of elements.

In *M. sibirica* the filaments are not only attached to the petals behind them but, almost at the same height, are united with each other to form a short filament tube. The adjacent bases of petals are not united so that the filament tube is notched on the outside between the petals. In *M. perfoliata*, the bases of the petals are laterally fused, too. In *M. fontana* only three stamens are initiated: the two posterior lateral ones and the anterior median one, all of them attached to the superposed petal at the base. The three petal-stamen groups are united by the bases of the remaining two petals into a distinct petal-stamen tube, slit open nearly to the base at the adaxial side. No rudimentary stamens for the origin of the anterior lateral single petals can be found, the relevant primordia must therefore develop in a different way.

Molluginaceae. In principle, the Molluginaceae share the possession of two bracteoles and a calyx in the 2/5 pattern of development (Plate 6.2,15) with sepal 2 in an adaxial position, but they differ in the remaining floral patterns. In agreement with this, developmental differences can be described as well.

Pattern 1: In *Corbichonia rubriviolacea* (unpublished SEM photographs by Bittrich) the youngest stages show five alternisepalous staminal primordia at the edge of the floral apex which (compared with the sepals) is bulged but flattened on top. Five smaller episepalous primordia occur below (similar to Plate 6.2,16). Alternating with these first ten primordia appear another ten. The centrifugal primordial activity is not exhausted after that, but becomes more irregular. Whether the primordia will differentiate into stamens or into petals cannot be decided in the early stages available. In serial sections of *C. decumbens* the elements initiated after the two series of 5 + 5 develop into narrow petals decreasing in size centrifugally. The five alternisepalous and the five episepalous stamens of *C. decumbens* insert in one whorl like the petals in an outer whorl. *Glinus lotoides* follows the same pattern (Plate 6.2,15–17). In the sample of Bittrich from Australia the third whorl of stamens is followed by about 10–20 further primordia growing up to narrow, long, deeply cleft petals (Plate 6.2,17; Plate 6.6,57). Material from Israel has fewer petals which can mostly be related in pairs to the alternisepalous radii (Hofmann 1973: p. 267).

Pattern 2: *Hypertelis salsoloides* (unpublished SEM photographs by Bittrich) shows an innermost pentamerous alternisepalous whorl, a medium episepalous one with primordia of almost equal size, and an outermost whorl with five alternisepalous smaller primordia (Plate 6.6,58). This points to a centrifugal development. As in *Corbichonia* and *Glinus*, the inner alternisepalous stamens are the longest of all. All primordia develop in *H. salsoloides* into stamens. Therefore, this pattern of initiation corresponds to that of *Trichostigma*. In the mature flower the ten inner stamens insert in a single whorl, even forming a short filament tube. The glandular tissue of the nectary is situated on the inner side of the tube and only in the case of the alternisepalous stamens reaches onto the free filaments. The outer alternisepalous stamens insert individually at the base of the filament tube and lack glandular tissue.

The genera *Macarthuria* (Plate 6.6,59) and *Limeum* (Plate 6.6,55) probably develop according to pattern 2, but in some of their species the outermost five alternisepalous primordia are differentiated into petals, whereas they are absent in the other species. In *Macarthuria australis* and *M. neocambrica* only the three episepalous stamens located before the sepals 3–5, in *Limeum* only the two in front of the sepals 4 and 5, are developed. This arrangement is strictly observed. In the mature flower the two or three episepalous stamens are longer than the five alternisepalous ones. All eight or seven stamens insert at equal distances in one whorl and form a short filament tube. The nectary lines the interior of the tube reaching up to the bases of the free filaments. The five alternisepalous stamens of *Pharnaceum* and *Psammotropha* have no nectary tissue at the base of the filaments. Instead, they insert outside a short tube-like disc with nectary tissue inside. Therefore, the five stamens may be homologized with the outer alternisepalous ones of *Hypertelis salsoloides*; the disc can be understood as the rest of a filament tube formed originally by the primordia of the inner ten stamens, which do not develop further. The only five alternisepalous stamens of *Mollugo cerviana* (Plate 6.6,60), *Hypertelis bowkeriana*, *Coelanthum semiquinquefidum*, and *Adenogramma myriantha* bear nectary tissue at the bases; therefore they probably correspond to the inner alternisepalous stamens which always emerge first in the examined Molluginaceae. In this case, the pattern breaks off after the first step.

The flower diagram of *Mollugo verticillata* (Plate 6.6,56) can only be interpreted as being formed by reduction. Sepals 1–3 are initiated in quick succession, but sepals 4 and 5 only after a break. The only three stamens alternate with the sepals 1–3 and the three carpels alternate with the stamens. In *M. verticillata* the change in the number of elements per whorl takes place at the transition from sepals to stamens and not from stamens to carpels as in most plants.

Patterns 1 and 2 coincide in the first two steps and only diverge later on. They can be interpreted as variants of a basic scheme as well.

Caryophyllaceae. Many Caryophyllaceae (Alsinoideae and Silenoideae) are similar in flower formula to some Molluginaceae, but their ontogeny differs

Plate. 6.4. Development of the androecium in Caryophyllaceae. *24–26*, *Stellaria graminea*; *27–29*, *Stellaria media*; *30* and *31*, *Cerastium vulgatum*. *1–5*, Sepals in successive appearance; •, episepalous stamen; *, alternisepalous stamen; ⧫, petal; *white arrowheads*, top of carpel; *gear*, flower primordium; *t*, transversal bracteole; *o*, ovule; *bar* denotes 0.1 mm

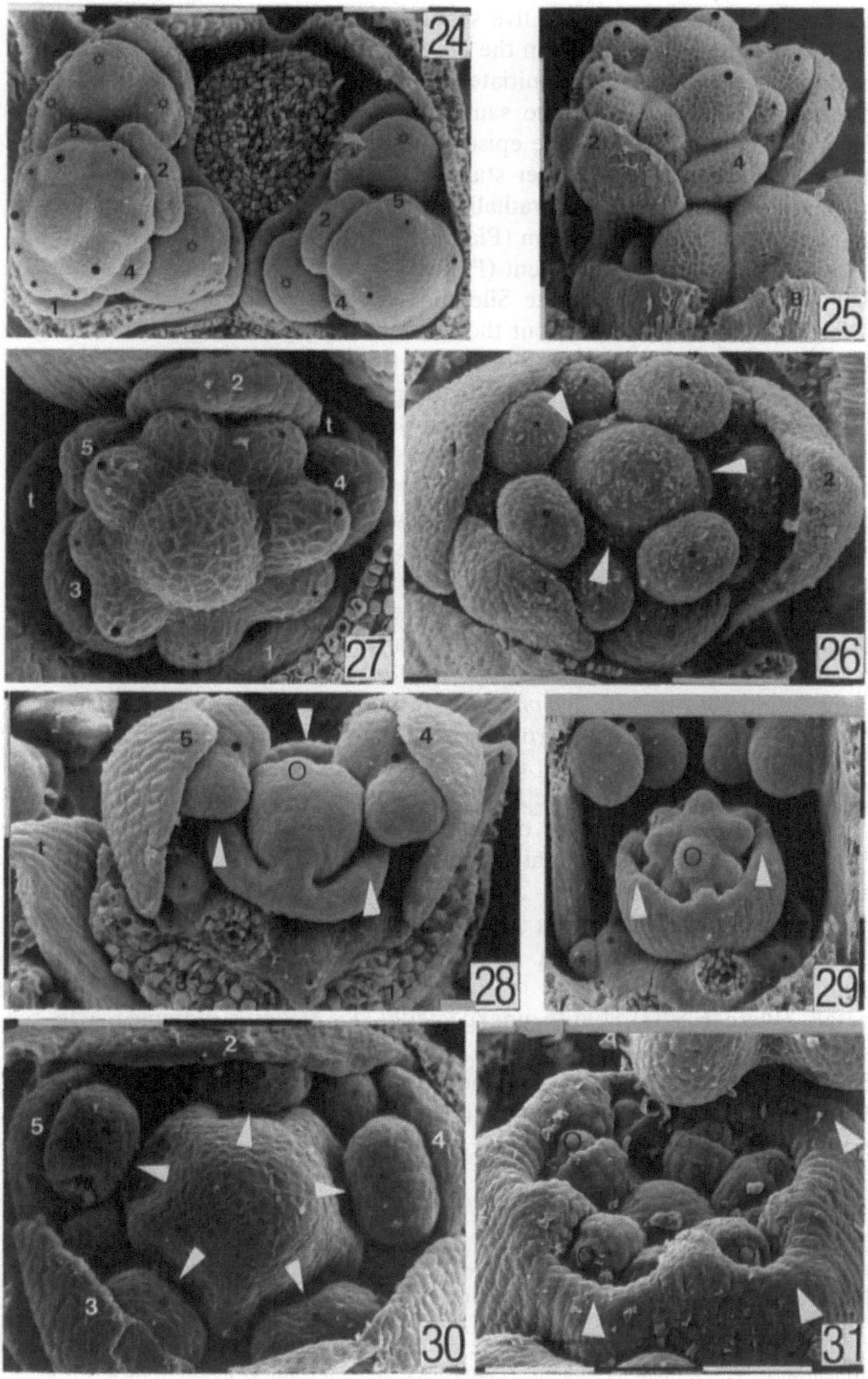
24
25
27
26
28
29
30
31

considerably. In flowers with five sepals, five petals and ten stamens, after the appearance of five sepals in the 2/5 pattern and with sepal 2 in an adaxial position, ten primordia are initiated simultaneously at the highly elevated floral apex and at almost the same distance to the centre (Plate 6.4,24; Plate 6.5,34 and 35). The five episepalous primordia (individual primordia) quickly develop into the larger stamens, but the alternisepalous primordia (primary primordia) extend radially and subdivide into an inner staminal and outer petalous primordium (Plate 6.4,25; Plate 6.6,62), which for a long time lags behind in development (Plates 6.4,28 and 29; Plate 6.5,37 and 39). This subdivision occurs in the Silenoideae earlier than in the Alsinoideae (Plate 6.4,25; Plate 6.5,36), but the elements of a pair are of the same age. It is therefore not possible to decide between a centrifugal or centripetal pattern. However, it is obvious that the petals originate in connection with the androecium, as already emphasized by Rohweder (1967, 1970).

Within the Caryophyllaceae, there exist several deviations from this basic pattern. In *Telephium imperati*, *Corrigiola litoralis* and *Drypis spinosa* (all with only five episepalous stamens in the mature flower) ten primordia arise simultaneously, or in *Corrigiola* the episepalous ones perhaps slightly earlier. But the alternisepalous ones are placed more to the outside right from the beginning. They also grow much slower than the episepalous primordia and develop directly into petals (Plate 6.6,64). Unpublished SEM photographs by Rutishauser of *Illecebrum verticillatum* show a situation comparable with that of *Telephium*, but the number of episepalous stamens is reduced to two placed in front of sepals 4 and 5 (Plate 6.5,40).

Of the genus *Stellaria*, *S. graminea* with ten stamens and *S. media* with reduced petals and only three fertile stamens were examined. In the flowers of *S. media* eight primordia of unequal size appear simultaneously (Plate 6.4,27). The three individual primordia of the episepalous stamens are always in front of sepals 3–5 and much larger than the alternisepalous primary primordia. The latter subdivide very late into a petal and a staminal primordium. The primordia of the epipetalous stamens remain minute humps (Plate 6.4,28), those of the petals grow later into small bipartite petals (Plate 6.6,63). Quite unexpectedly, in *Stellaria graminea* with ten simultaneously initiated primordia, the three primordia in front of sepals 3–5 were distinctly larger than the two in front of sepals 1 and 2 (Plate 6.4,24 and 25), which were about the same size as the alternisepalous primordia. In the further course of growth the differences in size disappear.

→

Plate. 6.5. Development of the androecium in Caryophyllaceae. *32* and *33*, *Lepyrodiclis holosteoides*; *34–38*, *Lychnis coronaria*; *39*, *Silene dioica*, female flower; *40*, *Illecebrum verticillatum* (courtesy of R.Rutishauser); *41*, *Corrigiola litoralis*. *1–5*, sepals in successive appearance; ●, episepalous stamen; *, alternisepalous stamen, ♦, petal; *white arrowheads*, top of carpel; *gear*, flower primordium; *L*, leaf; *o*, ovule; *S*, septum; *bar* denotes 0.1 mm

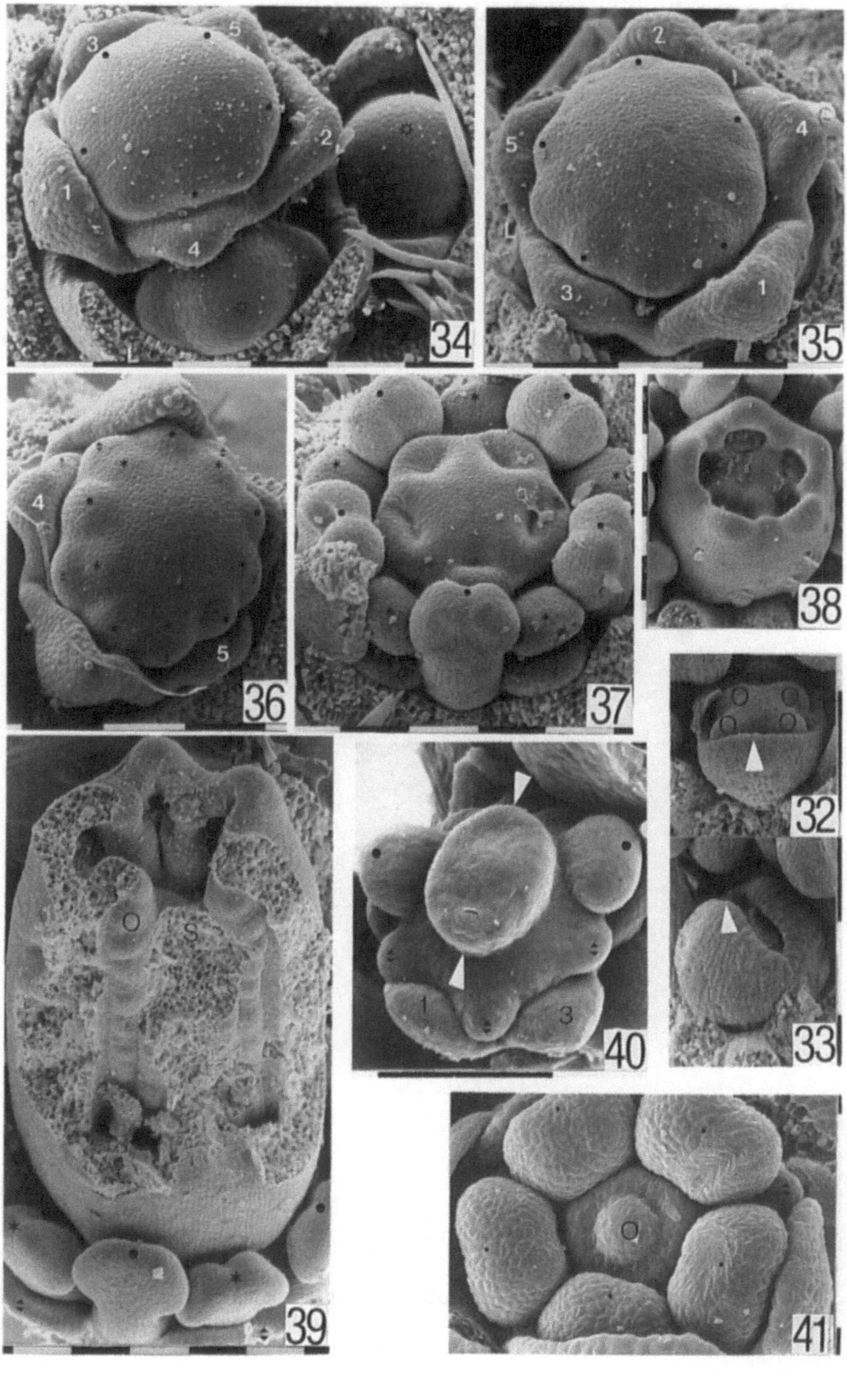
34
35
36
37
38
39
40
32
33
41

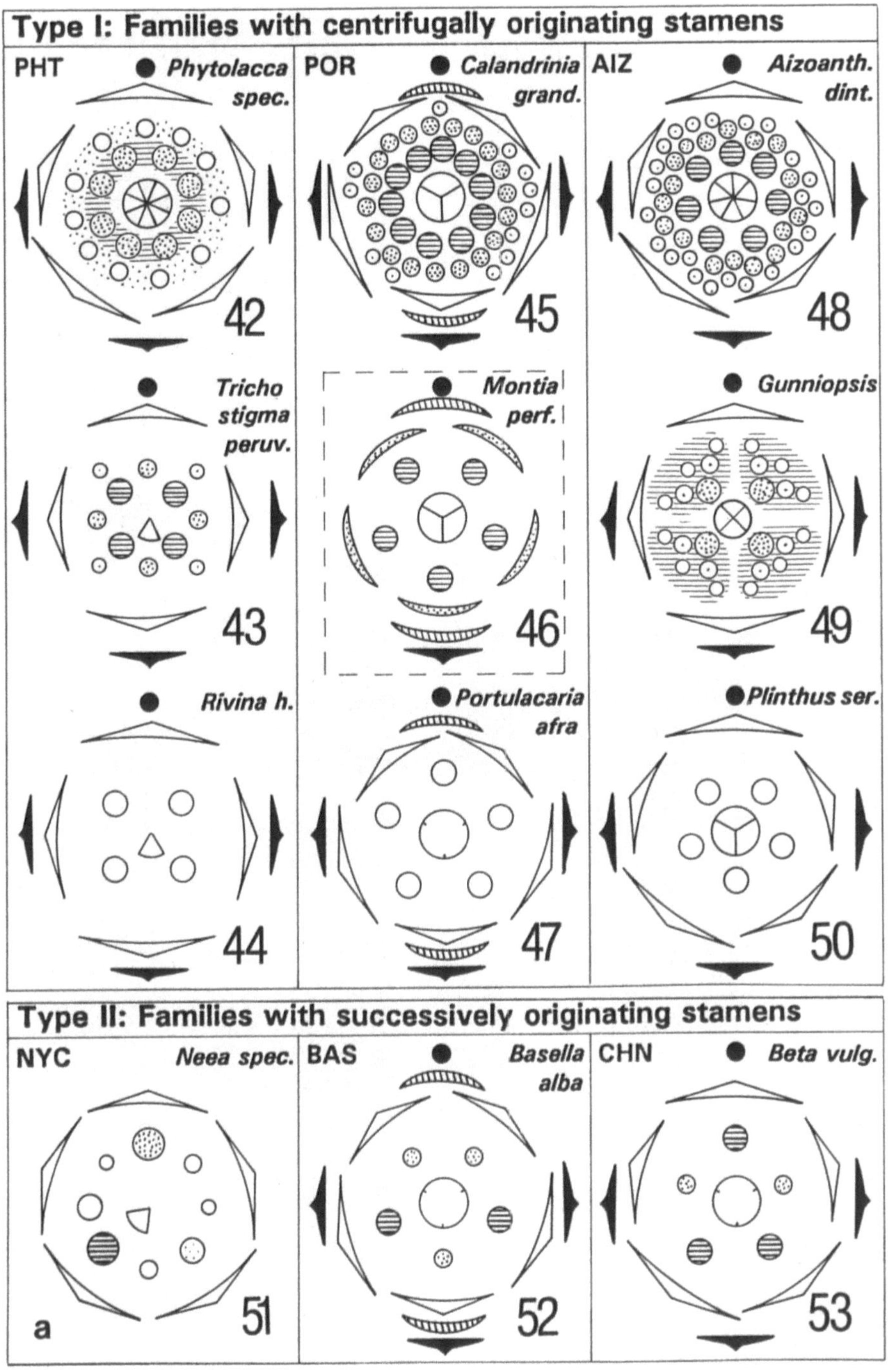

Plate. 6.6. Flower diagrams arranged according to physiognomical similarity, showing the diversity within the families (probably caused by reduction: *vertical columns*). *Hatching* shows the observed ontogeny within the androecium and petals.

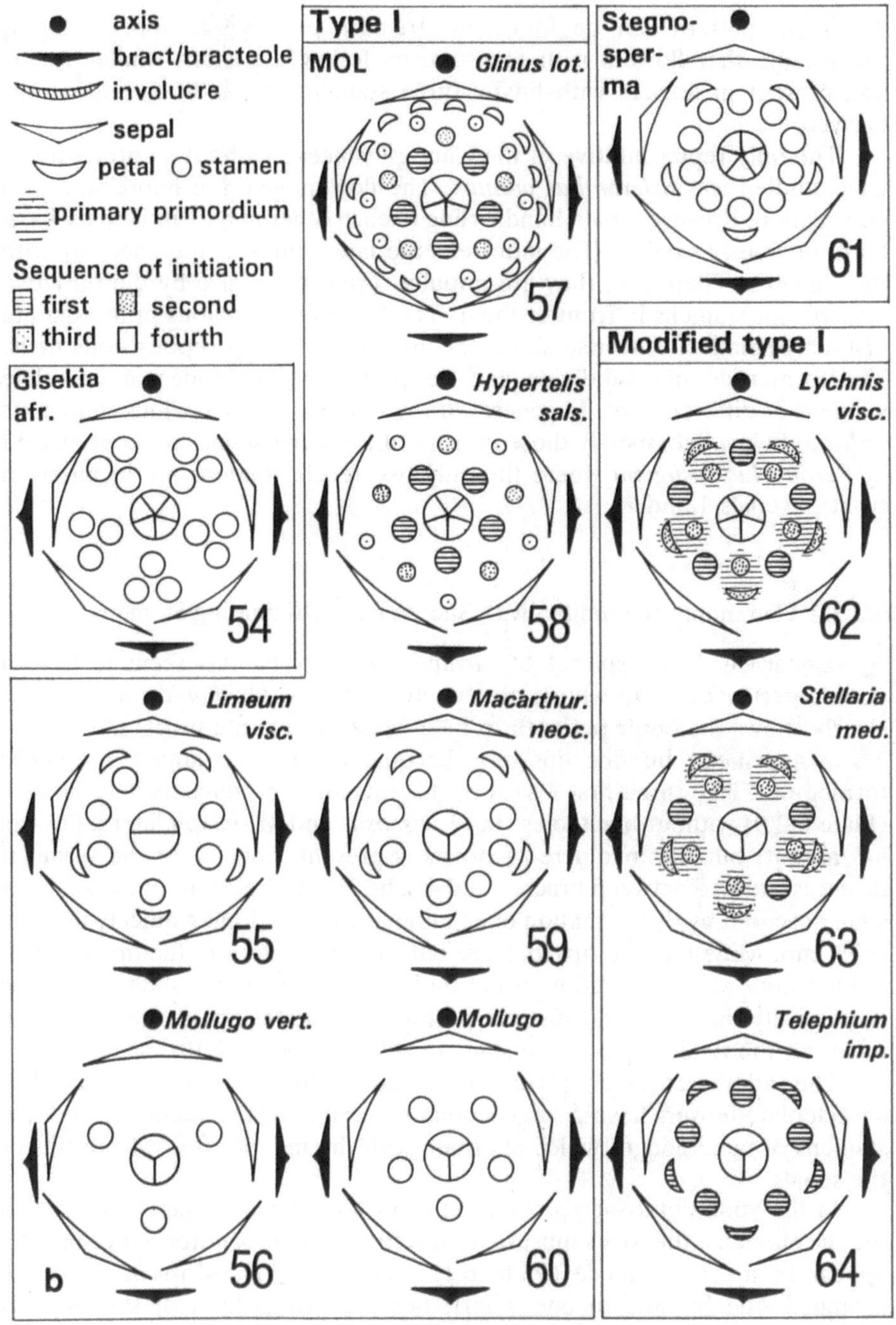

Plate. 6.6 (*Continued*) *42–50* and *55–60*, Type I, families with centrifugally originating stamens; *51–53*, type II, families with successively originating stamens; *54* and *61*, ontogeny of stamens unknown; *62–64*, modified type I

A similar pattern can be found in *Arenaria procera*. It appears highly remarkable that flowers with ten stamens follow a pattern similar to the one present in flowers with finally three stamens, a condition supposedly derived.

The only representative of the Caryophyllaceae without a corolla that I have studied was *Scleranthus annuus*. The flowers give the impression of a decussation, because sepals 1 and 3 cling to each other in bud and sepals 4 and 5 are obviously smaller. The latter emerge late, almost simultaneously with the stamens. Therefore, they are initially surpassed in size by the only two episepalous stamens in front of them. The five alternisepalous primordia are distinctly smaller, but arise simultaneously with the two episepalous ones. The former do not subdivide and develop into staminodes or complete stamens of different size. In general, if the number of episepalous stamens is reduced below five, mostly those in front of the inner sepals are kept, except in *Paronychia fastigiata* where the only two fertile stamens are localized in front of sepals 1 and 2.

6.2.2.2 Comments on Families with Successively Originating Stamens

Nyctaginaceae. Some species of *Mirabilis* and *Oxybaphus* seem to have a double perianth of two whorls of connate elements. *Oxybaphus nyctagineus* clearly shows the outer perianth to be a whorl of connate bracts initiated in 2/5 arrangement, but the tips soon become valvately arranged, the edges turning out. The three first leaves of this involucre produce axillary flowers (Plate 6.2,9) without bracteoles. In *O. viscosus* and *M. jalapa* lateral flowers are absent, but the involucre of bracts is present. Sepal 1 of the terminal flower is situated between bracts 1 and 4. The 2/5 arrangement of the sepals is soon obscured by the formation of an annular rim developing quickly into the calyx cup, while the free tips increase only a little. Later on, the tips show an induplicative valvate position in the bud. The three to four stamens appear successively and show no distinct positional relationship to the sepals. In addition, the final length of the stamens differs (Plate 6.2,10).

The inflorescences of *Abronia* are heads without terminal flowers. The ebracteolate flowers have a single connate perianth of five members and five stamens of equal length which alternate with the induplicate, valvate tips of the sepals.

In the young buds of *Neea* sp. the calyx lobes become valvate early. To see the stamens, the lobes must be removed. The eight androecial primordia appear in a 3/8 sequence (Plate 6.1,7; Plate 6.6,51). Most of them are arranged side by side in one whorl, but the primordium of stamen 6 is mostly placed outside stamens 1 and 3 (Plate 6.1,8). As a result of this peculiar arrangement, it cannot be decided with certainty whether the spiral proceeds inwards or outwards.

Sections of *Pisonia umbellifera* show eight (to 11) stamens of different length inserting in one whorl, as in *Bougainvillea* (Rohweder and Huber

1974). In these cases the different length agrees well with a 3/8 arrangement, as described by Sattler and Perlin (1982).

Basellaceae. This family is often interpreted as an appendage of the Portulacaceae. Both families share a two-leaved median involucre (somewhat ambiguously called bracteoles by Cronquist 1981) placed at a right angle to the bracteoles. In addition, the flowers of *Basella*, with five elements in the perianth and five superposed stamens, are reminiscent of *Montia*. However, the inflorescences of *Basella alba* are double racemes with terminal flowers, each preceded by three involucral phyllomes continuing the spiral of the bracts of the lateral flowers.

My results confirm the statements of Lacroix and Sattler (1988) regarding the ontogeny. In principle, the stamens would be superposed to the sepals in a 2/5 arrangement in a continuous undisturbed helical initiation. In *Basella* the origin of primordia does not occur at regular intervals, but rhythmically, so that the flower can be interpreted as decussately organized in whorls of two (rarely three) elements. The transversal pair of sepals 1 and 2 follows the decussate bracteoles and involucral phyllomes. Subsequently, a whorl of three members arises, consisting of sepals 4 (anterior median), and 3 and 5 (posterior lateral). Stamens 1 and 2 (Plate 6.3,22) as well as 3–5 repeat this pattern (Plate 6.3,23; Plate 6.6,52). The carpels alternate with the last three stamens. This pattern of initiation in *Basella* differs distinctly from all Portulacaceae under study.

Chenopodiaceae. In *Beta vulgaris* and *B. trigyna* the five sepals originate in a normal 2/5 arrangement after the bracteoles, which often produce lateral flowers in their axils. Sepal 2 is oriented to the axis (Plate 6.2,13). The ontogenetic spiral seems to be continued by five stamens with the same divergence angle placing the stamens in front of the sepals (Plate 6.2,14; Plate 6.6,53). This pattern of stamen initiation is very similar to that in *Basella alba*, but the resemblance is obscured by the thick red or white connate bracteoles, involucral phyllomes and sepals in *Basella* in contrast to the free, thin, and green perianth elements with somewhat membranous margins in the Chenopodiaceae.

6.2.2.3 Flower Ontogeny of Gyrostemonaceae

Until now, only *Gyrostemon racemiger* and *Codonocarpus cotinifolius* had been ontogenetically studied. All genera of the Gyrostemonaceae are considered to be unisexual. The ebracteolate flowers are found in the axils of bracts which are provided with awl-shaped stipules and form an open, sometimes few-flowered, raceme. The perianth consists of few (six to nine) scaly phyllomes, basally fused to some extent. The transversal ones are often larger and appear first. They may represent bracteoles integrated into the perianth. The subsequent initiation seems to progress, as in *Lewisia*, anterior and posterior to the median plane; the aestivation of the sepals is very variable. The floral apex of the female flower becomes a low, flat disc developing an edge later. This edge divides into 27–30 carpels, each at first consisting of a free stylar hunch. In front of it a minute depression soon appears, which later develops into the locule. Subsequently, inside this ring of depressions a second annular edge is formed, but this one is not subdivided surrounding a very large, flat remnant of the floral apex. These conditions are reminiscent of the Malvaceae, the only other family with a very high number of carpels in circular arrangement which I have already studied with SEM. In multilocular Aizoaceae, a similar situation is found: a central sterile region (columella) is retained in the gynoecium (Hartmann 1978). The similarity is perhaps functional: carpellary primordia need a fixed minimum size. To install a high number of primordia in a single whorl a very broad floral apex is required. Inevitably a large residual part remains in the centre.

In male flowers the androecial development starts similarly. After the emergence of the sepals in *Gyrostemon racemiger* an edge is formed along the margin of the flat, broad floral apex which simultaneously divides into ca. 16–20 staminal primordia. The rest of the floral apex forms a second inner edge which again divides into many (13–15) staminal primordia. It is followed by a third edge at which seven to nine primordia arise. The small central remnant splits according to its size into one to three staminal primordia. In this case the edges develop successively and centripetally and may be interpreted as ring-shaped primary primordia. In several attempts they divide into as many elements as can be placed on the primary primordium. Since there are no fixed numerical relations between the whorls, no pattern of positional relation can be recognized. Only a part of the genus *Gyrostemon* establishes several whorls of stamens (besides *G. racemiger*, also *G. ramulosus*, *G. subnudus*, *G. sheatii* and *G. sessilis*). The other species develop a single whorl only. The anthers are very long, latrorse and nearly sessile as in *Sarcobatus*. The broad central remnant of the floral apex does not develop further.

The initiation of stamens in *Codonocarpus cotinifolius* is similar to that in *Gyrostemon* species showing one whorl of stamens. However, surprisingly, an inner edge develops and divides into many carpellary primordia, but this growth stops rather early. The flowers are functionally male. It is interesting that this imitation of a hermaphrodite flower appears in a genus in which monoecious plants occur as well.

6.2.3 Gynoecium

6.2.3.1 Types of Gynoecia

Contrary to former opinions (Buxbaum 1961; Cronquist 1981), in the Caryophyllales there exist no examples of really free carpels. Species of

Phytolacca with ovaries appearing apocarpous to the naked eye have short, basal, distinctly fused carpellary sections. In *Gisekia* and in the Phytolacceae *Ercilla* and especially *Anisomeria*, the basic syncarpy is even more difficult to perceive. In these four genera elongated apical zones exist, which are really apocarpous, whereas at flowering time the symplicate and synascidiate zones contribute only minimally to the ovary height. The "apocarpous" appearance is supported by the fact that, in contrast to all other Caryophyllales, adjacent carpels are only joined at their margins and not radially over the entire flanks, i.e. the connate portions are near the centre and hidden. The ontogeny of similar forms of mature carpels in *Limnanthes* (Hofmann and Ludewig 1985) and in *Malope* (Malvaceae; Mampell, pers. comm.) has been studied with SEM. Unfortunately, I do not know the early ontogeny of *Ercilla*, and *Anisomeria*, or *Gisekia*. My opinion is that in all these cases it is not a matter of primitive but of highly advanced, carpellary form (Rohweder 1965b; Hofmann 1977).

I consider pentamerous flowers with five carpels to be primitive in Caryophyllales. In some taxa tetramerous flowers with four carpels occur, in others higher numbers of perianth members and carpels are found (in particular in the Aizoaceae, Ihlenfeldt 1960; Bittrich 1986; Hartmann 1993). I interpret these as parallel or analogous forms of the pentamerous flowers and do not see the necessity or possibility of deriving the tetramerous from the pentamerous ones.

As known from many other groups, in some families ovaries with three, two and one carpel (reduction series) occur beside those composed of isomerous five carpels. In some genera with penta- or tetramerous perianths higher numbers of carpels than elements of the perianth can be found (Plate 6.1,2). In the Cactaceae with an undetermined polymerous perianth, more than five carpels may occur, but their number is always lower than that in the perianth. I suppose that pleiomery of the gynoecium is another direction of derivation, because the carpels, as shown for *Phytolacca*, arise simultaneously instead of successively (the latter being reported for the many carpels of primitive Magnoliideae or Rosidae). Outside the Caryophyllales, families with more than five carpels in one whorl are much rarer (e.g. Malvaceae, Actinidiaceae and Gyrostemonaceae) than those with oligomerous gynoecia.

Cactaceae generally have carpels with many ovules. In the Caryophyllaceae, Molluginaceae, Portulacaceae, and Aizoaceae many genera with numerous ovules can be found (my ovary type A). This seems to be the most primitive condition, but other taxa are known to have one-ovuled carpels (ovary type B). As a rule, the only ovule is arranged in the median plane of the carpel at the "Querzone". This type is the only one in the Phytolaccaceae and Nyctaginaceae. In Caryophyllaceae and Portulacaceae genera with three- or two-carpellate gynoecia occur, containing only a single basal ovule each (ovary type C). This type of ovary, not divided into compartments, is the only one found in Chenopodiaceae, Basellaceae and Didiereaceae, and the dominant one in Amaranthaceae. It also occurs in the genera *Microtea*, *Lophiocarpus* and *Dysphania*, as well as in *Sarcobatus* and *Halophytum*.

The different numbers of ovules per carpel and the different proportions of ovary zones can be correlated. In the Molluginaceae and Caryophyllaceae the synascidiate zone will become short if the number of ovules per carpel is low. In these two families the majority of ovules insert in the synascidiate zone, whereas in Portulacaceae they predominantly insert in the symplicate zone. In those genera of Molluginaceae, Caryophyllaceae and Portulacaceae that possess single ovules per carpel, and in *Phytolacca*, *Agdestis*, *Barbeuia*, and *Stegnosperma*, the solitary apo- or epitropous ovules are placed nearly basally. In the remaining members of the Phytolacceae, Rivineae and Nyctaginaceae, and in *Gisekia*, the single ovules are strictly basal because in the mature gynoecium the synascidiate zone hardly exists any more. In contrast, in the few Aizoaceae with solitary apotropous ovules, these hang from the apex of the ovary into the well-developed locule formed completely by the synascidiate zone. In this family the symplicate zone is reduced even in many-ovuled representatives. In *Acrosanthes*, however, the symplicate zone is rather long. The very small septa do not join in the middle. The one ovule in each locule inserts laterally, not in the centre of the ventral ovary wall. The ovary therefore belongs to a reduced type A, not to ovary type B.

Additional characteristic properties (presumably derivations) occur in some taxa: (1) vanished septa (Portulacaceae, Silenoideae, Alsinoideae, multiovuled Paronychioideae and *Stegnosperma*; perhaps indicated in *Agdestis*, *Barbeuia*, *Sesuvium* and *Zaleya*); (2) formation of apical septa or walls (Phytolaccaceae, Nyctaginaceae, *Gisekia*, and *Agdestis*; more faintly in *Stegnosperma*, some Molluginaceae and Aizoaceae); (3) free or joined styles sunk into the thickened top of the ovary (Silenoideae, Alsinoideae and some Aizoaceae).

The flowers are mainly hypogynous, but the ovaries can be placed in a flower cup, yet free from it (flowers perigynous), or the flowers may be from half to completely epigynous with the ovary firmly connate to the outer flower cup (e.g. in Aizoaceae, Cactaceae, *Agdestis* and *Portulaca*). Table 6.2 summarizes the features of the ovaries.

6.2.3.2 Ontogeny of Carpels

I do not want to bring up the controversy regarding phyllospory and stachyospory again, but I believe that this problem cannot be solved with the help of anatomical facts. The argumentation of Hofmann and Ludewig (1985) concerning this problem can also be applied to the Caryophyllales. In my opinion an unstructured growing apex is not a pure product of the stem but should be interpreted as a future shoot, later organizing the two categories of organs (leaf and stem) which are difficult to delimit anatomically. It is a very laborious venture to investigate the early ontogeny of carpels by serial sections. Examinations by SEM make the task a great deal easier, although

Table 6.2. Survey of ovary characters

Taxon			No. carpels						Ovary type	Position of ovule		Peculiarities				
Cactaceae			>5	5					A					stj	fc	fe
Aizoaceae	Mesembry-anthemoideae		>5	5					A						fc	fe
	Ruschioideae			5					A							
	Tetragonioideae	*Tetragonia*	>5	5		3			B s	apo	ap	aps			fcll	fe
	Aizooideae	*Aizoanthemum*	>5	5					A							
		Aizoon		5es	4				A			aps			fcs	fe
		Gunniopsis			4				A			aps			fcss	–
		Galenia				3			B s	apo	ap				fcl	–
		Plinthus				3			B s	apo	ap				fcll	–
		Acrosanthes					2		A r						fcl	fe
	Sesuvioideae	*Sesuvium*				3			A			aps			fcl	(fe)
		Zaleya					2		A			aps			fcs	(fe)
		Trianthema						1	A						fcll	fe
Portulacaceae		*Portulaca*	>5	5					A				so	stj	fcl	fe
		Lewisia			4	3			A				so	stj	fcs	
		Calandrinia				3			A				so	stj	fcs	
		Anacampseros				3			A				so	stj	fcs	
		Talinum				3			A				so	stj		
		Montia				3			B s	apo	ba		so	stj		
		Portulacaria, *Ceraria*				3			C		ba			stj	fcs	
Phytolaccaceae	Phytolacceae	*Phytolacca americana*	>5						B s	apo	ba	aps				
		Phytolacca acinosa	>5						B pa	apo	ba	aps				
		Anisomeria	>5	5					B pa	apo	ba	aps				
		Ercilla		5					B pa	apo	ba	aps				
Gisekia				5es					B pa	apo	ba	aps				
Stegnosperma				5as					B s	epi	ba	aps	so			
Agdestis					4				B s	apo	ba	aps		stj	fcl	fe
Barbeuia							2		B s	(apo)	ba	aps		stj		

Table 6.2. *Continued*

Taxon			No. carpels	Ovary type	Position of ovule	Peculiarities
Molluginaceae		*Corbichonia*	5es	A		
		Hypertelis	5es	A		aps
		Glinus	5es	A		stj
		Pharnaceum	3	A		
		Mollugo, *Glischrothamnus*	3	A		aps
		Coelanthum	3	A		aps fc
		Limeum	2	A		aps
		Psammotropha	5es	B s	epi	stj
		Macarthuria	3	B s	apo	aps stj
		Adenogramma	1	B	epi	
Polpoda			2	B s	epi	stj
Caryophyllaceae						
	Silenoideae	*Lychnis*, *Silene dioica* (L.) Clairv.	5es	A		so
		Uebelinia spp.	5as 3	A		so
		Agrostemma githago L.	3	A		so
		Silene vulgaris (Moench) *Garcke*	3	A		so
		Cucubalus baccifer L.	3	A		so
		Saponaria ocymoides L.	2	A		so
		Dianthus carthusianorum L.	2	A		so
		Vaccaria pyramidata Medicus	2	A		so
		Gypsophila repens L.	2	A		so
	Alsinoideae	*Cerastium vulgatum*	5es	A		so fcss
		Moenchia mantica (L.) Bartl.	5es	A		so fcss
		Myosoton	5as	A		(aps) so fcss
		Sagina	4	A		so fcss
		Stellaria, *Arenaria*	3	A		so fcss
		Moehringia trinervia (L.) Clairv.	3	A		so fcss
		Minuartia laricifolia (L.) Schinz & Tell.	3	A		so fcss

		Lepyrodiclis			2		A						so		fcss	
		Buffonia paniculata F. Dubois			2			B	s	epi			so		fcs	
		Scleranthus			2				C		ba				fcl	
	Paronychioideae	*Spergula*	5as				A						so		fcss	
		Polycarpaea teneriffae Lam.		3			A						so	stj		
		Loeflingia hispanica L.		3			A						so	stj		
		Drymaria cordata (L.) Willd.		3			A						so	stj		
		Pollichia campestris Ait.			2			B	s	epi			so	stj	fc	
		Dicheranthus plocamioides Webb		3					C		ba			stj	fcl	
		Gymnocarpos decander Aschers. & Schweinf.		3					C		ba			stj	fcl	(fe)
		Illecebrum verticillatum L.			2				C		ba			stj	fc	
		Herniaria glabra L.			2				C		ba			stj	fc	
		Paronychia argentea Lam.			2				C		ba			stj	fc	
		Pterantus dichotomus Forsk.			2				C		ba			stj	fc	
Telephium				3			A						so		fcss	
Geocarpon				3			A						so		fcs	
Corrigiola				3					C		ba			stj	fcl	
Phytolaccaceae Rivinoideae						1		B		apo	ba	aps				
Nyctaginaceae						1		B		apo	ba	aps				
Amaranthaceae		*Celosia*		3	2?		A?				ba			stj		
		Amaranthus		3	2				C		ba					
Chenopodiaceae				3	2				C		ba			stj		
Didiereaceae				3					C	apo	ba					
Basellaceae				3					C		ba					
Achatocarpaceae					2				C		ba					
Microtea					2				C		ba					
Lophiocarpus					2				C		ba					

Families and genera arranged according to decreasing number of carpels and predominante ovary type. 5es, five episepalous carpels; 5as, five alternisepalous (epipetalous) carpels; A, ∞ ovules/carpel; B, one ovule/carpel; C, one ovule/ovary of two to three carpels; B s, ovary syncarpous; B pa, ovary pseudoapocarpous; apo, ovule apotropous; epi, ovule epitropous; ba, ovule inserted basally; ap, ovule inserted apically; aps, apical septum, apical wall; so, septa obliterating; stj, styles joined; fc, flower cup (ss, very short; s, short; l, long; ll, very long); fe, flower epigynous, ovary (partly) joined with the flower cup; r, reduced; ( ), character weak they chiefly only confirm results of the light microscopy methods. SEM pictures also do not settle the question whether in the centre of an ovary there is still a remnant of the flower axis, or whether in the vegetation cone fused carpel parts below the distinct primordial hunches of the carpel apices already exist. Those fused carpel parts emerge by intercalary extension growth.

The syncarpous ovaries of many dicotyledons are similar just in the beginning. In top view a very early stage of development often appears at its circumference as many projections, as the carpels are initiated. In side view, near the flattened apex an edge can be found (Plate 6.5,40), which develops as many small outer hunches as carpels, each accompanied by a small depression (Plate 6.1,2; 6.3,18; and 6.5,37: plug stage). Carpel primordia of this form can be found in the Phytolaccaceae, Portulacaceae, Molluginaceae and some Caryophyllaceae, e.g. Silenoideae.

Because the mature gynoecia of Silenoideae and Alsinoideae principally show the same zones in the same arrangement, it has been rather surprising to detect different developments. In Silenoideae, the primarily ascending margins of the septa between the depressions quickly change into a horizontal position (Plate 6.5,37) and soon incline towards the centre by raising the outer rim. Quite early, the outer ovary wall surpasses the apex of the later placental column (Plate 6.5,38), the carpel tips approach each other, the apical pore closes, and the styles elongate. At the same time, the synascidiate zone expands by intercalary growth. Not before the septa disappear in the closing ovary, the ovules, arranged in two rows per carpel, emerge from the central column successively downwards in the synascidiate zone (Plate 6.5,39). In the Alsinoideae, e.g. in *Cerastium*, *Myosoton* or *Stellaria*, the rest of the floral apex (after the androecium has been formed) is a high paraboloid cone (Plate 6.4,25 and 27). The developing outer walls of the carpels hang deep below the apex on the thick centre of the ovary, which changes later into the placental column (Plate 6.4,26 and 30). Between the flat depressions, the first indications of the locules, the septa appear rather hesitantly bending towards the centre of the placental column. The outer wall of the ovary lags behind in growth, so that for a long time the septa slope from the centre to the periphery (Plate 6.4,28). In the long run the upper edge of the outer ovary wall gradually grows up to the level of the apex of the placental column, where the first and uppermost ovules already originate (Plate 6.4,28 and 29). In *Cerastium vulgatum* the outer integument is already visible before the upper, and almost even, outer ovary wall definitely surpasses the centre (Plate 6.4,31). *Lepyrodiclis* with only four ovules is an exception in the Alsinoideae by closing the ovary rather early (Plate 6.5,32 and 33).

In the Portulacaceae, the outer part of the ovary forming the symplicate zone develops rather rapidly. The styles elongate very early in comparison with the Caryophyllaceae (Plate 6.3,20).

My SEM pictures confirm the results on the ontogeny of the monomerous ovary in Nyctaginaceae by Rohweder and Huber (1974) and Sattler and Perlin (1982). At first, the ovule primordium and the rim of the dorsal ovary wall lie side by side (Plate 6.1,8; Plate 6.2,10). Finally the faster growing wall begins to enclose the ovule (Plate 6.2,11 and 12), but previous to this, the young ovule closes the ventral cleft. The ventral cleft is rather short and neither elongates nor extends to the style since the ovary grows further by forming an apical wall.

It seems that the ovules in uniovulate ovaries appear earlier than the first ones in a multiovulate ovary (Plate 6.5,41).

6.2.3.3 Stigmas, Styles, Pollen Tube Transmitting Tissue and "Free-Central Placenta"

In my opinion stigma and style need not be recognized as distinct regions, since the stigma is a differentiation of the apex and the inner side of the style. The stigmatic surface is formed by papillae, but in the Mirabileae and *Agdestis* there are bundles of connate hairs instead. In the Phytolacceae, Aizoaceae (with the exception of *Argyroderma*, Hartmann 1978), Silenoideae, Alsinoideae and some Molluginaceae the styles and stigmas are free. They are basally fused in most Paronychioideae and form a rather long common style in the Portulacaceae and Cactaceae. Small-celled pollen tube transmitting tissue is located below the stigmatic papillae and adjacent to the ventral sides or ventral clefts if these are indicated on styles. If the styles are free down to the base, it is necessary to block the apical pore with papillae, which in Caryophyllaceae often protrude between the stylar bases and look like stigmatic papillae. Bogle et al. (1971) state for *Geocarpon* small, sessile stigmatic recurved lobes inserted between the tips of the later fruit valves. True styles are said to be absent in this genus.

With the exception of the Phytolaccaceae, young stages of all multicarpellous and more than one-seeded ovaries are regularly divided into compartments by septa, which do not necessarily meet in the centre. Mostly their margins are covered with papillae, probably indicating the path for the pollen tubes. Papillae are rarely absent at the margins, and the margins alone consist of small-celled pollen tube transmitting tissue (Rohweder 1965a, 1967, 1970; Hofmann 1973, 1977). The smooth central margins of Ruschioideae need a reinvestigation, keeping in mind that the papillate placenta is found in a parietal-basal position. Except for in some Portulacaceae, the greatest portion of ovules insert in the synascidiate zone. Therefore, the bands of papillae are continuous and mostly placed between rows of ovules and septa.

Phytolacca differs from the described typical form because of its pseudoapocarpous ovary (Rohweder 1965b), in which the adjacent carpel flanks are connate only in their most proximal portions, thus not exhibiting typical septa. In addition, a central, rather wide channel is formed by the apical walls or apical septa. A shallow ventral cleft with adjacent pollen tube transmitting tissue runs from the apical walls up to the stigma; it transgrades downwards into the short symplicate zone. The channel is basally filled with pollen tube transmitting papillae mainly inserted on a conical elevation in the centre of the carpels. Papillae cover also the margins of the very small radial septa in the short symplicate zone. This pseudoparenchyma of papillae presumably serves as compitum where transgression of pollen tubes between carpels is possible. From here individual accesses lead as "ventral tubes" (Hofmann 1973) to each locule, either horizontally or diagonally downwards (e.g. in *Phytolacca americana*). These tubes presumably present the pathways for the pollen tubes. The structure of the carpels in *Gisekia*, *Ercilla*, and *Anisomeria* is principally the same, but at flowering time the ovules of *Anisomeria* are inserted basally, slightly parietal and not at the "Querzone". Perhaps this is caused by a shifting of the placenta, resembling the processes occurring in Cactaceae and Aizoaceae. From the insertion of the ovule, the ventral tubes descend further to the funnel-shaped centre of the flower, joining there the pollen tube transmitting tissue placed under the epidermis in the median plane of the apical wall (Fig. 8 in Hofmann 1977).

The solitary carpels of Rivineae and Nyctaginaceae are very similar to the individual carpels of *Gisekia* or Phytolacceae. However, the pollen tube transmitting tissue does not lie under the surface of the median plane of the apical wall, but by continual divisions of the epidermal cells is shifted deeper into the wall (Rohweder and Huber 1974). Slightly forked at the inner end of the ventral tube, it leads into the locule just above the point of insertion of the apotropous ovule. On the funicle, an anulus of papillae permits the growth of the pollen tube to the micropyle.

The unilocular ovaries of Paronychioideae with one basal campylotropous ovule possess no septa, but the two comissural parts of the ovary wall are lined with pollen tube conducting papillae in a strand reaching as far down as the micropyle. In ovaries with several basal seeds these strands are detached.

In *Corrigiola* and *Scleranthus* the one basally inserted ovule has a long curved funicle, thus placing the micropyle precisely below the apical pore of the ovary. Therefore, there are no strands of pollen tube conducting tissue in these ovaries. Similarly, a shortening of the pollen tube conducting tissue is found in *Gymnocarpos* and *Paronychia* (Rohweder and Urmi-König 1975).

The "free-central placenta" of the Caryophyllaceae, Portulacaceae and *Stegnosperma* is the result of disproportionate growth. In the course of differentiation the small cells of the mesophyll of the peripheral regions of the septa become separated by enlarging intercellular spaces, while the epidermis grows normally. The mesophyll disintegrates early leaving the epidermis intact for some time, but finally the outer portion of the septum disappears leaving marginal portions in the centre of the ovary around the central column. These portions bear ovules in the lower parts of the ovary forming the placental column, but are sterile above, connecting the apical pore with the top of the placental column. This sterile region remains alive at least until flowering time when the pollen tubes have to seek their path to the ovules. Later, they may break off the top of the ovary, while the symplicate zone is often inflated to make room for the enlarging seeds.

6.3 Conclusions

Conventional treatment distinguishes rather large families in the Caryophyllales (e.g. Cronquist 1981). In this study deviating genera have been treated separately, yet Nakai's conclusion (1942) to establish all these at the family rank has not been followed, because the problem is only shifted and not solved in this way. Taxonomy includes separation and unification, and both have to be considered. Until my investigations have reached equal levels in all groups, I prefer to keep the disputed genera as such, preferably as *taxa incertae sedis*, in order to work unbiased.

Yet, some conclusions can be drawn. The fundamental subdivision of the Caryophyllales can probably best be based on some chemical characters, like betalains or anthocyanins (e.g. Mabry 1976; see also Chaps. 1 and 10), and on some ultrastructural characters, like sieve-element plastids (see also Chap. 5). In contrast to this, flower morphological characters can be used to separate units at a lower taxonomic level. This is due to the high adaptive value of the flower, resulting in many parallel or analogous evolutionary lines. On the other hand, for many character expressions the adaptive value is yet unknown. The obliterating septa represent an example: the detached margins of the septa in some Caryophyllaceae and Portulacaceae look alike and seem to represent homologous events. I am sceptical, however, whether this rather rare character is a synapomorphy of these families, which do not share the same type of flower pigments.

It seems to me a problem to decide whether character states are primitive or advanced, or, even more, whether character expressions occur once or several times. The Rivineae and Nyctaginaceae share a peculiar monocarpellate, one-seeded ovary with a specially placed pollen tube transmitting tissue, but the Rivineae have free sepals and racemes, like most of the Phytolacceae, and not panicles or head-like aggregations of flowers in addition to solitary terminal flowers on modules like the Nyctaginaceae. Only Nyctaginaceae show anthocarps, perhaps a synapomorphy for the family.

Portulacaceae, Basellaceae and Didiereaceae share an involucre decussated to the two bracteoles, but in most Portulacaceae there are numerous stamens in centrifugal appearance or few, occurring simultaneously, as in *Montia*. In contrast, in *Basella* the stamens emerge successively. This character expression links Basellaceae to Chenopodiaceae and Nyctaginaceae, suggesting another possible grouping. In this way suggested groups overlap when groupings based on other characters are considered dominant. Nevertheless, some statements can be made.

Table 6.3. Survey of some characters

		Growth form	Succulent parts	S. gr.	Spiny parts	Phyllotaxis	Stipules	Inflorescence	Modular growth		Sex
Phytolaccaceae											
Phytolacceae		t h	(sho)	an		alt	–	Thyrse› raceme		(E)	he/f:m
Rivineae		t, sh, h			br	alt	–	Double raceme/raceme		(E)	he/f:m
Agdestis		h				alt	–	Thyrse		E	he
Barbeuia		Liana				alt	–	Umbell-like raceme		E	he
Stegnosperma		sh		n		alt	–	Raceme		–	he
Microtea		h				alt	–	Raceme			he
Lophiocarpus		h	l	n		alt	–	Thyrse		–	he
Achatocarpaceae	Lit.	sh		n	sho	alt	–	Short lateral raceme		E?	f:m
Nyctaginaceae											
Mirabileae		h			sho	op	–	Head-like raceme/E	M	E/–	he
Pisonieae		sh				op	–	Panicle		E	he/f:m
Aizoaceae s.s.											
Mesembryanthemoideae		sh, dsh, h	l sho	an	l	op, alt	–	Dichasia, monochasia		E/–	he
Ruschioideae		sh, dsh, h	l sho	an	sho	op, alt	–	Dichasia, monochasia, E	(M)	E/–	he
Aizooideae		h	l			op, alt	–	Thyrse		E	he/f:m
Tetragonioideae		sh, h	(l)		Fruit	alt	–	Raceme		E	he/f:m
Sesuvioideae		h	(l)			op	–/(+)?	Dichasia/E	M	(E)	he
Didiereaceae	Lit.	t, sh	sho	n	br/l	alt	–	Thyrse		E	f:m
Cactaceae	Lit.	t h	sho/(l)	n	l	alt	–	Thyrse/E		E	he
Chenopodiaceae	Lit.	t, sh, h	shol/–	an		alt/op	–	Thyrse/raceme		E/–	he/f:m
Halophytum	Lit.	h						Spike			f:m
Sarcobatus		sh	–			alt	–	m, raceme; f, axillary		m–/fE’	f:m
Dysphania	Lit.	h	–			alt	–	Dichasia			he + f
Amaranthaceae	Lit.	h	–	an		alt/op	–				he/f:m

Portulacaceae										
Portulacoideae		sh, h	(sho)/l	n	alt/op	+/ha/–	Dichasia ›monochasia		E	he
Montioideae		h	–		alt	–	Monochasia		E	he
Basellaceae		h	(l)	n	alt/op	–	Double raceme		E	he
Molluginaceae										
Corbichonieae		dsh?	–		alt	–	Thyrse		E	he
Limeeae		h	–		alt	–	Dichasia	(M)	E	he
Mollugineae		h	(l)		alt	–/+	Dichasia	M	E	he/f:m
Gisekia		h	–		alt	–	Dichasia	M	E	he
Polpoda		dsh	–	n	alt	+	?		E?	he
Caryophyllaceae										
Paronychioideae		dsh, h	–	n	op	+	Dichasia		E	he
Alsinoideae		h	–	n	op	–	Dichasia	(M)	E	he
Silenoideae		h	–	n	op	–	Thyrse/dichasia		E	he/f:m
Telephium		dsh	–	n	alt	+	Monochasium?		E	he
Corrigiola		h	–	n	alt	+	Monochasium	(M)	E	he
Hectorella	Lit.	dsh	–		alt	–	E		E	he
Lyallia	Lit.	dsh			alt	–	Solitary in axils			he
Geocarpon		h	–		op	–	Monochasium		E	he

+, Present; –, absent; no record given, not known (though a record mentioned does not mean that all members of the group show that character)
Growth form: t, tree; sh, shrub; dsh, dwarf shrub; h, herb
Succulent parts: sho, shoot; l, leaf; ( ) weak
S. gr., secondary growth: an, anomalous; n, normal
Spiny parts: br, bracts; l, leaf; sho, shoot
Phyllotaxis: alt, leaves alternate; op, leaves opposite
Stipules: ha, stipules represented by hairs
Inflorescence: E, terminal flower present; (E), terminal flower sometimes present
Modular growth: M, distinct number of leaves; (M), indistinct number of leaves
Sex: he, hermaphrodite; f, female; m, male
Lit, taken from literature

Table 6.3. *Continued*

	Invo-lucre	Showy parts	Calyx	Petals	Stamens	Dev. andr	Carpels	Ovary type	Fruit	Oxalate
Phytolaccaceae										
Phytolacceae	–	s	>5 5	–	∞	cf	>5 5	B	Berry	raph
Rivineae	–	s	5 4	–	∞–12	cf	1	B	Berry/achene	raph
Agdestis	–	s	4	–	∞		4	B	Achene	d
Barbeuia	–	s	5	–	∞		2	B	One-seeded	–
Stegnosperma	–	s+p	5	5	5es + 5ep		5ep	B	capsule	d
Microtea	–		5 4	–	4es + 5-3as		2	C	Capsule	–
Lophiocarpus	–		5	–	1es + 3as		2	C	Achene	d
Achatocarpaceae Lit.	–		5 4		∞		2	C	Achene Berry	d
Nyctaginaceae										
Mirabileae	⟨5–3⟩	br/s	(5)	–	8–<5	sc	1	B	Achene+anthocarp	+
Pisonieae	–	(s)	(5)	–	8–11	sc	1	B	Achene+anthocarp	+
Aizoaceae s.s.										
Mesembryanthemoideae	–	p	5 4	∞	∞	cf	>5 5 4	A	Capsule/nut	raph
Ruschioideae	⟨2⟩	p	>5 5 4	∞	∞	cf	>5 5 4 3	A	Capsule/berry	raph
Aizooideae	–	s	5 4	–	∞–5	cf	>5 5 4 2	A	Capsule	d
Tetragonioideae		s	5 4 3	–	∞		>5 3	B	Achene/nut	d
Sesuvioideae	⟨2⟩	s	5	–	∞ as (+es)		3 2 1	A	Pyxidium	d
Didiereaceae Lit.	2	s	4	–	11-8-6(es, as)		3	C	Achene	d
Cactaceae Lit.	–	s	∞	–	∞	cf	>5–3	A	Berry	d/raph
Chenopodiaceae Lit.	–		5(–1)	–	5es	sc	3 2	C	Achene	+
Halophytum Lit.			f0/m4	–			?	C	Achene	
Sarcobatus	–		f0/m0	–	???		f2	C	Achene	
Dysphania Lit.	–		3 2	–	he 2es		he2 f2	C	Achene	
Amaranthaceae Lit.	–	s+app	5 (3)	–	5es		3 2	A C	Capsule/achene	+, d

Portulacaceae																
Portulacoideae		2	s	>5	5	–	∞-5as	cf	>5–3			A		C	Capsule	d
Montioideae		2	p	–		5	5ep			3			B			d
Basellaceae		2	s	5		–	5es	sc		3				C	Berry/nut	d
Molluginaceae																
Corbichonieae		–	s+p	5		∞	∞	cf	5es			A			Capsule	d
Limeeae		–	s+p	5		5/0	3-2es + 5ep	cf		3	2	A			Capsule	d
Mollugineae		–	s+p	5		–/∞	∞–15 es+as/5as	cf	5	3		A	B		Capsule	d
Gisekia		–	s	5		–	5x 3as/1as		5es						Mericarps ?	raph
Polpoda		–	–		4	–	4as				2		B			
Caryophyllaceae																
Paronychioideae		–	–/(p)	5	4	5/0	5-1es+ 5/r ep	mcf	5	3	2	A	B	C	Capsule/achene	d
Alsinoideae		–	p	5	4	5/0	5-3es+ 5/r ep	mcf	5	3	2	A	B		Capsule	d
Silenoideae		–	p	5		5	5es + 5/0 ep	mcf	5	3	2	A			Capsule	
Telephium		–	p	5		5	5es	mcf		3		A			Capsule	
Corrigiola		–	p	5		5	5es	mcf		3				C	Nut	
Hectorella	Lit.	2	s	5		–	5as				2	A			Achene?	
Lyallia	Lit.	2=br?	s		4	–	3				2	A			Capsule?	
Geocarpon		–	(s)	5	4	–	4es+4r			3		A			Capsule	

Involucre: 2, two intercalated floral phyllomes; ⟨ ⟩, empty bracts; br, bracteoles
Showy parts: s, sepals; p, petals (stapetals); app, petal-like appendages of the filament tube
Calyx: 5, five free sepals; (5), five joined sepals
Stamens: es, episepalous; as, alternisepalous; ep, epipetalous; r, reduced (staminodes)
Dev. andr., developement of the androecium: cf, occurrence of stamina (and petals) centrifugally (type I); sc, occurrence of stamina successively (spirally, type II); mcf, modified type I
Ovary type: A, ∞ ovule/carpel; B, one ovule/carpel; C, one ovule/ovary
Oxalate: +, present (into parenchyma); d, druses; raph, raphides
Lit, taken from literature

Gisekia, with betalains and a pseudoapocarpous gynoecium, appears better placed near the Phytolaccaceae. In growth it agrees well with the Molluginaceae, but the modular construction of the plant body and the dichasiale inflorescence have also been found in other groups (Table 6.3), so that the character expression is not as unique as was thought earlier (Hofmann 1973). It was not possible to study *Gisekia* ontogenetically, but the arrangement of the androecium in ripe flowers fits neither of the two molluginaceous patterns nor that of the Phytolaccaceae (Plate 6.6,54).

In the Aizoaceae, I suggest that Tetragonioideae is grouped with Aizooideae because the reduction series in the gynoecium in *Plinthus* and *Galenia* is carried further in *Tetragonia* (Hofmann 1973). In a similar way, the inflorescences form transitional series ending in *Tetragonia*. Bittrich (1990), by naming epidermal bladder hairs as a synapomorphy for both subfamilies, stressing that they form a monophylum, supports this view. Clearly, the separation of *Tetragonia* as a distinct family can no longer be considered since the group presents the highest developed branch of the here newly circumscribed Aizooideae.

Gilbert (1987) points out that *Telephium* and *Corrigiola* share two characters (alternate leaves and cordate stipules) otherwise not known in the Caryophyllaceae. He places them tentatively in the Molluginaceae, from which they differ in detached septa (*Telephium*) or in uniovulate ovaries (*Corrigiola*). The ontogeny of the androecium clearly represents a variant of the caryophyllaceous pattern, similar to that of *Drypis*. Therefore, *Telephium* and *Corrigiola* belong to the Caryophyllaceae.

In conclusion, I repeat that flower morphological characters are more numerous than often anticipated. A wealth of data is still hidden and awaiting discovery. It seems important, though, that data are compiled and stored sensibly to make comparisons and conclusions retraceable. Certainly, the computer operates the character matrix absolutely objectively but, perhaps, the subjectivity arises again in selecting the characters included in a cladistic analysis (Carolin 1987).

Acknowledgements. Critical-point drying was carried out by WD Kötting, Max-Planck-Institut für Biophysikalische Chemie, Göttingen. K Wehr from the Forestry department assisted me with the Phillips 515 SEM studies and S Hourticolon of our institute performed all the photographic work. V Bittrich (Hamburg), K Huber and R Rutishauser (Zürich) provided unpublished SEM pictures. In addition to several students, J Clement (Austin) and HEK Hartmann (Hamburg) helped me with the English version of this paper. To all those mentioned I am most grateful.

References

Behnke H-D (1975) *Hectorella caespitosa*: Ultrastructural evidence against its inclusion into Caryophyllaceae. Plant Syst Evol 124:31–34

Bittrich V (1986) Untersuchungen zu Merkmalsbestand, Gliederung und Abgrenzung der Unterfamilie Mesembryanthemoideae (Mesembryanthemaceae Fenzl). Mitt Inst Allg Bot Hamb 21:5–116
Bittrich V (1990) Systematic studies in Aizoaceae. Mitt Inst Allg Bot Hamb 23b:491–507
Bittrich V (1993) Caryophyllaceae. In: Kubitzki K, Rohwer J, Bittrich V (eds) Families and genera of vascular plants, vol 2. Springer, Berlin Heidelberg New York, pp 206–236
Bittrich V, Hartmann H (1988) The Aizoaceae – a new approach. Bot J Linn Soc 97:239–254
Bogle AL, Swain T, Thomas RD, Kohn ED (1971) *Geocarpon*: Aizoaceae or Caryophyllaceae? Taxon 20:473–477
Buxbaum F (1961) Vorläufige Untersuchungen über Umfang, systematische Stellung und Gliederung der Caryophyllales (Centrospermae). Beitr Biol Pflanz 36:1–56
Carolin R (1987) A review of the family Portulacaceae. Aust J Bot 35:383–412
Cronquist A (1981) An integrated system of classification of flowering plants. Columbia University Press, New York
Dahlgren R (1980) A revised system of classification of the angiosperms. Bot J Linn Soc 80:91–124
Dahlgren R (1983) General aspects of angiosperm evolution and macrosystematics. Nord J Bot 3:119–49
Eckardt T (1964) 13. Reihe Centrospermae. In: Melchior H (ed) A. Engler's Syllabus der Pflanzenfamilien, 12th edn. Borntraeger, Berlin, pp 99–101
Eckardt T (1967a) Blütenbau und Blütenentwicklung von *Dysphania myriocephala* Benth. Bot Jahrb Syst 86:20–37
Eckardt T (1967b) Vergleich von *Dysphania* mit *Chenopodium* und mit Illecebraceae. Bauhinia 3:327–344
Eckardt T (1968) Zur Blütenmorphologie von *Dysphania plantaginella* F.v.M. Phytomorphology 17:165–172
Eckardt T (1969) Morphology and systematic position of *Dysphania*. XIth Int Bot Congr Abstr 51, Seattle, Washington
Eckardt T (1974) Vom Blütenbau der Centrospermen-Gattung *Lophiocarpus* Turcz. Phyton (Austria) 16:13–27
Engler A (ed) (1934) Die natürlichen Pflanzenfamilien, 2nd edn, vol 16c. Engelmann, Leipzig
Gilbert MG (1987) The taxonomic position of the genera *Telephium* and *Corrigiola*. Taxon 36:47–49
Hartmann HEK (1978) Monographie der Gattung *Argyroderma* N. E. Br. (Mesembryanthemaceae Fenzl). Mitt Inst Allg Bot Hamb 15:121–235
Hartmann HEK (1991) Mesembryanthema. Contrib Bolus Herb 13:75–157
Hartmann HEK (1993) Aizoaceae. In: Kubitzki K, Rohwer J, Bittrich V (eds) Families and genera of vascular plants, vol 2. Springer, Berlin Heidelberg New York, pp 37–69
Hartmann HEK, Bittrich V (1991) Typification of suprageneric names – some nomenclatural changes in Aizoaceae. S Afr J Bot 57:73–75
Heimerl A (1934) Phytolaccaceae. In: Engler A (ed) Die natürlichen Pflanzenfamilien, 2nd edn, vol 16c. Engelmann, Leipzig, pp 135–164
Hershkovitz M (1991) Phylogenetic assessment and revised circumsription of *Cistanthe* Spach (Portulacacaeae). Ann M Bot Gard 78:1009–1021
Hofmann U (1973) Centrospermenstudien. 6. Morphologische Untersuchungen zur Umgrenzung und Gliederung der Aizoaceen. Bot Jahrb Syst 93:247–324
Hofmann U (1977) Centrospermenstudien. 9. Die Stellung von *Stegnosperma* innerhalb der Centrospermen. Ber Dtsch Bot Ges 90:39–52
Hofmann U, Göttmann J (1990) *Morina* L. und *Triplostegia* Wall. ex DC im Vergleich mit Valerianaceae und Dipsacaceae. Bot Jahrb Syst 111:499–553
Hofmann U, Ludewig J (1985) Morphologie und systematische Stellung von *Limnanthes douglasii* R. Brown, einem repräsentativen Vertreter der Limnanthaceae. Bot Jahrb Syst 105:401–431

Ihlenfeldt H-D (1960) Entwicklungsgeschichtliche, morphologische und systematische Untersuchungen an Mesembryanthemen. Feddes Repert 63:1–104
Lacroix C, Sattler R (1988) Phyllotaxis theories and sepal-stamen superposition in *Basella rubra*. Am J Bot 75:906–917
Leins P (1964) Die frühe Blütenentwicklung von *Hypericum hookerianum* Wight et Arn. und *H. aegypticum* L. Ber Dtsch Bot Ges 77:112–123
Leins P (1975) Die Beziehungen zwischen multiplen und einfachen Androeceen. Bot Jahrb Syst 96:231–237
Leins P, Schwitalla S (1986) Studien an Cactaceen-Blüten. I. Einige Bemerkungen zur Blütenentwicklung von Pereskia. Beitr Biol Pflanz 60:313–323
Mabry TJ (1976) Pigment dichotomy and DNA-RNA hybridization data for centrospermous families. Plant Syst Evol 126:79–94
Mabry TJ, Neumann P, Philipson WR (1978) *Hectorella*: A member of the betalain-suborder Chenopodiineae of the order Centrospermae. Plant Syst Evol 130:163–165
Nakai T (1942) Notulae ad plantas Asiae Orientalis (XVIII). J Jpn Bot 18:91–120
Pax F, Hoffmann K (1934) Caryophyllaceae. In: Engler A (ed) Die natürlichen Pflanzenfamilien, 2nd edn, vol 16c. Engelmann, Leipzig, pp 275–364
Payer J-B (1857) Traité d'organogénie comparée de la fleur. Masson, Paris
Rauh W, Schölch H-F (1965) Weitere Untersuchungen an Didiereaceen. 2. Teil Infloreszenz-, blütenmorphologische und embryologische Untersuchungen mit Ausblick auf die systematische Stellung der Didiereaceen. Sitzungsber Heidelberger Akad Wiss Math-Naturwiss Kl 11:221–434
Rohweder O (1965a) Centrospermenstudien. 1. Der Blütenbau bei *Uebelinia kiwuensis* T.C.E. Fries. Bot Jahrb Syst 83:406–418
Rohweder O (1965b) Centrospermenstudien. 2. Entwicklung und morphologische Deutung des Gynöciums bei *Phytolacca*. Bot Jahrb Syst 84:509–526
Rohweder O (1967) Centrospermenstudien. 3. Blütenentwicklung und Blütenbau bei Silenoideen. Bot Jahrb Syst 86:130–185
Rohweder O (1970) Centrospermenstudien. 4. Morphologie und Anatomie der Blüten, Früchte und Samen der Alsinoideen und Paronychioideen. Bot Jahrb Syst 90:201–271
Rohweder O, Huber K (1974) Centrospermenstudien. 7. Beobachtungen und Anmerkungen zur Morphologie und Entwicklungsgeschichte einiger Nyctaginaceae. Bot Jahrb Syst 94:327–359
Rohweder O, König K (1971) Centrospermenstudien. 5. Bau der Blüten, Früchte und Samen von *Pterantus dichotomus* Forsk. (Caryophyllaceae). Bot Jahrb Syst 90: 447–468
Rohweder O, Urmi E (1978) Centrospermenstudien. 10. Untersuchungen über den Bau der Blüten und Früchte von *Cucubalus baccifer* L. und *Drypis spinosa* L. (Caryophyllaceae-Silenoideae). Bot Jahrb Syst 100:1–25
Rohweder O, Urmi-König K (1975) Centrospermenstudien. 8. Beiträge zur Morphologie, Anatomie und systematischen Stellung von *Gymnocarpos* Forsk. und *Paronychia argentea* Lam. (Caryophyllaceae). Bot Jahrb Syst 96:375–409
Rutishauser R (1981) Blattstellung und Sprossentwicklung bei Blütenpflanzen unter besonderer Berücksichtigung der Nelkengewächse (Caryophyllaceae s.l.). Cramer, Vaduz (Dissertationes Botanicae vol 62)
Rutishauser R (1985) Blattquirle, Stipeln und Kolleteren bei den Rubieae (Rubiaceae) im Vergleich mit anderen Angiospermen. Beitr Biol Pflanz 59:375–424
Sattler R, Perlin L (1982) Floral development of *Bougainvillea spectabilis* Willd., *Boerhaavia diffusa* L. and *Mirabilis jalapa* (Nyctaginaceae). Bot J Linn Soc 84: 161–182
Ulbrich E (1934) Chenopodiaceae. In: Engler A (ed) Die natürlichen Pflanzenfamilien, 2nd edn, vol 16c. Engelmann, Leipzig, pp 379–584

7 Pollen Morphology and Exine Ultrastructure

JOAN W. NOWICKE

7.1 Introduction

The pollen characteristic that links all the betalain families (Mabry 1976), Aizoaceae, Amaranthaceae, Basellaceae, Cactaceae, Chenopodiaceae (including *Dysphania*), Didiereaceae, Halophytaceae, Nyctaginaceae, Phytolaccaceae s.l., Portulacaceae and Stegnospermataceae with the anthocyanin-pigmented Caryophyllaceae and Molluginaceae is a tectum that is spinulose with punctae and/or annular perforations (Nowicke 1975; Skvarla and Nowicke 1976; Nowicke and Skvarla 1977; Nowicke and Skvarla 1980). Within the Caryophyllales there are only simple apertures: 3-colpate, pantoporate and pantocolpate are common; 6- to 8-zonocolpate is more restricted. Specialized pollen, i.e., nonspinulose-punctate/annular perforate, is usually coarsely to finely reticulate.

The Achatocarpaceae, a small family of two genera and as many as 17 species, have a scabrate tectum (Skvarla and Nowicke 1982) and thus are the only family in Caryophyllales to lack the typical one described above. Even the apertures, four to six large, poorly defined pores, have no counterpart in the remaining Caryophyllales. The pigment condition is unknown, but Achatocarpaceae do have the P-type sieve-element plastid (see Chap. 5), characteristic of the order.

Other families have been associated with or assigned to Caryophyllales. Earlier studies have shown that pollen data do not support a close relationship between the Caryophyllales and three associated families, Plumbaginaceae, Polygonaceae, and Primulaceae (Nowicke and Skvarla 1977); nor do pollen data support a relationship between Batidaceae and/or Gyrostemonaceae with the Caryophyllales. In fact, the thick solid exines found in the latter two families (Nowicke and Skvarla 1980, Figs. 65–70) are so rare in the dicotyledons that no relationships to any other families are suggested. Pollen morphology does not support a relationship between Theligonaceae and Caryophyllales (Nowicke 1975).

The very distinction of Caryophyllales makes their ancestry speculative. However, both Takhtajan (1987) and Cronquist (1988) consider

Ranunculidae and Caryophyllidae to have a common lineage, and many Ranunculaceae (Nowicke and Skvarla 1981, Figs. 1–18) have pollen similar to Caryophyllales. In a study of 160 species of Ranunculaceae, Nowicke and Skvarla (in press) document a range of aperture types that overlap those in the Caryophyllales: 3-colpate, pantoporate, pantocolpate, and even 6- to 8-zonocolpate. The majority of Ranunculaceae have a spinulose and punctate/perforate tectum. In fact, in scanning electron microscopy (SEM) many grains in Ranunculaceae are indistinguishable from their apertural counterparts in Caryophyllales. However, Ranunculaceae differ from Caryophyllales in the absence of annular perforations and the presence of large columellae and usually well-developed apertural endexines (Nowicke and Skvarla 1981, Figs. 115–121; Nowicke and Skvarla, in press).

The thrust of this effort is to expand the knowledge of exine ultrastructure in Caryophyllales, and not to provide an exhaustive review of previous palynological publications. Yet, it would be impossible, given the size of Caryophyllales, to fully document the range of pollen morphology/ultrastructure in this publication. I have, therefore, made reference to previous studies when a pollen morphology or distinctive characteristic is not illustrated here, e.g., finely reticulate exines in Nyctaginaceae (*Reichenbachia* and *Neea*) and in Portulacaceae (*Mona meridensis*). Palynological studies of Caryophyllales and their scope and methodology include: Nowicke (1975) (the order using SEM), Skvarla and Nowicke (1976) [the order using transmission electron microscopy (TEM)], Nowicke and Skvarla (1977) (the order plus Plumbaginaceae, Polygonaceae and Primulaceae using SEM and TEM), Nowicke and Skvarla (1980) (the order plus Batidaceae, Gyrostemonaceae, and Plumbaginaceae, Polygonaceae, and Primulaceae in SEM and TEM). The Russian palynologist A. N. Sladkov has published a series of familial treatments (in Russian), but only the most recent one, Nyctaginaceae (1990), is cited, on the assumption that the bibliography will list his earlier studies for those readers who are interested. More restricted studies include those of Bittrich (1986) on Mesembryanthemoideae (SEM), Eliasson (1988) on certain Amaranthaceae (SEM), Straka (1965a,b) on Didiereaceae [light microscopy (LM) and TEM], Nilsson (1966, 1967) on Portulacaceae (line drawings), Nowicke (1970) and Nowicke and Luikart (1971) on Nyctaginaceae (SEM), Bortenschlager (1973) on Phytolaccaceae, Leuenberger (1976) on Cactaceae (SEM), and Bortenschlager et al. (1972) and Skvarla and Nowicke (1982) on Achatocarpaceae (SEM/TEM). Pollen studies limited to genera and/or species are not cited here.

The selection of taxa to be examined in TEM was based on: general interest, e.g., *Barbeuia*, *Sarcobatus*, *Pereskia*, and *Lophiocarpus*; confirmation of TEM information based on a single species in problematical genera, e.g., *Stegnosperma* and *Microtea*; a dearth of TEM information for a palynologically diverse family, e.g., Cactaceae; and documentation of a unique exine structure, e.g., the endexine in Didiereaceae. Because Achatocarpaceae were the subject of an earlier SEM/TEM study (Skvarla and Nowicke 1982), only two SEMs are included here. Two families, Basellaceae with a greater-than-expected range of pollen variation, and Didiereaceae with an unusual exine structure, will be the subject of a separate publication (Nowicke and Begle 1993). The absence of new Molluginaceae pollen data is not intentional and is due mostly to the poor circumscription of the family (Gilbert 1987; Bittrich 1990). Very few genera are known with regard to pigment type (*Gisekia* was found to have betacyanins and not the expected anthocyanins) and Molluginaceae may be the only centrospermous family to have sieve-element plastids with two types of crystals, globular or angular (see Chap. 5).

To the best of the author's knowledge, none of the species illustrated here in TEM have been illustrated elsewhere; however, in a few cases, species previously illustrated in SEM, viz., Achatocarpaceae and Didiereaceae, have been included for the purposes of representation, comparison and the convenience of the reader.

7.2 Materials and Methods

Anthers were removed from herbarium specimens and pollen for all three preparations, LM, SEM, and TEM, was first acetolyzed according to Erdtman (1966). Pollen for LM was mounted in glycerin jelly and sealed with paraffin. For SEM, pollen was coated with carbon and then gold-palladium, and examined and photographed with a Hitachi 570 SEM. For TEM, pollen was incorporated into agar, fixed with osmium tetroxide, stained with uranyl acetate and embedded in L. R. White acrylic resin. After sectioning with a diamond knife, the sections were stained with lead citrate and examined and photographed in a JEOL 1200EX TEM. All glass slides and electron micrographs are deposited at the Palynological Laboratory, National Museum of Natural History.

The species examined, voucher data, aperture type, tectum morphology, and figure numbers are given in Table 7.1. For the most part, the species names in Table 7.1 are taken from the collection label or the most recent annotation. The 18 plates are organized alphabetically by family, except for Chenopodiaceae in Plate 7.2 with Amaranthaceae, Didiereaceae in Plate 7.3 with Basellaceae, and Achatocarpaceae in Plate 7.15 with Phytolaccaceae.

7.3 Results

The pollen of 63 of the 77 species examined (Table 7.1) in this study is illustrated in Plates 7.1–7.18, comprising 123 SEMs and TEMs. Representatives of Aizoaceae are shown in Plate 7.1, Amaranthaceae and Chenopodia-

Table 7.1. Species examined, voucher data

Species	Voucher data	Aperture	Tectum	Plate	SEM	TEM
Achatocarpaceae						
Achatocarpus mollis H Walter	Eggers 15503 Ecuador	6 PP, IRREG	SC	7.15,5	SEM	–
A. nigricans Triana	Allen & Van Severen 6891 El Salvador	6 PP, IRREG	SC	7.15,6	SEM	–
Aizoaceae						
Drosanthemum striatum (Haw.) Schwantes	UC Bot. Gard. Berkeley 50.812 South Africa	3 C	MRET	7.1,1 & 2	SEM	TEM
Gisekia miltus Fenzl	Seydel 4111 SW Africa	3 C	SP-PT	7.1,3	SEM	TEM
Mesembryanthemum nodiflorum L.	Wolf 662 California	3–4 ZCP, 6 PC	MRET		SEM	–
Psammotropha myriantha Sond.	Bayliss s.n. S Africa	3 C	SP-PT	7.1,4 & 5	SEM	TEM
Sesuvium portulacastrum L.	Stoddart & Head 9102 Jamaica	3 C	SP/?		SEM	–
Tetragonia arbuscula Fenzl	Bayliss 2171 S Africa	3 C	SP-PT		SEM	–
T. fruticosa L.	Buys s.n. S Africa	3 C	SP-PT	7.1,6 & 7	SEM	TEM
Trianthema monogyna L.	Pringle 9420 Mexico	3 C	SP-PT	7.1,8	SEM	–
Amaranthanceae						
Pfaffia lanata (Poir.) Gib.	Anderson et al. 36582 Brazil	PP	RET		SEM	–
Philoxerus vermicularis (L.) R Br	Stoddart 107 Belize	PP	RET	7.2,1 & 2	SEM	TEM

Basellaceae						
Anredera baselloides (HBK) Baillon	Campinas 365 Brazil	6, 12 PC	SP-PT		SEM	TEM
A. vesicaria (Lam.) Gaertner	Hinton 6903 Mexico	6 PC, PP	SP/?	7.3,1 & 2	SEM	TEM
Basella excavata S Elliot	Croat 32007 MO Madagascar	6 PC	PT	7.3,3	SEM	–
B. madagascariensis Boiv. ex H Perr.	Bernardi 12010 Madagascar	6 PC	PT		SEM	TEM
Tournonia hookeriana Moquin-Tandon	Asplund 9755 Ecuador	7–10 PP	SP-PT		SEM	–
	Holmgren 475 Ecuador	7–10 PP	SP-PT		SEM	TEM
B. paniculata Volkers	Raven & Raven 26116 S Africa	12-PC	SP-PT		SEM	TEM
Cactaceae						
Aporocactus flagriformis Lem.	Sokoi s.n. Mexico	3 C	SP-AP	7.5,1 & 2	SEM	TEM
Ariocarpus agavioides (Castan.) EF Anders.	Anderson 1616 Mexico	3 C, 6 PC	SP-PT	7.5,7	SEM	TEM
Carnegiea gigantea (Engelm.) Britt. & Rose	Toumey s.n. Arizona	3 C	SP-AP	7.5,3	SEM	TEM
Cereus peruvianus (L.) Mill.	US Botanic Garden District of Columbia	3 C	SP-AP	7.5,4–6	SEM	TEM
Cleistocactus candelilla Cardenas	Ritter 357 Bolivia (UCB Bot. Gard.)	3 C	SP-AP		SEM	–
Disocactus himantocladus (Rol.-Goss) Kimn.	Lankester s.n. Costa Rica	3 C	SP?-AP	7.4,1–4	SEM	TEM
D. macranthus (Alex.) Kimn. & Hutchis.	MacDougal A197 Mexico	6 PC	SP-AP	7.4,5	SEM	–
Echinocereus pacificus (Engelm.) Britt. & Rose	Moran 15085 Mexico	3 C	SP-PT		SEM	TEM

Table 7.1. *Continued*

Species	Voucher data	Aperture	Tectum	Plate	SEM	TEM
Mammillaria meiacantha Engelm.	Palmer 573 Mexico	3 C	SP-PT		SEM	TEM
Nopalea auberi (Pfeiff.) Salm-Dyck	Rose & Rose 11309 Mexico	12 PP	SP-PT		SEM	–
Opuntia erinacea Engelm. & Bigel	Bentley et al. 13403 Nevada	12–15 PC	RET	7.6,1 & 2	SEM	TEM
O. macbridei Britt. & Rose	Wurdack 1790 Peru	6 PP, 6 PC	PT-MRET	7.6,3 & 4	SEM	TEM
O. macrorhiza Engelm.	Lathrop 421 Kansas	PP	RET		SEM	TEM
O. microdasys (Lehmann) Pfeiffer	Nelson 6145 Mexico	10–12 PP	RET	7.6,5 & 6	SEM	TEM
O. triacantha (Willd.) Sweet	Britton et al. 1565 Puerto Rico	IRREG PP	PT	7.4,6 & 7	SEM	TEM
Parodia aureispina Backeb.	UCB Bot. Gard. 49.1758-1 Argentina	3 C	SP-PT		SEM	–
Pereskia aculeata Miller	Hatschbach 47785 Brazil	6–12 PC	SP-AP	7.7,1–3	SEM	TEM
P. guamacho FAC Weber	Pittier 6179 Venezuela	12 PC (6, 15 PC)	SP-AP		SEM	TEM
Rebutia kupperiana Bödeker	UCB Botanical Garden Bolivia	12 PC IRREG	SP-AP	7.7,6	SEM	–
Rhipsalidopsis gaertneri (Reg.) Moran	Cordeiro 169 Brazil	12 PC	SP-AP	7.7,4 & 5	SEM	TEM
Caryophyllaceae						
Acanthophyllum microcephalum Boiss.	Rechinger 751 Iran	PP	SP-PT	7.8,1 & 2	SEM	TEM

Agrostemma githago L.	McDougall s.n., 1914 Tennessee	PP	SP-PT	7.8,3 & 4	SEM	TEM
Arenaria capillaris Poir.	Hodgdon 8359 Alaska	PP	SP-PT	7.10,7 & 8	SEM	TEM
A. macradenia S Watson	Atwood 4745 Utah	PP	SP-PT	7.8,5 & 6	SEM	TEM
Lychnis floscuculi L.	Fosberg 55527 Massachusetts	PP	SP-PT	7.9,6 & 7	SEM	TEM
Malachium aquaticum (L) Fries	Blaszczyk et al. 526 Poland	PP	SP-PT	7.9,4 & 5	SEM	TEM
Melandrium apricum Rohrb.	Hatusima s.n. Japan	PP	FRET	7.10,3 & 4	SEM	TEM
Saponaria officinalis L.	Thomas 11423 Montana	PP	SP-AP	7.9,1–3	SEM	TEM
Silene aperta Greene	Ferris & Lorraine 10870 California	PP	FRET	7.10,1 & 2	SEM	TEM
Spergularia macrotheca (Britt.) Robinson	Keck & Stockwell 3350 California	3 C	SP-PT	7.10,5 & 6	SEM	TEM
Chenopodiaceae						
Salsola kali L.	Barrell 187-60 Colorado	PP	SP-PT	7.2,7 & 8	SEM	TEM
Sarcobatus vermiculatus Torrey	Moodie 12 Canada	10–15 PP	SP-PT	7.2,5 & 6	SEM	TEM
Suaeda australis Moquin-Tandon	Stoddart 4643 Australia	PP	SP-PT	7.2,3 & 4	SEM	TEM
Didiereaceae						
Alluaudia procera Drake	Humbert & Swingle 5702 Madagascar	6–8 ZCP	SP-AP	7.3,4	SEM	TEM
Alluaudiopsis fiherenensis Humbert & Choux	Lorence 1937 MO Madagascar	6–8 ZCP	SP-AP	7.3,6	SEM	TEM
Decaryia madagascariensis Choux	Humbert 20318 Madagascar	6–8 ZCP	SP-AP	7.3,5	SEM	–

Table 7.1. *Continued*

Species	Voucher uata	Aperture	Tectum	Plate	SEM	TEM
Nyctaginaceae						
Acleisanthes greggii Standley	Pennell 18022 Mexico	PP	SP-AP	7.11,4	SEM	–
Anulocaulis leiosolenus Standley	Stearns 448 Texas	PP	SP/AP	7.11,1 & 2	SEM	TEM
Guapira costaricana (Standley) Woodson	Bro. Maurice 777 Panama	6? PC, IRREG.	SP-PT	7.12,1 & 2	SEM	TEM
Mirabilis albida (H Walter) Heimerl	Waterfall 12295 Oklahoma	PP	SP-AP		SEM	–
M. tenuiloba S Watson	Moran & Reveal 19585 Mexico	PP	SP-AP	7.11,3	SEM	TEM
Pisonia ambigua Heimerl	Venturi 998 Argentina	3 C	SP-PT	7.12,3 & 4	SEM	TEM
P. grandis R Br	Fosberg 49420 Aldabra	3 C	SP-PT	7.12,5–7	SEM	TEM
Tripterocalyx micranthus Standley	Holmgren & Hansen 3404 Utah	3 C	CRET	7.11,5–7	SEM	TEM
Phytolaccaceae						
Agdestis clematidea Moc. & Sesse	Small et al. 10694 Florida	3 C	SP-PT	7.16,7	SEM	–
Barbeuia madagascariensis Steudel	Miller & Randrianasoles 6220 MO Madagascar	3 C	SP-PT	7.15,1–4	SEM	TEM
Ercilla spicata (Bertero) Moquin-Tandon	Werdermann 1922 Chile	3 C	SP-PT, IRREG.	7.13,1 & 2	SEM	TEM
Lophiocarpus polystachyus Turcz.	Seydel 4438 SW Africa	3 C	SP-PT	7.13,3–5	SEM	TEM
Microtea maypurensis (Kunth) G Don	Silva & Hage 606 Brazil	PP	SP-PT	7.16,4–6	SEM	TEM

M. scabrida Urb.	Hatschbach et al. 13304 Brazil	PP	SP-PT	7.16,1–3	SEM	TEM
Petiveria alliacea L.	Curtiss 5520 Florida	PP, IRREG	SP-PT, IRREG	7.13,6 & 7	SEM	TEM
Phytolacca dioica L.	Jorgensen 1995 Argentina	3 C	SP-PT	7.14,3	SEM	TEM
P. pruinosa Fenzl	Haradjian 3164 Syria	3 C	SP-PT	7.14,1 & 2	SEM	TEM
Seguieria aculeata Jacq.	Dugand 5190 Colombia	3 C	SP-PT	7.14,6–8	SEM	TEM
S. langsdorffii Moquin-Tandon	Hatschbach 8363 Brazil	3 C	SP-SC	7.14,4 & 5	SEM	TEM
Portulacaceae						
Claytonia sarmentosa CA Mey.	Spetzman 784 Alaska	3 C	SP-PT	7.17,3 & 4	SEM	TEM
Montiastrum lineare (Dougl.) Rydb.	Thompson 14628 Washington	28–35 PC	TH	7.17,5 & 6	SEM	TEM
Talinella sp.	Croat 31953 MO Madagascar	PC	SP-AP		SEM	–
	Dorr 4097 MO Madagascar	PC	SP-AP	7.17,7	SEM	–
	Humbert 18878 P Madagascar	PC	SP-AP		SEM	–
Talinum patens (Jacq.) Willd.	Venturi 4981 Argentina	15 PC	SP-PT	7.17,1 & 2	SEM	TEM
Stegnospermataceae						
Stegnosperma cubense A Rich.	Solis 22 Mexico	3 C	SP-PT/AP	7.18,1 & 2	SEM	TEM
S. watsonii DJ Rogers	Wiggins 6247 Mexico	3 C	SP-PT/AP	7.18,3–5	SEM	TEM

(All US National Hebarium unless otherwise noted), aperture types and tectum types [3 C, 3-colpate; PP, pantoporate, pores numerous unless stated otherwise; PC, pantocolpate; ZCP, zonocolpate; TH, tholate; SP, spinulose; PT, punctate; AP, annular perforate; RET, reticulate; MRET, microreticulate; FRET, finely reticulate; CRET, coarsely reticulate; SC, ± scabrate; IRREG, irregular] ceae in Plate 7.2, Basellaceae and Didiereaceae in Plate 7.3, Cactaceae in Plates 7.4–7.7, Caryophyllaceae in Plates 7.8–7.10, Nyctaginaceae in Plates 7.11 and 7.12, Phytolaccaceae in Plates 7.13–7.16, Portulacaceae in Plate 7.17, and Stegnospermataceae in Plate 7.18.

7.3.1 Caryophyllales Pollen Description (Plates 7.1–7.18)

Pollen is shed as monads, only very rarely in tetrads; spheroidal, subprolate to suboblate in shape, very rarely cuboidal; apertures 3-colpate (3-zonocolpate), pantoporate, pantocolpate, rarely 5- to 8-zonocolpate, irregularly porate, or 4-colpate (4-zonocolpate); aperture membranes/covers are usually present and well developed; tecta complete (continuous) and mostly spinulose and punctate or annular perforate, occasionally tecta incomplete and coarsely reticulate to microreticulate, or, more rarely, scabrate; in thin section endexine is usually very thin to even absent in nonapertural areas, and thin and loosely lamellate or fibrous beneath the apertures, or, more rarely, consisting of vertical strands, or appearing solid; the foot layer ranges from threadlike to very thick; the columellae are short (less than one-third the exine thickness) to long (more than one-third); the tectum is usually well developed, sometimes thickened.

General pollen features – tectum, apertures, and exine structure – are discussed first and followed by pollen descriptions for each family.

7.3.2 Tectum

The pollen character that unites the centrospermous families (Achatocarpaceae excepted) is a tectum that is spinulose (SP) and has, to some degree, openings that have been designated punctae (PT) if without a thickened or raised margin, e.g., Plates 7.1,3, 7.5,7, 7.12,3 and 7.17,4, or annular perforations (AP) if crassimarginate or with a raised margin, e.g., Plates 7.3,5, 7.5,2, 7.7,3 and 4, and 7.11,1 and 4. Sometimes PT are more obvious in TEM than in SEM, e.g., Plate 7.1,4 versus Plate 7.1,5. Annular perforations are distinguishable in TEM, e.g., Plates 7.5,3, 7.7,5 and 7.11,2 and 3.

Plate. 7.1. SEM and TEM of Aizoaceae s.l. pollen. *1* and *2*, *Drosanthemum striatum*: *1*, microreticulate tectum (SEM); *2*, mostly radial section through mesocolpus (TEM). *3*, *Gisekia miltus*: tectum (SEM). *4* and *5*, *Psammotropha myriantha*: *4*, slightly oblique radial sections of two grains; the endexine is threadlike in the mesocolpus but loosely lamellate-granular beneath the colpus; the foot layer is about half as thick as the complete tectum (TEM); *5*, colpus and adjacent tectum (SEM). *6* and *7*, *Tetragonia fruticosa*: *6*, radial section through mesocolpus; note very thick tectum, short columellae and foot layer half as thick as the tectum (TEM); *7*, polar view (SEM). *8*, *Trianthema monogyna*: note prominent colpal membrane (SEM). Unless otherwise indicated, *bars* denote 1 μm

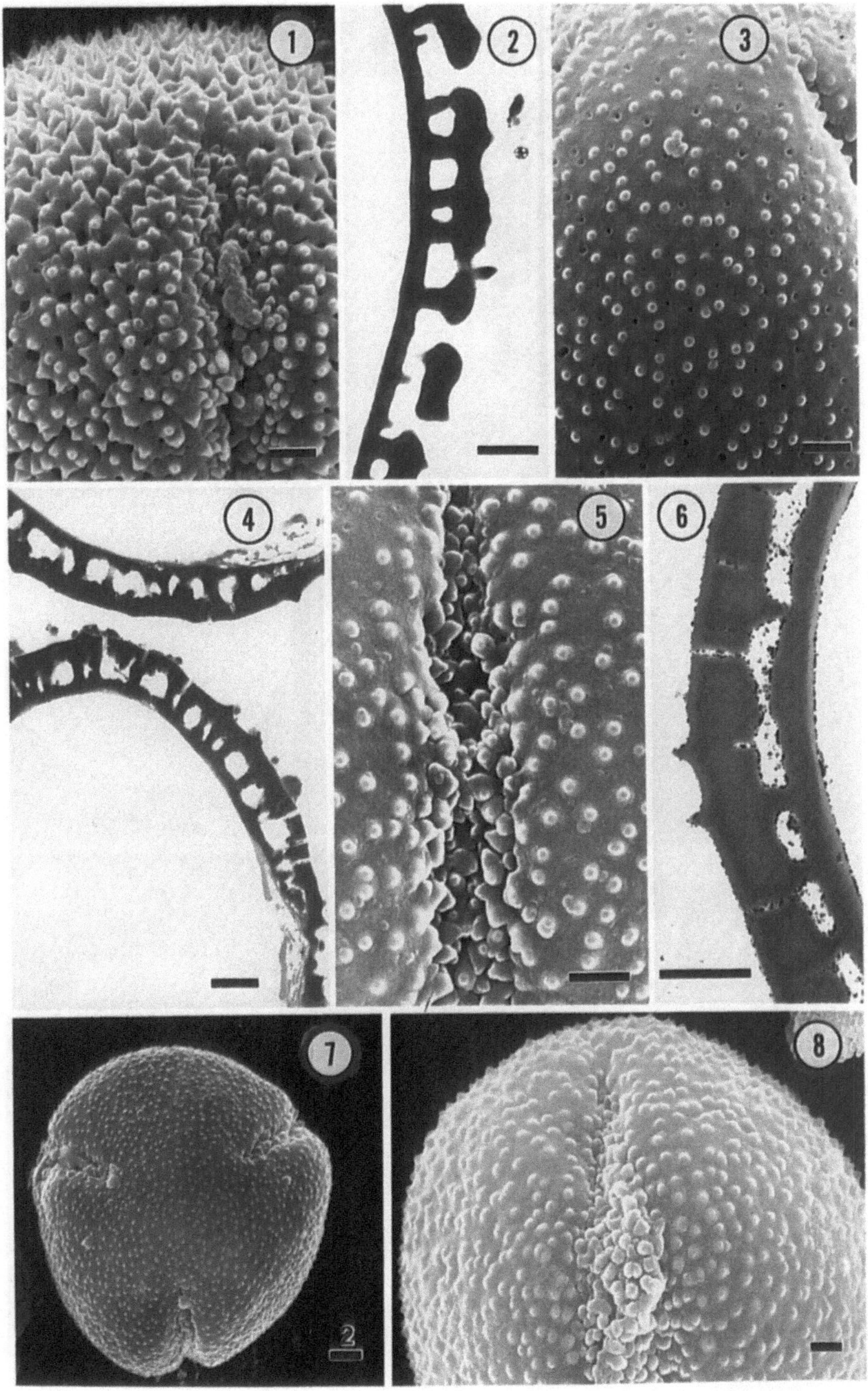
1
2
3
4
5
6
7
8

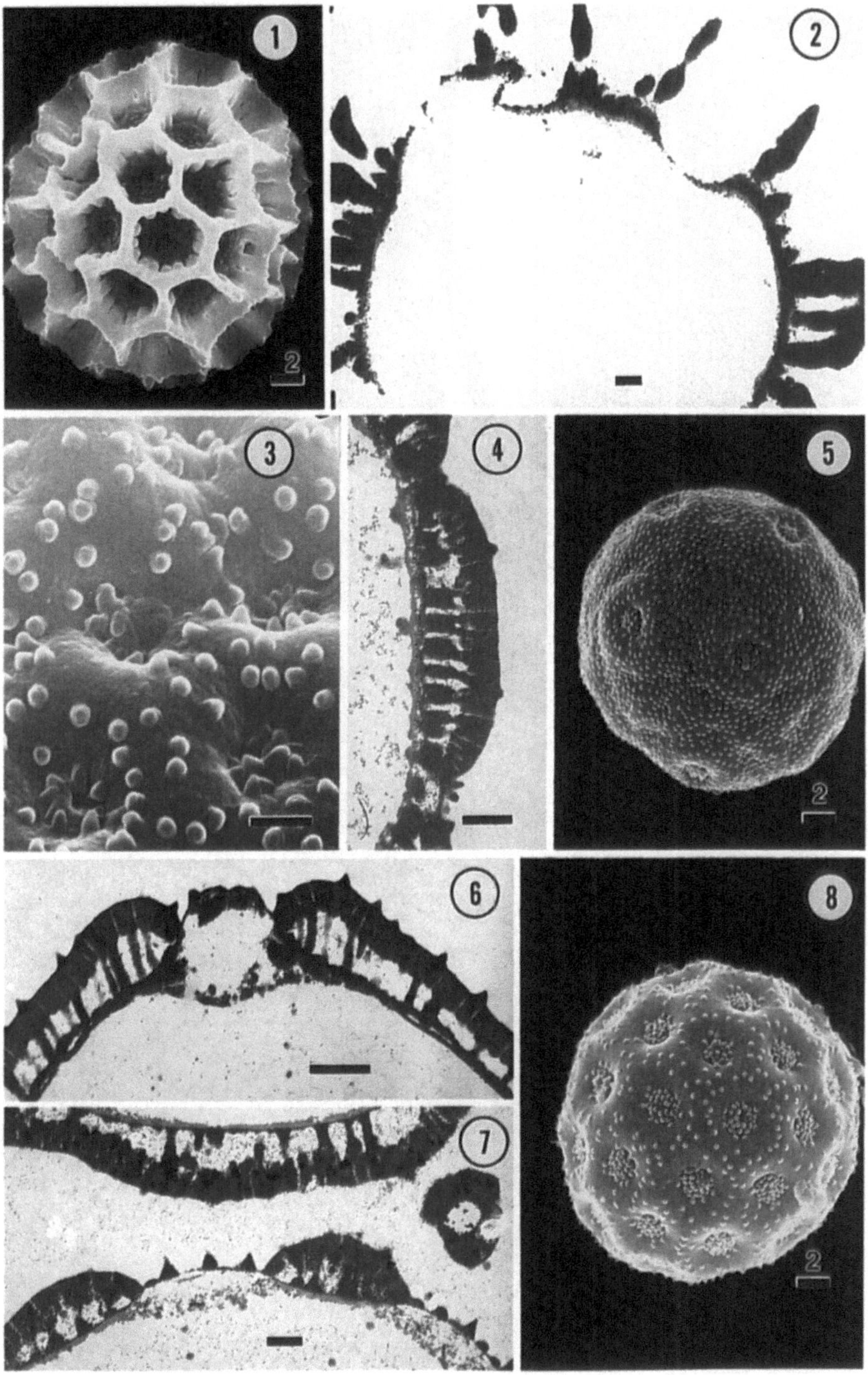
1
2
3
4
5
6
7
8

Plate. 7.2. SEM and TEM of Amaranthaceae (*1* and *2*) and Chenopodiaceae (*3–8*) pollen. *1* and *2*, *Philoxerus vermicularis*: *1*, Gomphrena-type Amaranthaceae; pantoporate grain with pores separated by muri consisting of a single row of columellae which fuse distally to form an incomplete, reticulate tectum (SEM); *2*, section of grain; at the *top*, where the section passed through three pores, note paucity of endexine (TEM). *3* and *4*, *Suaeda australis*: *3*, tectum (SEM); *4*, radial section; both the endexine and foot layer are thin, irregular and not always in contact; the columellae are stout, sometimes appearing to narrow or bifurcate at the thick tectum (TEM). *5* and *6*, *Sarcobatus vermiculatus*: *5*, pantoporate grain showing pores with "raised" margins (SEM); *6*, radial section across pore; note that raised margins are from elongated columellae, not a locally thickened tectum (TEM). *7* and *8*, *Salsola kali*: *7*, slightly oblique radial sections of two grains; there is a thin endexine, which becomes loosely and sparsely granular under the pores, a very thin foot layer, and a thick tectum (TEM); *8*, note the uniformity of pores, both in shape and distribution (SEM). Unless otherwise indicated, *bars* denote 1 μm

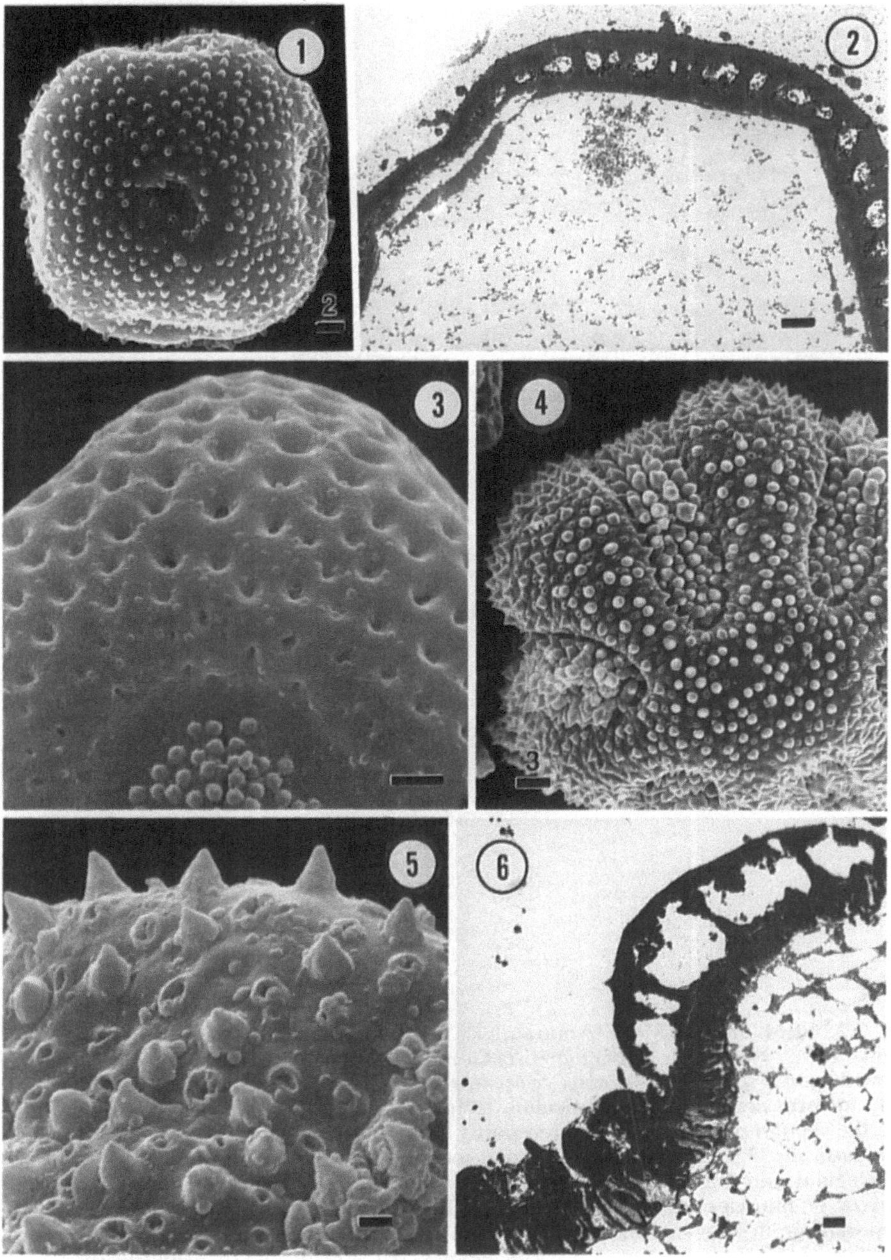
1
2
2
3
4
3
5
6

Plate. 7.3. SEM and TEM of pollen in Basellaceae (*1–3*) and Didiereaceae (*4–6*). *1* and *2*, *Anredera vesicaria*: *1*, polar or equatorial view; the colpi are very short in this 12-pantocolpate grain (SEM); *2*, radial section through mesocolpus; *Anredera* species have a solid, inner apertural endexine, usually more prominent than the layer at *upper left* (TEM). *3*, *Basella excavata*: the grains of most *Basella* species are cuboidal with a colpus in the center of each face (SEM). *4*, *Alluaudia procera*: polar view of 6-zonocolpate grain; note very prominent aperture (SEM). *5*, *Decaryia madagascariensis*: tectum showing prominent spines and prominent annular perforations (SEM). *6*, *Alluaudiopsis fiherenensis*: section across colpus and mesocolpus; note distinctive endexine consisting of vertical strands and expansion of columellae nearer the tectum (TEM). Unless otherwise indicated, *bars* denote 1 μm

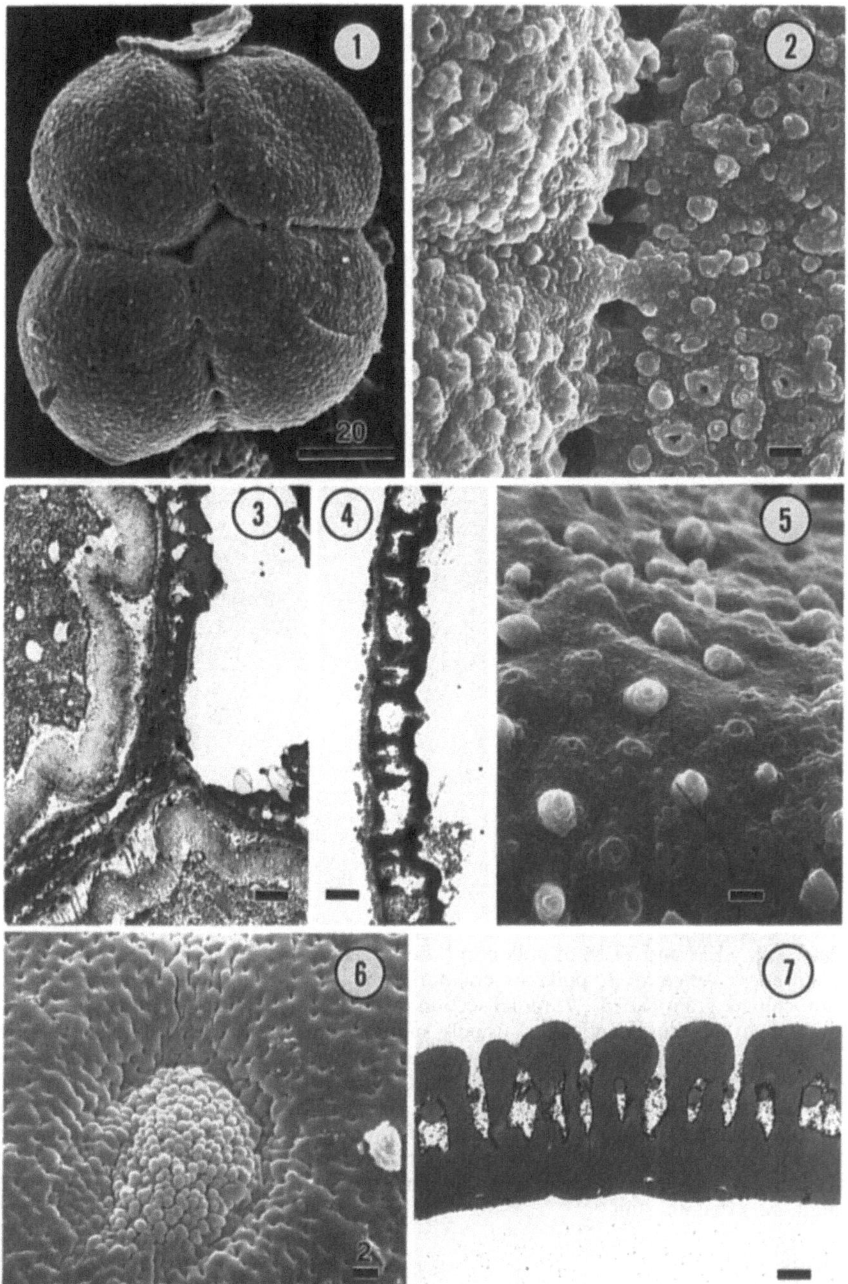
1
20
2
3
4
5
6
2
7

Plate. 7.4. SEM and TEM of Cactaceae pollen. *1–4*, *Disocactus himantocladus*: *1*, rhomboidal tetrad; tetrad members are 3-colpate, but only *lower left* tetrad member shows half of colpus; most tetrads were in a tetrahedral configuration (SEM); *2*, exine bridges between tetrad members; note highly granular surface and prominent annular perforations (SEM); *3*, section through partially acetolyzed tetrad; note prominent, less electron dense intine and common tectum between adjacent tetrad members (*lower far left*) (TEM); *4*, radial section of exterior wall of tetrad member; the irregular undulate tectum agrees with that portrayed by SEM in *2* (TEM). *5*, *Disocactus macranthus*: this species has pollen in monads and is ± 6-pantocolpate; the tectum has a granular appearance similar to that in *2* (SEM). *6* and *7*, *Opuntia triacantha*: *6*, pore in finely punctate tectum; this species is 15–20 porate but the pores are irregular in shape, size and distribution (SEM); *7*, section through mesoporus; note massive foot layer, well-developed columellae, and thick tectum with gaps corresponding to the punctae in *6* (TEM). Unless otherwise indicated, *bars* denote 1 μm

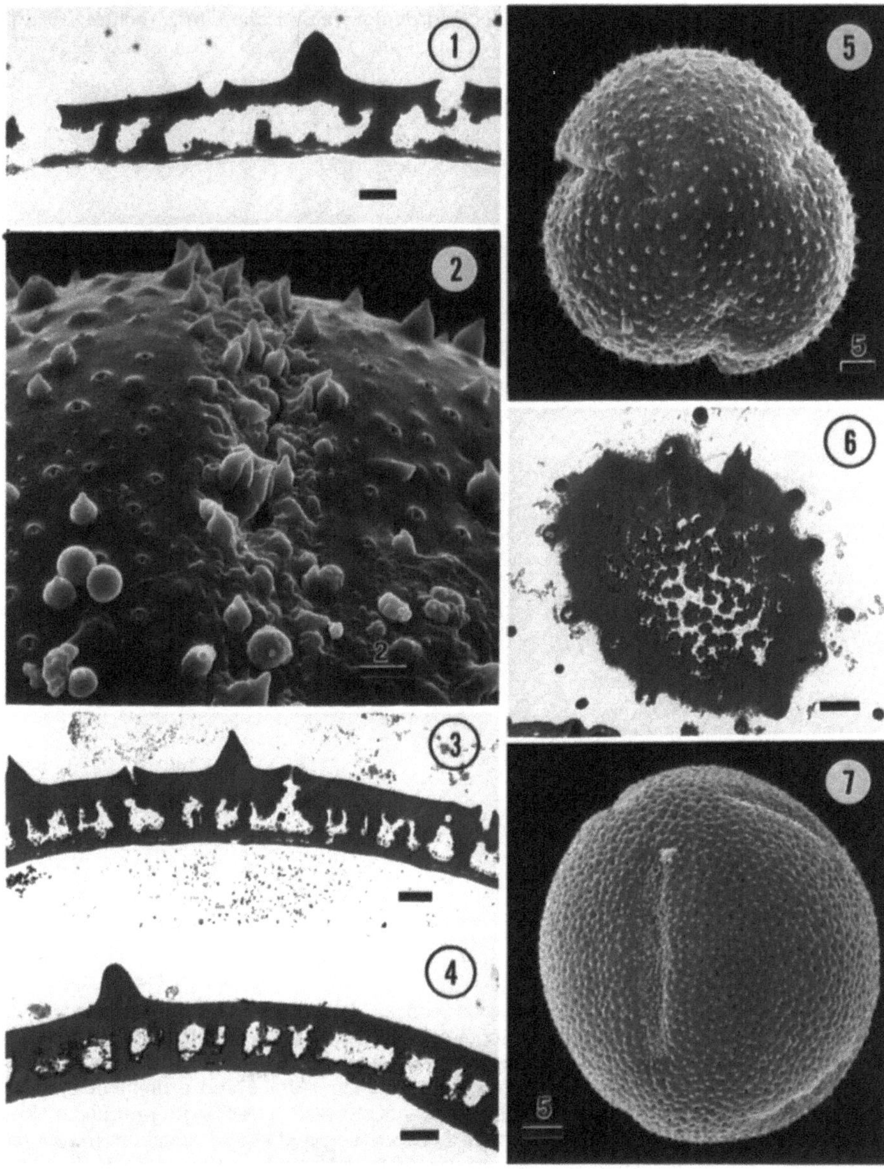

Plate. 7.5. SEM and TEM of Cactaceae pollen. *1* and *2*, *Aporocactus flagriformis*: *1*, oblique radial section; all three layers (foot layer, columellae and tectum) appear irregular; the tectal perforations agree with those depicted in SEM (see *2*) (TEM); *2*, tectum showing one of three colpi (SEM). *3*, *Carnegiea gigantea*: radial section; although the tectum is better defined than in *1*, this exine also has irregular columellae and foot layer (TEM). *4–6*, *Cereus peruvianus*: *4*, radial section (TEM); *5*, equatorial view (SEM); *6*, tangential section (TEM). *7*, *Ariocarpus agavioides*: a 6-pantocolpate grain, but the sample also has some 3-colpate grains (SEM). Unless otherwise indicated, *bars* denote 1 µm

Plate. 7.6. SEM and TEM of Cactaceae pollen, *Opuntia*. *1* and *2*, *O. erinacea*: *1*, in these large grains, thin-walled areas are separated by ridges of reticulate ektexine; note that thin area at *upper right* is not an aperture, and that given the orientation of three colpi the

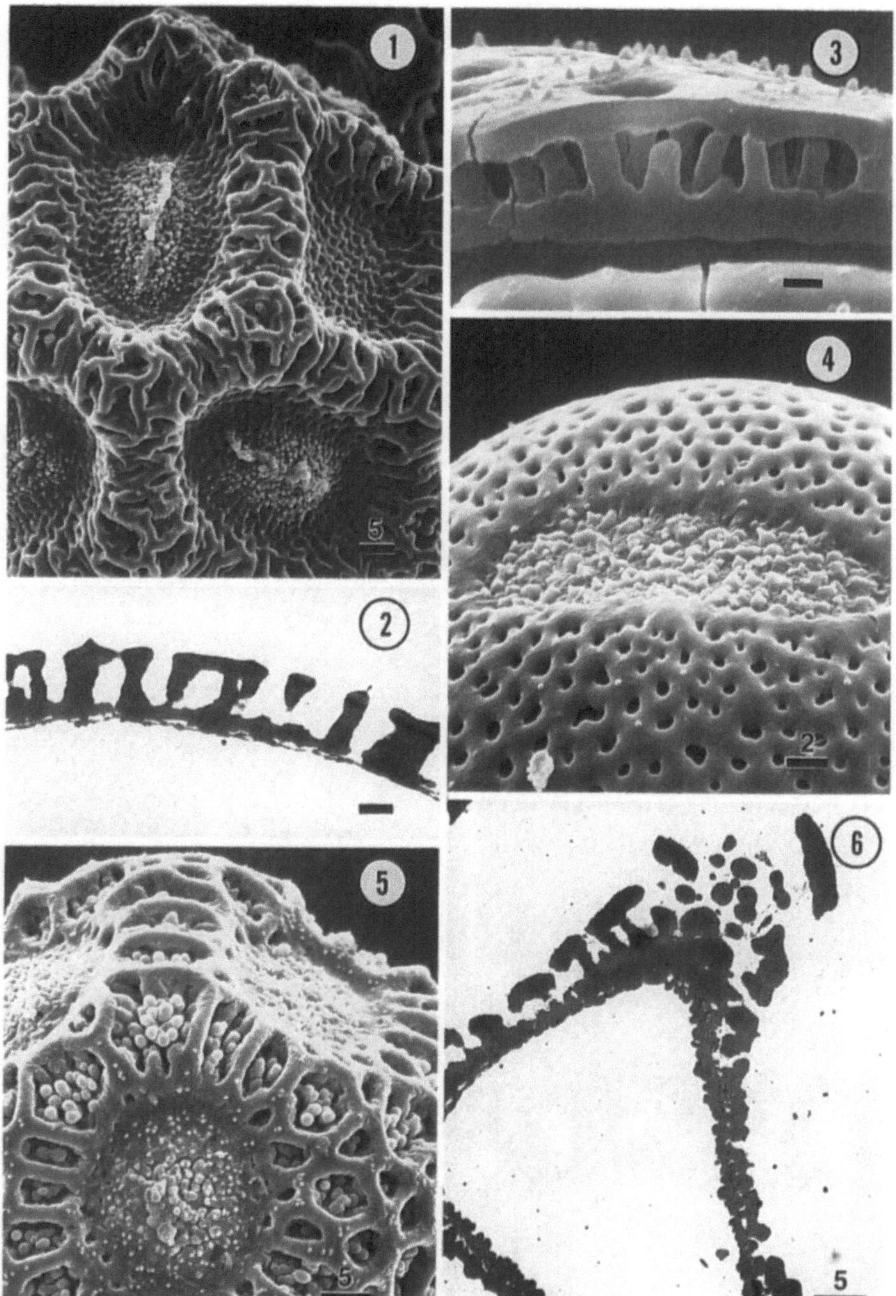

grain is pantocolpate (SEM); *2*, section through reticulate ridge (TEM). *3* and *4*, *O. macbridei*: *3*, fracture; note vestigial (?) spinules and thick foot layer, the latter confirmed in TEM (SEM); *4*, aperture shape in this species varies, some are porate, others brevicolpate; the spinules are very reduced in size and number (SEM). *5* and *6*, *O. microdasys*: *5*, in the reticulate ridges the lumina are larger than in *1* and have many free columellae (SEM); *6*, section through a reticulate ridge (*upper right*), then along edge or ridge (*upper left*) and across a pore (*right*) (TEM). Unless otherwise indicated, *bars* denote 1 μm

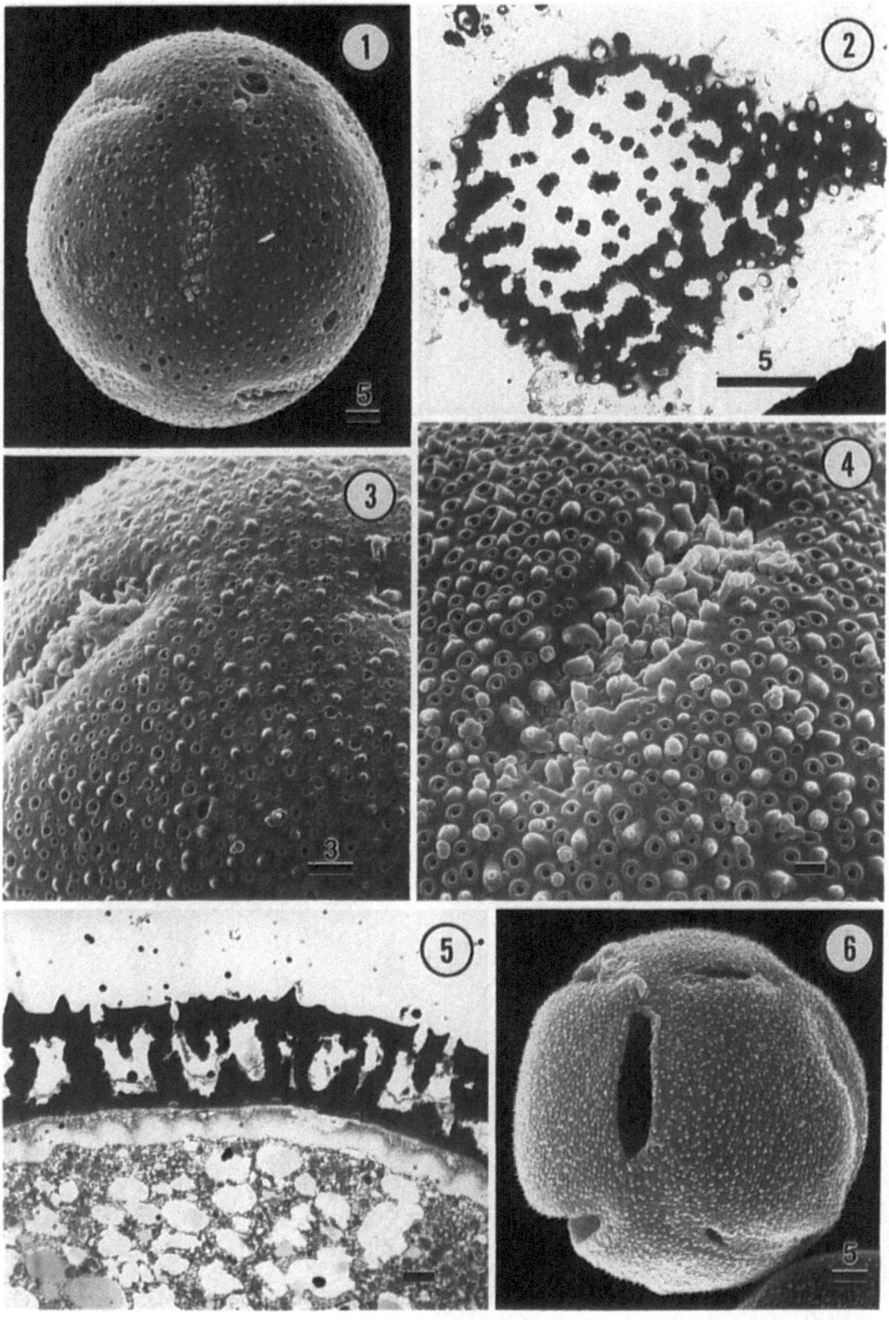
1
5
2
5
3
3
4
5
6
5

Plate. 7.7. SEM and TEM of Cactaceae pollen. *1–3*, *Pereskia aculeata*: *1*, a 12-pantocolpate grain; note irregularity of tectal openings (SEM); *2*, tangential section; the columellae are irregular in shape and surface (TEM); *3*, tectum (SEM); *4* and *5*, *Rhipsalidopsis gaertneri*: *4*, one colpus of a 12-pantocolpate grain (SEM); *5*, radial section of partially acetolyzed grain; note thick intine and irregular columellae with granular surfaces; in tangential section (unpubl. data) the columellae of this species are as irregular as those portrayed in *2* (TEM). *6*, *Rebutia kupperiana*: the grains in this sample are irregularly pantocolpate and, although not apparent at this low magnification, the tectum is annular perforate (SEM). Unless otherwise indicated, *bars* denote 1 μm

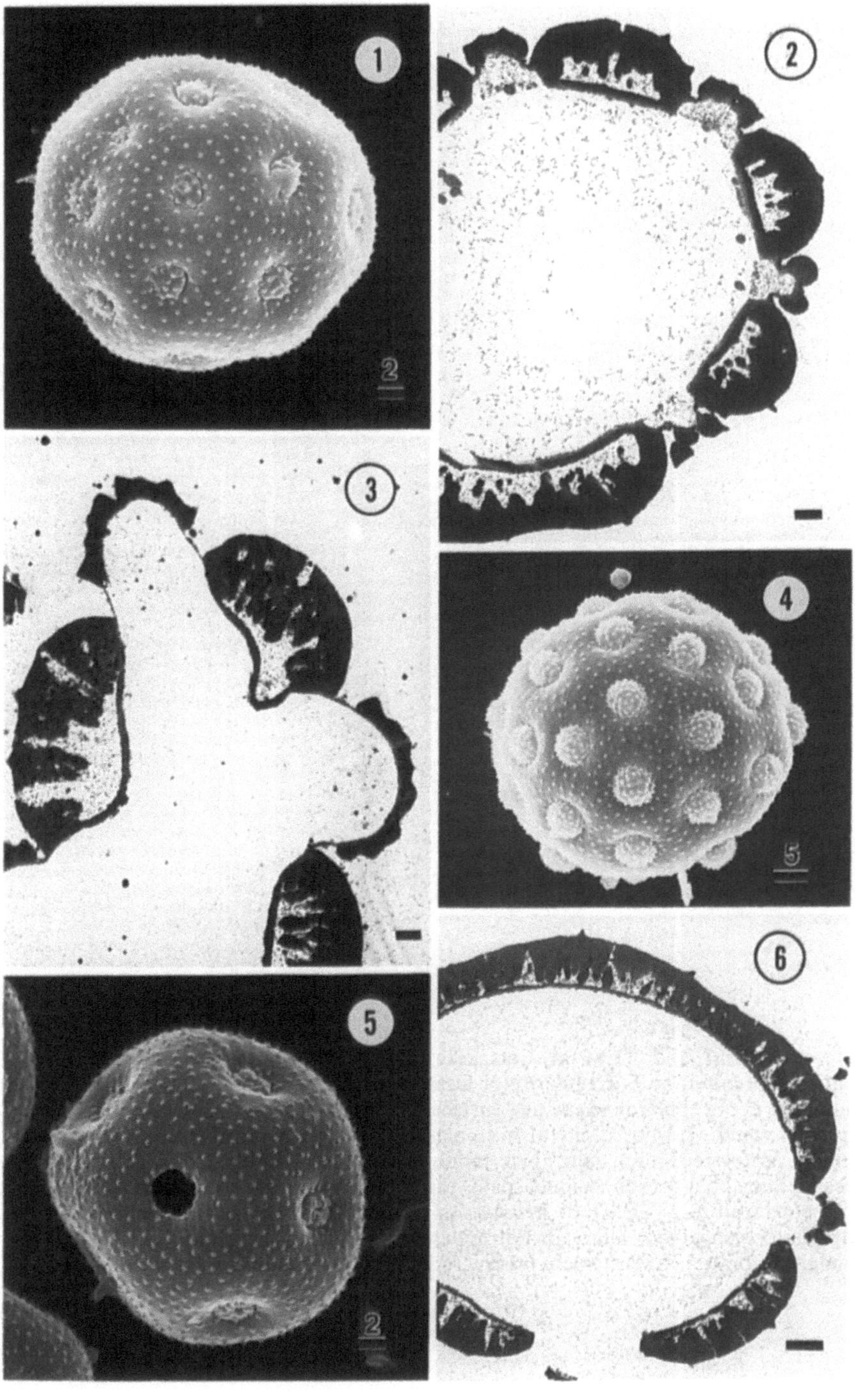
1
2
2
3
4
5
5
2
6

Plate. 7.8. SEM and TEM of Caryophyllaceae pollen. *1* and *2*, *Acanthophyllum microcephalum*: *1*, whole grain (SEM); *2*, somewhat oblique section; note thick tectum, thin foot layer, and very thin endexine, even under pores (TEM). *3* and *4*, *Agrostemma githago*: *3*, slightly oblique radial section; the pore cover or membrane is cuplike and separates intact; the endexine-foot layer is threadlike; none of the sections show columellae continuous with the foot layer, all were hanging and of the same length; the tectum is the predominant component (TEM); *4*, note pores that are regular in size and uniformly distributed (SEM). *5* and *6*, *Arenaria macradenia*: *5*, although in this grain the pores are uniform in size and evenly distributed, other grains had irregularities (SEM); *6*, radial section; the pantoporate exines in many Caryophyllaceae are characterized by a very thick tectum, columellae that narrow proximally, and a very thin, even threadlike, foot layer/endexine (TEM). Unless otherwise indicated, *bars* denote 1 μm

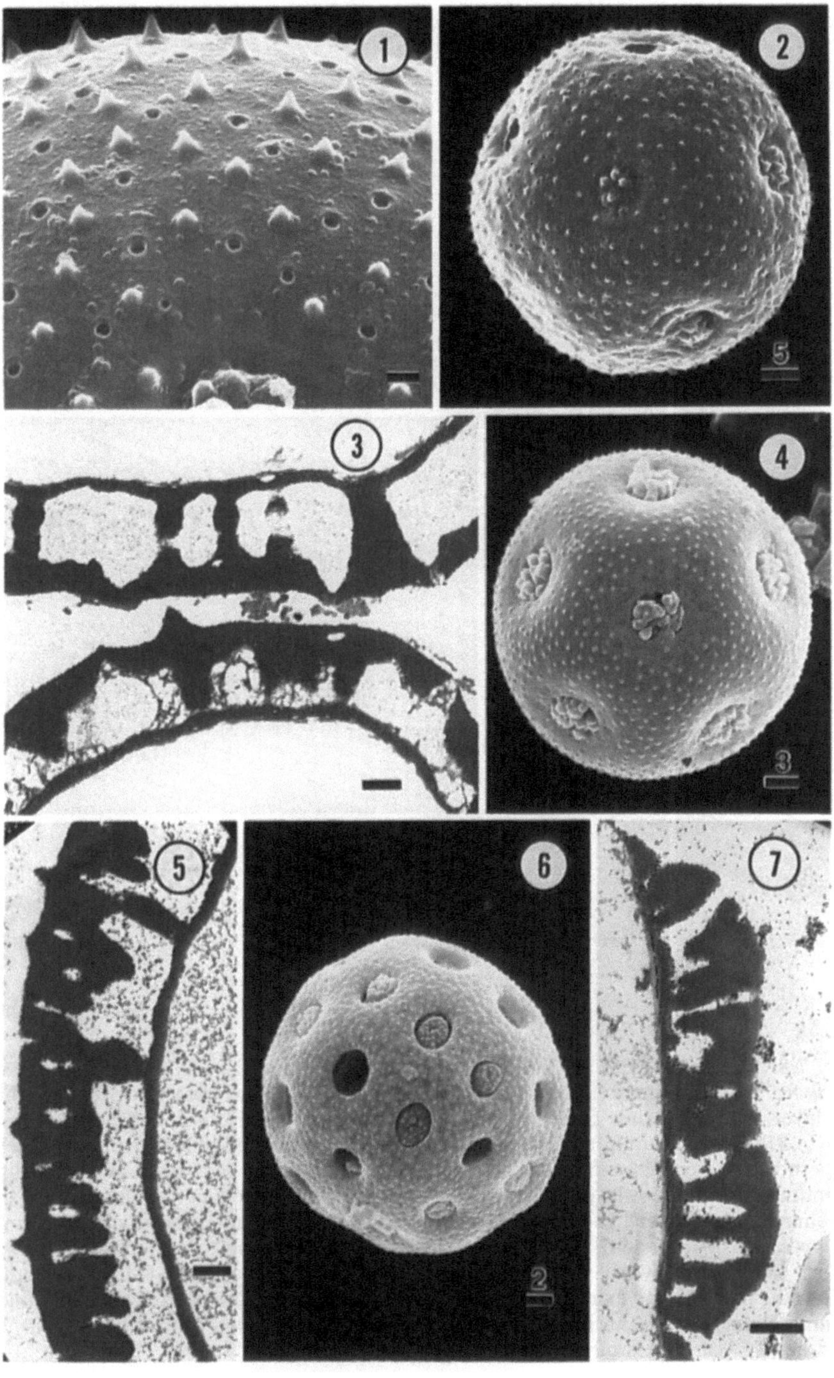
1
2
5
3
4
3
5
6
7
2

Plate. 7.9. SEM and TEM of Caryophyllaceae pollen. *1–3*, *Saponaria officinalis*: *1*, this is one of the few Caryophyllaceae that has annular, albeit weakly so, perforations (SEM); *2*, this grain has lost two aperture covers in acetolysis (SEM); *3* slightly oblique section through two grains; this pantoporate caryophyll has a better developed foot layer and mostly continuous columellae (TEM). *4* and *5*, *Malachium aquaticum*: *4*, whole grain (SEM); *5*, slightly oblique radial section (at *upper* and *lower parts* there are spines that end in a point, indicating that the knife was perpendicular to the grain); note thick tectum, only two continuous columellae, and very thin foot layer/endexine (TEM). *6* and *7*, *Lychnis flos-cuculi*: *6*, some disk-like aperture covers or membranes have been lost in acetolysis (SEM); *7*, radial section; there is a pore at each end and the solid concave area in center suggests proximity to a third one; note tapered columellae (TEM). Unless otherwise indicated, *bars* denote 1 μm

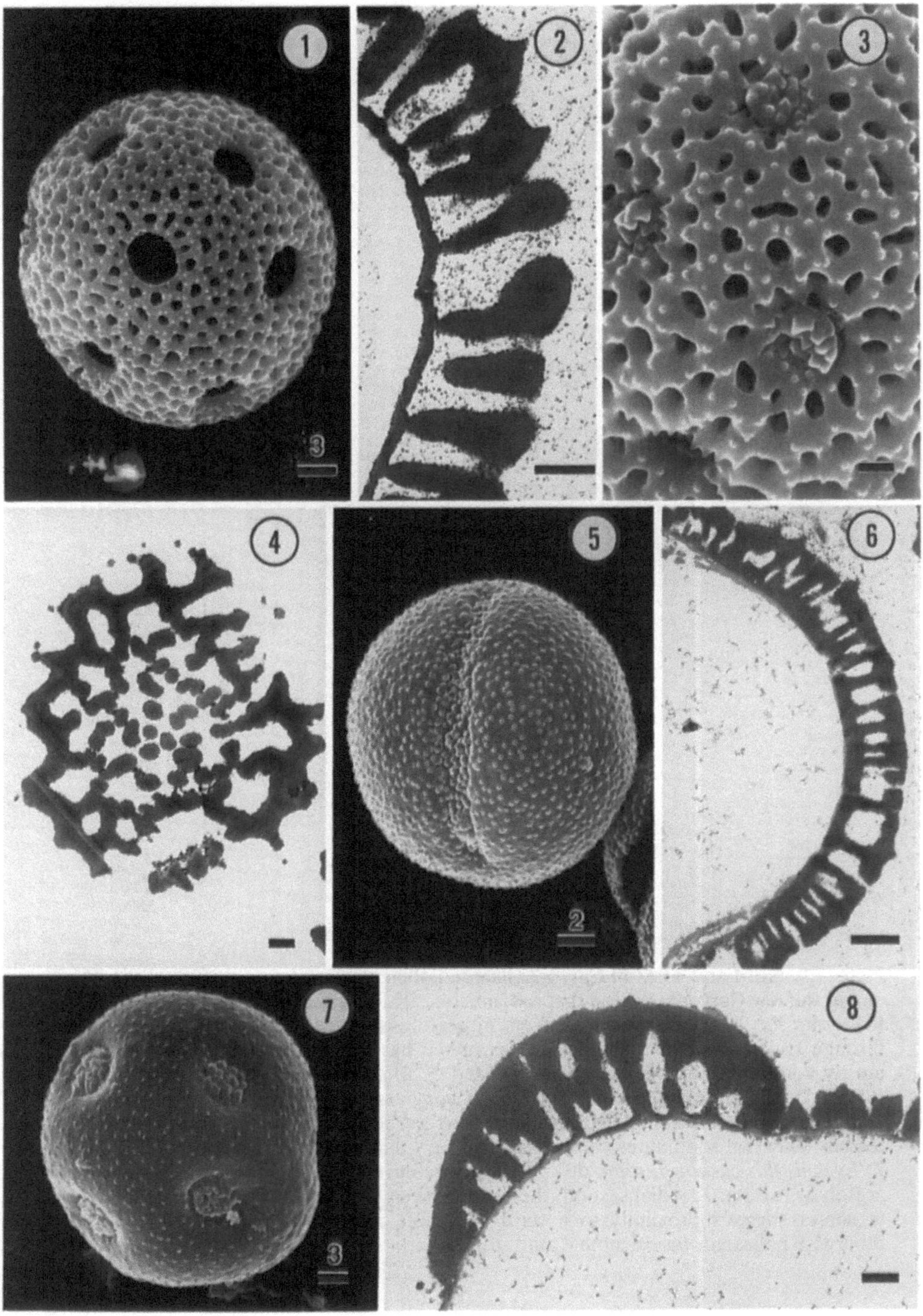
1
2
3
4
5
6
7
8
3
2
3

Plate. 7.10. SEM and TEM of Caryophyllaceae pollen. *1* and *2*, *Silene aperta*: *1*, in this grain all the aperture membranes, consisting of small islands of muri, were lost in preparation (SEM); *2*, radial section; note thin foot layer/endexine, tapered columellae (a fracture in SEM confirms this proximal narrowing), and discontinuous, but thick, tectum (TEM). *3* and *4*, *Melandrium apricum*: *3*, three aperture membranes are intact (SEM); *4*, tangential section (TEM). *5* and *6*, *Spergularia macrotheca*: *5*, equatorial view of 3-colpate grain (SEM); *6*, radial section through mesocolpus; contrast this 3-colpate exine structure with those of pantoporate SP-PT caryophylls: the tectum is much thinner and uniform; the columellae are slender with a uniform diameter; the foot layer is as thick as the tectum; and the endexine, while very thin in mesocolpus, becomes more prominent nearer the colpi (TEM). *7* and *8*, *Arenaria capillaris*: *7*, the pores are irregular in size and distribution (SEM); *8*, mostly radial section across mesoporus and adjacent pore; see also legend of Plate 7.9,6 (TEM). Unless otherwise indicated, *bars* denote 1 μm

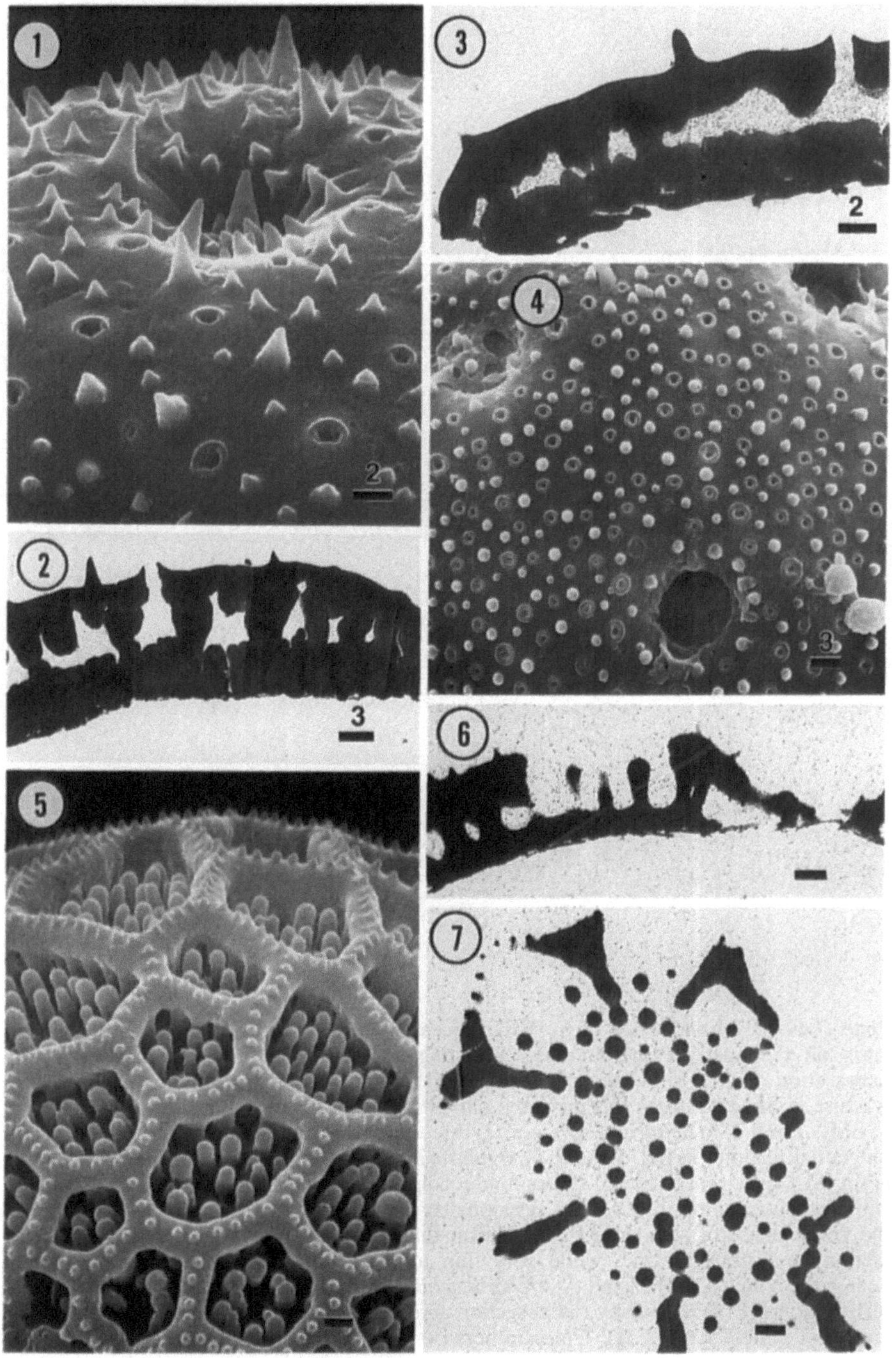
1
2
2
3
3
2
4
3
5
6
7

Plate. 7.11. SEM and TEM of Nyctaginaceae pollen. *1* and *2*, *Anulocaulis leiosolenus*: *1*, the grains illustrated in *1–4* are from the subtribe Boerhaviinae, characterized by very large, thick-walled grains with numerous, small pores, and a tectum that is spinose with prominent, annular perforations (SEM); *2*, radial section near aperture (*right*); the very thick, even massive, foot layer and irregular but thick tectum are connected by short, thick, and sometimes irregular, columellae (TEM). *3*, *Mirabilis tenuiloba*: see legend of *2* (TEM). *4*, *Acleisanthes greggi*: see legend of *1* (SEM). *5–7*, *Tripterocalyx micranthus*: *5*, reticulate exine with free columellae in lumina, and muri with small uniform spinules (SEM); *6*, radial section near colpus, the tectum (*left*) is thick and incomplete, the columellae forming the muri are short and thick, the foot layer is about two-thirds the thickness of the tectum, and the endexine is very thin (TEM); *7*, tangential section; the solid outer structures are the distal part of muri, the circular structures are either columellae in the muri or free columellae in the lumina, that have been cut at a right angle to their long axes (TEM). Unless otherwise indicated, *bars* denote 1 μm

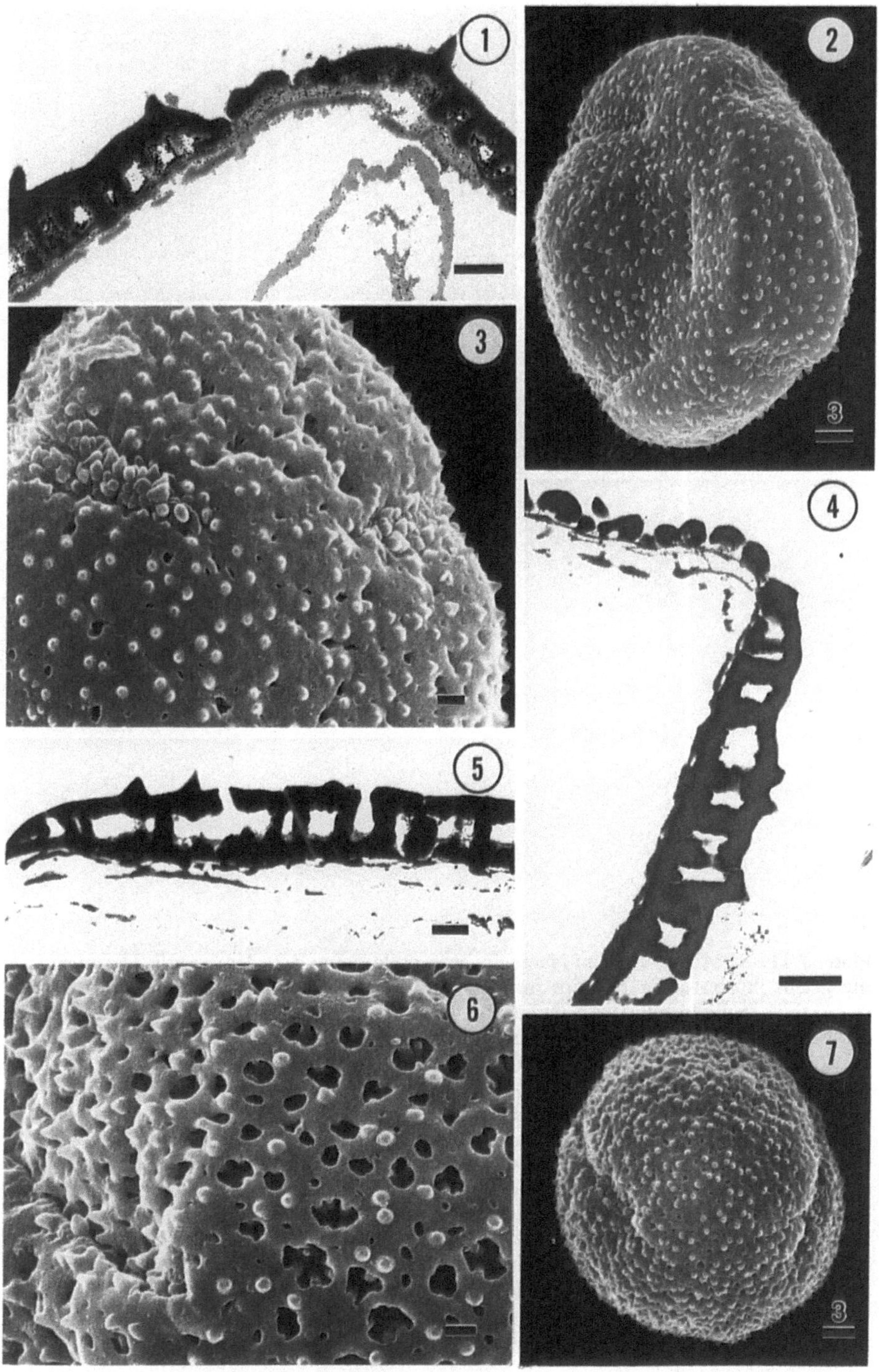
1
2
3
3
4
5
6
7
3

Plate. 7.12. SEM and TEM of Nyctaginaceae pollen. *1* and *2*, *Guapira costaricana*: *1*, radial section across colpus; note presence of less electron dense endexine under colpus and to each side, and unattached fragment (TEM); *2*, most pollen in this sample was collapsed, masking the apertures, but some appeared 12-(4-4-4-) pantocolpate (SEM). *3* and *4*, *Pisonia ambigua*: *3*, polar view illustrating irregularity of tectal openings (SEM); *4*, radial section including part of colpus; an exine typical of the dicotyledons: thin endexine, foot layer and tectum about equal in thickness, and slightly longer columellae (TEM). *5–7*, *Pisonia grandis*: *5*, see legend of *4* (TEM); *6*, in this particular grain the tectal openings are so large in the polar area that it is almost reticulate (SEM); *7*, oblique polar view (SEM). Unless otherwise indicated, *bars* denote 1 μm

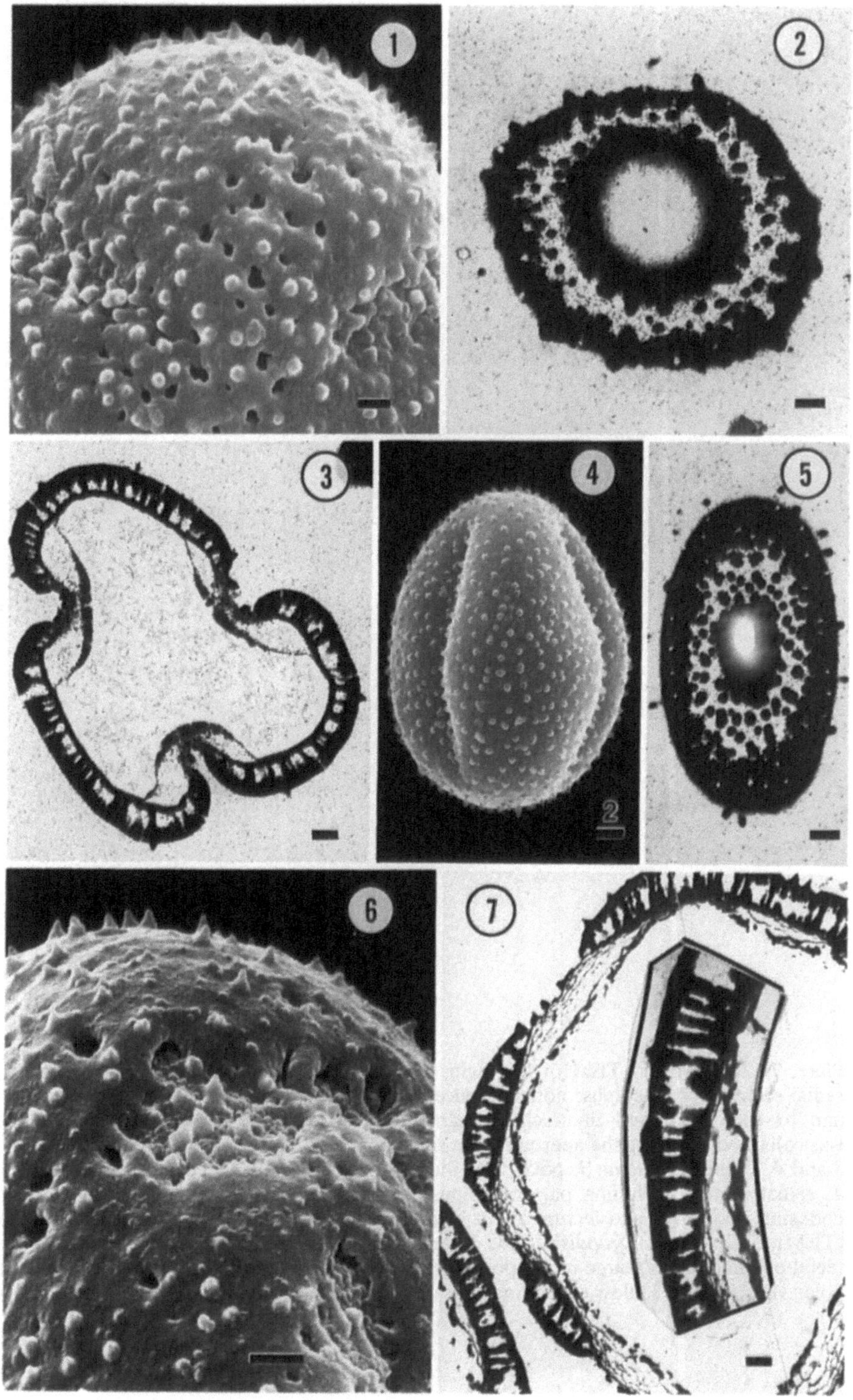
1
2
3
4
5
6
7

Plate. 7.13. SEM and TEM of Phytolaccaceae pollen. *1* and *2*, *Ercilla spicata*: *1*, in this particular grain the perforations at the pole are very large compared with the rest of the grain (SEM); *2*, tangential section; the outer ring corresponds to the tectum, the inner one to the foot layer, and both are equal in thickness; inbetween are the short, sparse columellae (TEM). *3–5*, *Lophiocarpus polystachyus*: *3*, radial section of whole grain; note, in contrast to most Caryophyllales, the increased endexine and its unusual intact separation (found in most grains) from the ektexine; the tectum is thick, the columellae well defined and the foot layer thin (TEM); *4*, equatorial view (SEM); *5*, tangential section (TEM). *6* and *7*, *Petiveria alliacea*: *6*, portion of grain illustrating irregular apertures and irregular tectal openings (SEM); *7*, radial section; in this study, *Petiveria* had the most distinctive endexine: under nonapertural regions it consists of very thin lamellae, but under the apertures it becomes much "looser" (which probably resulted, at least in part, from acetolysis) and forms a network, and on the inner boundary (apertural and nonapertural) the endexine is much more dense, perhaps because the lamellae remain adhered together; the foot layer can be irregular or sometimes equal to the tectum in thickness; the columellae are the most prominent component (TEM). *Inset* mostly radial section (TEM). Unless otherwise indicated, *bars* denote 1 μm

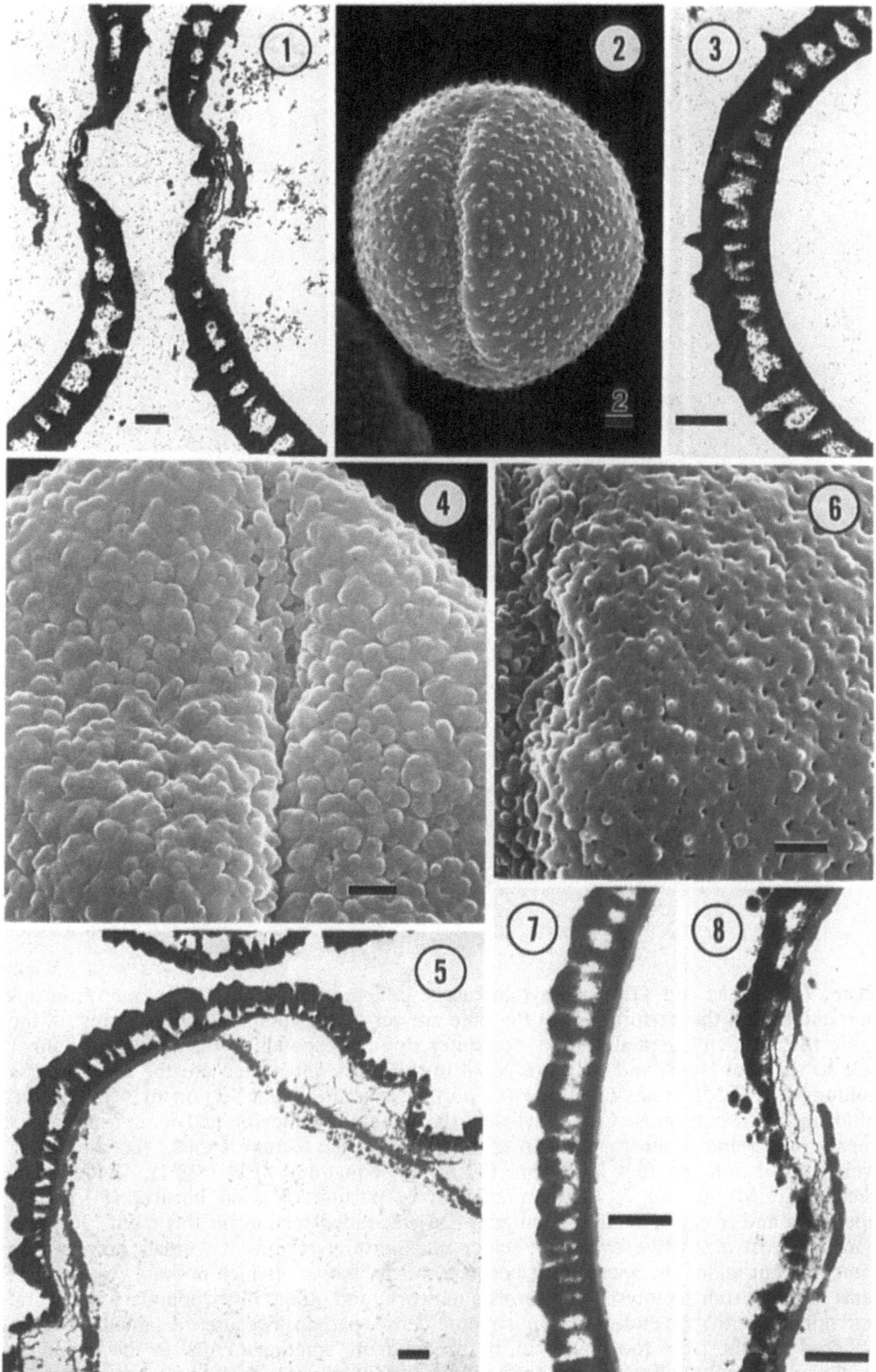
1
2
3
4
6
5
7
8

Plate. 7.14. SEM and TEM of Phytolaccaceae pollen. *1* and *2*, *Phytolacca pruinosa*: *1*, radial section; two grains with colpi showing sparse, slightly detached endexines (TEM); *2*, equatorial view (SEM). *3*, *Phytolacca dioica*: somewhat oblique radial section (TEM). *4* and *5*, *Seguieria langsdorffii*: *4*, granular-scabrate tectum; in both SEM and TEM (*5*) the tectum is composed of "elements" that, at least in part, are discrete or individual; the tectum in Achatocarpaceae (Plate 7.15,5 and 6) is somewhat similar (SEM); *5*, radial section through mesocolpus and adjacent colpi; note lamellate-fibrous endexine (TEM). *6–8*, *Seguieria aculeata*: *6*, tectum (SEM); *7*, radial section through mesocolpus; there is a very thin, less electron dense endexine (TEM); *8*, radial section across colpus; compare endexine with *Phytolacca* (*1*), *Seguieria langsdorffii* (*5*), *Petiveria* (Plate 7.13,7), *Barbeuia* (Plate 7.15,2 and 3), and *Stegnosperma* (Plate 7.18,2 and 5) (TEM). Unless otherwise indicated, *bars* denote 1 μm

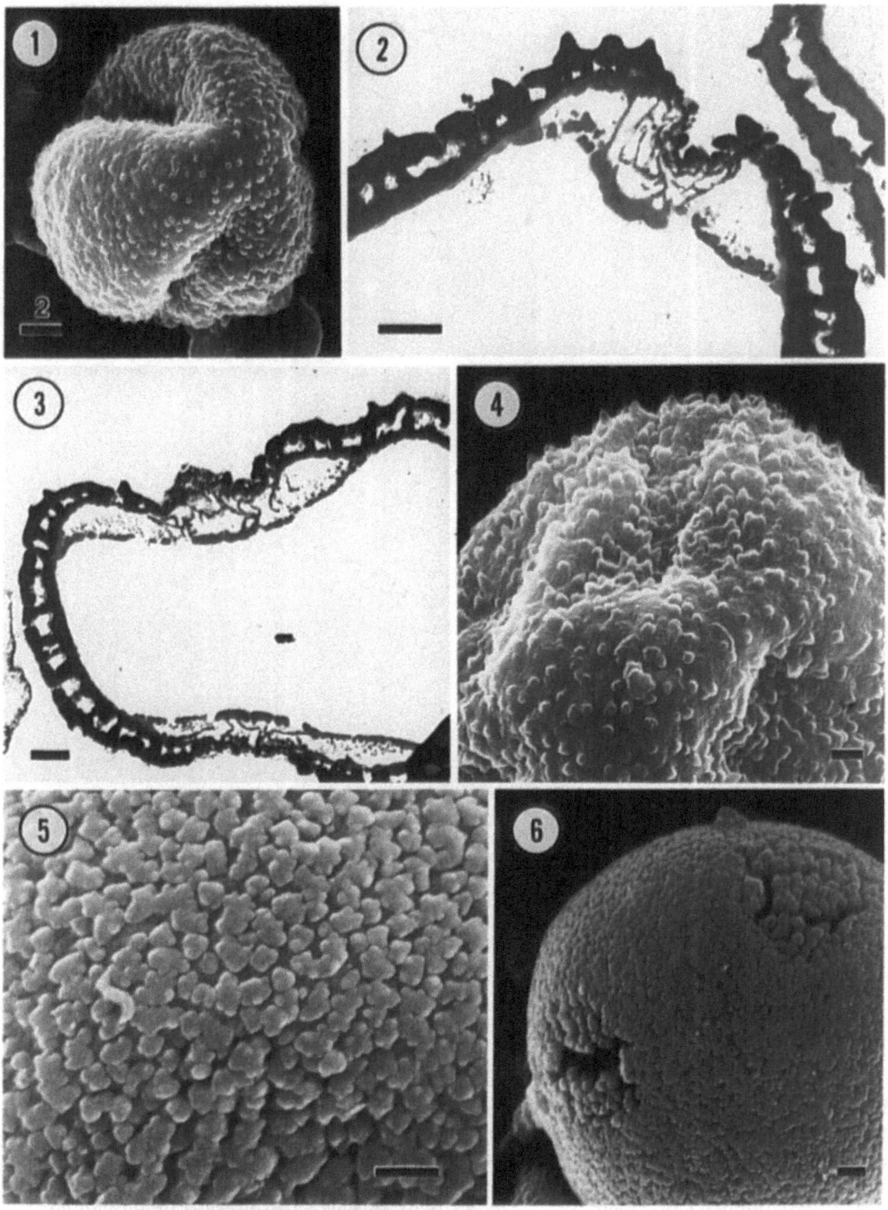

Plate. 7.15. SEM and TEM of Phytolaccaceae (*1–4*) and Achatocarpaceae (*5* and *6*) pollen. *1–4*, *Barbeuia madagascariensis*: *1*, oblique polar view (SEM); *2*, radial section through colpus; note similarity of endexine to that in *Petiveria* (Plate 7.14,7) and *Seguieria* (Plate 7.14,5): the separation from the ektexine, the thin, loose lamellae under the colpus, and the thicker, denser inner boundary (TEM); *3*, see legend of *2* (TEM); *4*, polar view; tectum with one colpus (*top*) visible (SEM). *5*, *Achatocarpus mollis*: scabrate tectum (SEM). *6*, *A. nigricans*: portion of grain including two poorly defined pores (SEM). The two species of *Achatocarpus*, part of an earlier SEM and TEM study of the family (see Skvarla and Nowicke 1982), were included for purposes of comparison. Unless otherwise indicated, *bars* denote 1 μm

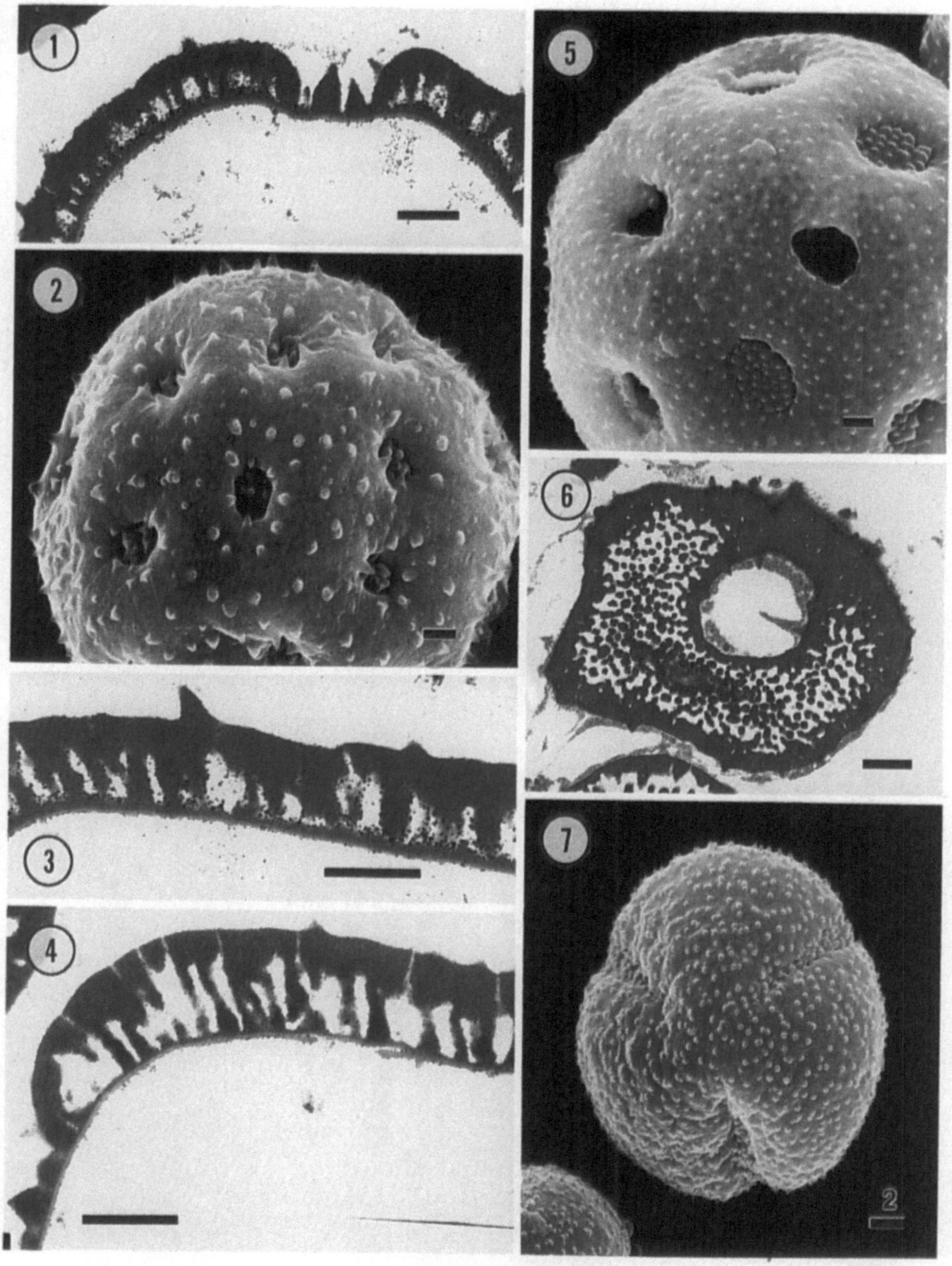

Plate. 7.16. SEM and TEM of Phytolaccaceae s.l. pollen. *1–3*, *Microtea scabrida*: *1*, radial section across aperture; note absence of lamellate-fibrous endexine found in many Phytolaccaceae (TEM); *2*, this grain had the most regular pores in the sample, similar in size and uniformly distributed (SEM); *3*, radial section through mesoporus (TEM). *4–6*, *Microtea maypurensis*: *4*, radial section near pore; note that all radial thin sections (*1*, *3* and *4*) show a thick tectum and a thin foot layer or endexine (TEM); *5*, portion of grain; note similarity of aperture covers to those of *Lychnis flos-cuculi*: (Plate 7.9,6) (SEM); *6*, tangential section; as seen in cross section, the columellae are very thin but circular (TEM). *7*, *Agdestis clematidea*: polar view (SEM). Unles otherwise indicated, *bars* denote 1 μm

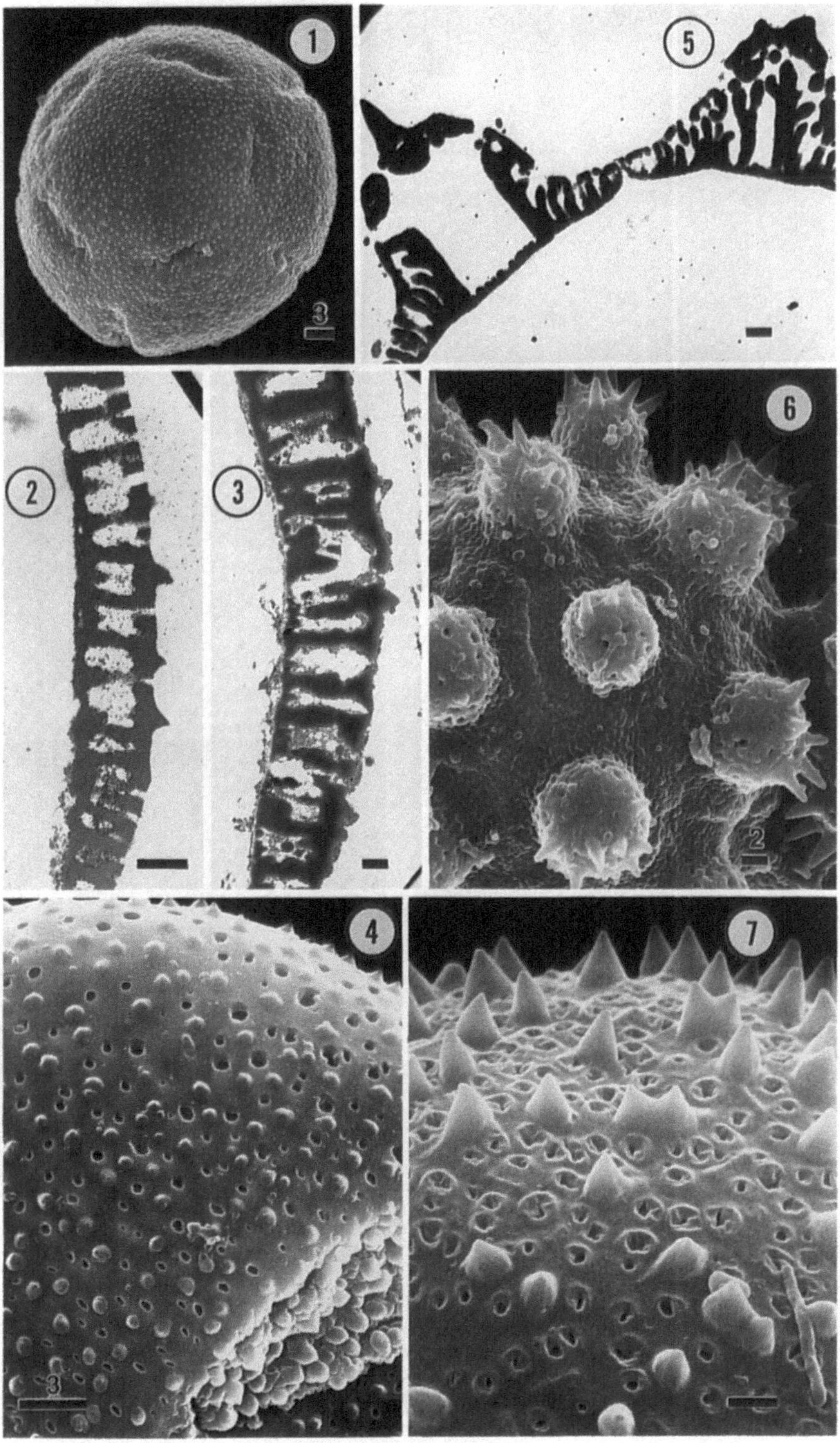
1
5
2
3
6
4
7
3
2
3

Plate. 7.17. SEM and TEM of Portulacaceae pollen. *1* and *2*, *Talinum patens*: *1*, polar view of 15-pantocolpate grain (SEM); *2*, radial section of mesocolpus near colpus (*bottom*); note paucity of endexine (TEM). *3* and *4*, *Claytonia sarmentosa*: *3*, radial section near colpus (*bottom*); the columellae in this species are large, irregular (confirmed in tangential section), and sometimes fused, e.g., a block of ektexine with two or three vertical, parallel, narrow slits (TEM); *4*, note different sized punctae (SEM). *5* and *6*, *Montiastrum lineare*: *5*, section through two tholi; the section passed through the middle of the left tholus and through the wall of the tholus on the right (TEM); *6*, portion of grain (SEM). *7*, *Talinella* sp.: note that some spinules are fused basally and that some annular perforations are partially occluded (SEM). Unless otherwise indicated, *bars* denote 1 μm

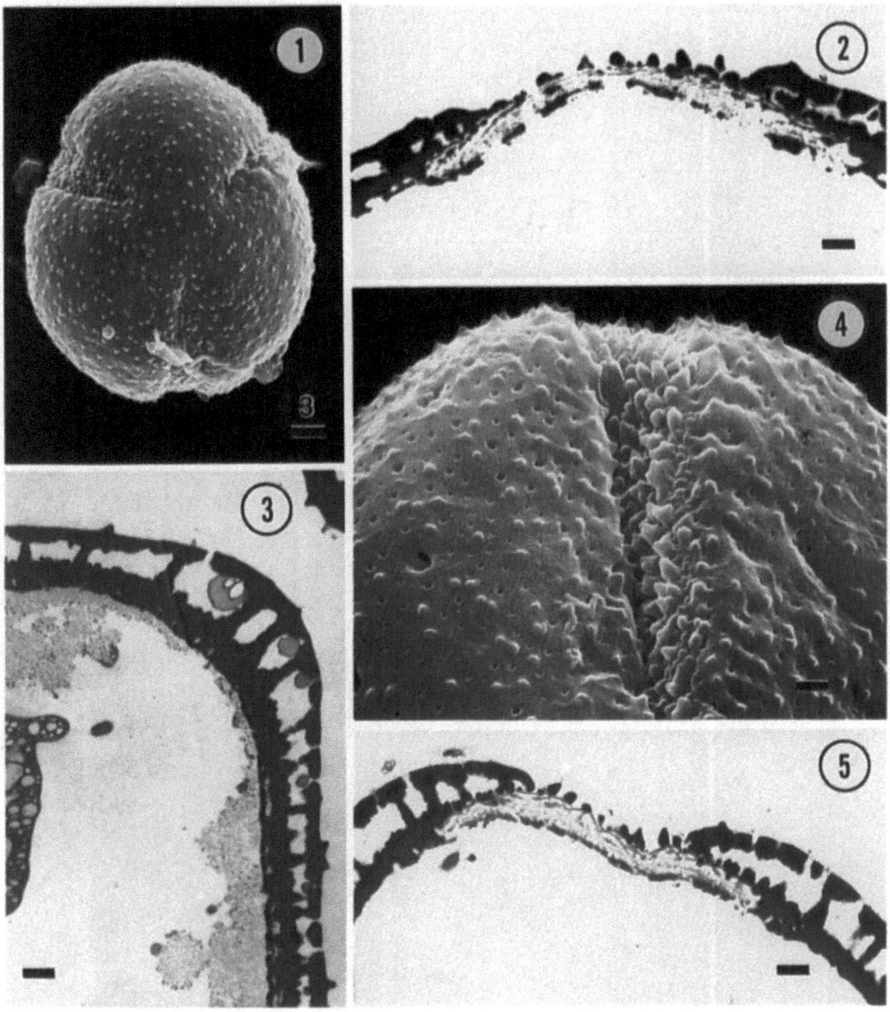

Plate. 7.18. SEM and TEM of pollen in Stegnospermataceae. *1* and *2*, *Stegnosperma cubense*: *1*, polar view (SEM); *2*, radial section across aperture; note similarity of endexine, lamellate proximally and netlike or loosely lamel-late distally, to that in *Petiveria alliacea* (Plate 7.13,7), *Barbeuia madagascariensis* (Plate 7.13, 2 and 3), *Seguieria aculeata* (Plate 7.14,7) and *S. langsdorffii* (Plate 7.14,5) (TEM). *3–5*, *S. watsonii*: *3*, slightly oblique radial section through pole; note elongated columellae at *curve* (pole), but many 3-colpate taxa show this structural modification, e.g., *Lophiocarpus*, some *Phytolacca*, some Ranunculaceae, and some Labiatae (TEM); *4*, polar view showing one colpus; note presence of occasional larger punctae (SEM); *5*, radial section across colpus; seen legend of *2* (TEM). Unless otherwise indicated, *bars* denote 1 μm

Some grains have areas of irregular or larger tectal openings, e.g., Cactaceae (Plate 7.7,1), Nyctaginaceae (Plate 7.12,3 and 6), Phytolaccaceae (Plate 7.13,1 and 6), and Stegnospermataceae (Plate 7.18,4).

Reticulate tecta are found in Aizoaceae, Amaranthaceae, Basellaceae, Cactaceae, Caryophyllaceae, Nyctaginaceae and Portulacaceae. Muri of all variants – microreticulate (Plate 7.1,1), finely (Plate 7.10,1 and 3), and coarsely reticulate (Plate 7.11,5) – frequently have spinules that may be vestiges from the common SP-PT/AP tecta. Seemingly, reticulate tecta could be easily derived from SP-PT/AP ones by enlargement of perforations, e.g., the polar region of some grains of *Pisonia grandis* (Plate 7.12,6). In *Anredera baselloides* which is SP-PT, the columellae (as seen in tangential section) are distributed or organized in rings or circles much like the columellae that form the muri in reticulate tecta. This suggests that reticulate and SP-PT/AP tecta are not as distinct as they appear, and that SP-PT/AP could be derived from reticulate forms. In the coarsely reticulate *Tripterocalyx*, the presence of free columellae in the lumina (Plate 7.11,5) obscures the organization of columellae in the muri (Plate 7.11,7). Free columellae in reticulate tecta could be evidence of a closer relationship to the SP-PT/AP form in which the columellae are (more) uniformly distributed.

Reticulate tecta occur in certain genera of Gomphrenoideae (Amaranthaceae), e.g., *Philoxerus* (Plate 7.2,1 and 2), where there is a pore in each lumen and the muri consist of a single row of columellae (see discussion of family).

Seguieria langsdorffii has, for Phytolaccaceae, a distinctive tectum (Plate 7.14,4): irregularly granular or scabrate with small blunt spines. This is somewhat similar to the scabrate tectum in Achatocarpaceae (Plate 7.15,5; and Skvarla and Nowicke 1982) and in *Sesuvium portulacastrum* (Nowicke, unpubl. data).

Bittrich (1986) illustrated two unreported tectum variations in Aizoaceae, rugulose and reticulate laevigate (see discussion of family).

7.3.3 Apertures

All Caryophyllales have simple apertures: 3-colpate (3-zonocolpate), pantoporate, pantocolpate, or, more rarely, 6- to 8-zonocolpate. Bittrich (1986) reported 4-colpate grains in *Mesembryanthemum nodiflorum*; a collection examined here (Table 7.1) is 3, 4-colpate and 6-pantocolpate.

Amaranthaceae and Chenopodiaceae are reported as having only pantoporate apertures, as does the large subtribe Boerhaviinae (Nyctaginaceae, Heimerl 1934a), and many Caryophyllaceae. Aizoaceae (Plate 7.1) are mostly 3-colpate, but this aperture is also common in Cactaceae, Nyctaginaceae, Phytolaccaceae (Plates 7.13–7.16), and Portulacaceae.

Within the pantoporate type there is considerable variation in pore size and number. Numerous small pores occur, e.g., in Amaranthaceae (Plate

7.2,1), Chenopodiaceae (Plate 7.2,3 and 8), many Caryophyllaceae (Plates 7.8,1–4 and 7.9,6), the subtribe Boerhaviinae (Plate 7.11,1 and 4), and sporadically elsewhere.

Large, poorly defined pores that are fewer in number occur in Achatocarpaceae (Plate 7.15,6) and in some species of Cactaceae (see family discussion).

Pantocolpate apertures can also vary in the number of colpi and can be 3-3-pantocolpate (Plate 7.5,7), 4-4-4-pantocolpate (Plates 7.7,1 and 7.12,2), 5-5-5-pantocolpate (Plate 7.17,1), or much higher in number as in some Portulacaceae (Nilsson 1967) and in the unique, tholate grains of *Montiastrum lineare* (Plate 7.17,6). Frequently, 12-pantocolpate and 15-pantocolpate apertures occur together within a sample (a collection).

Only Didiereaceae are 5-8-zonocolpate (Plate 7.3,4; Nowicke 1975, Fig. 66; Nowicke and Skvarla 1980, Figs. 11, 12 and 43).

Those Centrospermae with SP-PT/AP tecta frequently have conspicuous aperture membranes or covers consisting mostly of flecks of ektexine with spinulose processes. Examples include species in Aizoaceae (Plate 7.1,1 and 8), Chenopodiaceae (Plate 7.2,5 and 8), Cactaceae (Plate 7.7,1 and 4), Caryophyllaceae, (Plate 7.8,1–8; 7.9,2,4 and 6; 7.10,3,5,7 and 8), Didiereaceae (Plate 7.3,4 and 6), Nyctaginaceae (Plate, 7.12,3), Phytolaccaceae (Plate 7.13,1), Portulacaceae (Plate 7.17,4) and Stegnospermataceae (Plate 7.18,1). In some taxa, mostly Caryophyllaceae, Erdtman's (1966) definition of operculum as "a thickening of measureable bulk" could apply, but the boundary between operculum and aperture membrane would be difficult to establish and for this reason the term operculum has not been used. Perhaps the definition of operculum should depend on structure, i.e., the presence of columellae.

The widespread occurrence of well-developed aperture membranes or covers in the Caryophyllales may be correlated with the frequent paucity of apertural endexine (Didiereaceae excepted, Straka 1965a,b), i.e., the aperture covers may assume, in part, a supposed function of the endexine and help to reduce desiccation. Even beneath or near the apertures there is frequently only a thin endexine (see Sect. 7.3.4).

Many samples contained pollen with aberrant apertures, and some of these irregularities have been documented by Nilsson (1967, Fig. 3) for *Claytonia*, *Montia* and allied genera, and by Leuenberger for Cactaceae (1976, Fig. 4c).

7.3.4 Exine Structure

With the exception of thin or unusual endexines, the pollen of most Centrospermae is typical of dicotyledons generally: usually a well-defined foot layer, columellae about one-third the total exine thickness, and, in the common SP-PT/AP type, a complete tectum. Examples representing most families include Plates 7.1,4, 7.2,4, 7.3,2, 7.4,4, 7.5,3, 7.10,6, 7.12,5, 7.13,3, 7.14,1 and 3, 7.15,3, 7.16,3 and 7.17,2.

In Caryophyllales the apertural endexine ranges from seemingly absent, e.g., some pantoporate Caryophyllaceae and some Gomphrena types (Amaranthaceae); sparsely and loosely granular, e.g., some Chenopodiaceae and some Cactaceae; fibrous or netlike, e.g., some Phytolaccaceae; partially solid, e.g., some Basellaceae; to vertical strands, e.g., some Didiereaceae.

It is unusual for dicotyledons to have thin or seemingly absent endexines, but this character is widespread in the Caryophyllales (Plates 7.2,2, 6 and 7; 7.8,2 and 3; 7.10,8; 7.11,3 and 6; 7.12,4 and 5; 7.16,1 and 4; 7.17,2 and 3).

In thin sections across the large apertures of certain Cactaceae, e.g., *Opuntia* species (Plate 7.6,6; and unpubl. data), the granules or fragments have the same electron density as the nonapertural ektexine, suggesting that even though these elements have a morphology typical of endexines, i.e., granular, they may be ektexinous in origin.

Many Phytolaccaceae s.l. and Stegnospermataceae have distinctive endexines: directly beneath the apertures it is finely lamellate or fibrous and netlike, although the inner boundary is much denser (Plates 7.13,7; 7.14,5 and 8; 7.15,2 and 3; 7.18,2 and 5). The netlike condition and the denser inner boundary may result, at least in part, to a shrinking of the endexine.

Anredera (Basellaceae) has an unusual solid endexine (Nowicke and Begle 1993), better illustrated in Nowicke and Skvarla (1977, Fig. 21) than in Plate 7.3,2. *Lophiocarpus* (Phytolaccaceae) also has a solid endexine (Plate 7.13,3) that separates (at least after acetolysis) from the ektexine.

Some Didiereaceae have a unique endexine of densely spaced vertical strands (Plate 7.3,6) that also appear to have the same electron density as the ektexine (see family discussion).

More specialized ultrastructures include some pantoporate grains that have a thin endexine, thin foot layer and a thick tectum, and are found in some Amaranthaceae (Skvarla and Nowicke 1976, Fig. 14), many Chenopodiaceae (Plate 7.2,4 and 7; Skvarla and Nowicke 1976, Figs. 23–25), and some Caryophyllaceae (Plate 7.8,2, 3 and 6).

The subtribe Boerhaviinae is characterized by exines with massive foot layers, thick tecta and short columellae (Plate 7.11,2 and 3).

In thin section, the reticulate (Plate 7.2,1 and 2), finely reticulate (Plate 7.10,1–4) and coarsely reticulate (Plate 7.11,5–7) exines reflect the discontinuous tectum (see also Skvarla and Nowicke 1976, Figs. 15–17, 32, 33 and 43–45).

7.3.5 Pollen Descriptions of the Caryophyllales Families

For the most part I follow Cronquist's system (see Chap. 1), except for including *Gisekia* in Aizoaceae and recognizing Stegnospermataceae.

Achatocarpaceae (Plate 7.15,5 and 6). The pollen of five species of *Achatocarpus* and the monotypic *Phaulothamnus* was examined in SEM and TEM by Skvarla and Nowicke (1982). The apertures, large poorly defined pores (Plate 7.15,6), and the tectum, coarsely granular or scabrate (Plate 7.15,5), are unique in the order; even the ultrastructure, which lacks a foot layer, is unusual. Some Cactaceae, mostly assigned to the genus *Opuntia*, also have large pores but the similarity is probably superficial.

Aizoaceae (Plate 7.1). This family, which at one time included Molluginaceae (Pax and Hoffmann 1934a), was thought to have only 3-colpate apertures but Bittrich (1986) reported 4-colpate ones in *Mesembryanthemum nodiflorum* and a collection examined here has 3- to 4-colpate and 6-pantocolpate apertures. The most common tectum is SP-PT, but microreticulate does occur, e.g., *Drosanthemum striatum* (Plate 7.1,1 and 2) and *Delosperma ecklonis* (Nowicke and Skvarla 1980, Fig. 23). Bittrich (1986) illustrates in SEM (Fig. 48, p. 109) two tecta, "distinctly rugulose" in *Phyllobolus* and "reticulate, laevigate" in the *Mesembryanthemum nodiflorum* group, that are unique in the Caryophyllales but, as such, suggest no relationships, at least at this time, to other taxa. In TEM, *Tetragonia fruticosa* (Plate 7.1,6) has a well-developed foot layer, short columellae, and a very thick tectum – a sharp contrast with *Mesembryanthemum variabile* (Nowicke and Skvarla 1977, Fig. 19) which has an extremely thickened foot layer, short columellae and a thin tectum. Both *Gisekia miltus* (Nowicke, unpubl. data) and *Psammotropha myriantha* (Plate 7.1,4), have similar ektexines: thin foot layer, slender columellae, and a tectum about twice as thick as the foot layer. *Gisekia* (Plate 7.1,3), a betalain species formerly assigned to the anthocyanin-pigmented Molluginaceae, has pollen typical of the order and could be assigned to almost any other centrospermous family, except those where the 3-colpate condition is unknown: Achatocarpaceae, Amaranthaceae, Basellaceae, Chenopodiaceae, and Didiereaceae. However, the endexine of *Gisekia* (Nowicke, unpubl. data; Nowicke and Skvarla 1980, Fig. 37) suggests a relationship to the Phytolaccaceae.

Amaranthaceae and Chenopodiaceae (Plate 7.2). These two families have only pantoporate grains, and within a given species the pores are generally uniform in size and distribution. Eliasson (1988) does illustrate a few examples of irregular apertures in Amaranthaceae, but some grains are just collapsed. Most members of Amaranthaceae and Chenopodiaceae have thin foot layers/endexines and thick tecta (Plate 7.2,2, 4, 6 and 7); the crassimarginate pores in *Sarcobatus* (Plate 7.2,5 and 6) are the result of longer columellae, not a locally thickened tectum. Erdtman (1966) recognized an Amaranthus type with numerous small pores and a Gomphrena type with a pore in each lumen or luminoid depression. The latter type is illustrated here by *Philoxerus vermicularis* (Plate 7.2,1 and 2), in which each lumen has one pore and the muri consist of a single row of columellae. In some Gomphrena types the sides of the muri are covered by a very thin layer of ektexine forming, in a technical sense, a tectum (e.g., *Gomphrena elegans*, Skvarla and Nowicke 1976, Figs. 9 and 16–18). Stellate aperture covers are present in some Amaranthoideae (Nowicke 1975, Figs. 55 and 59; Eliasson 1988). The pollen of some Amaranthaceae and some Chenopodiaceae is very similar to certain Caryophyllaceae (Nowicke and Skvarla 1980, Figs. 25–36); especially noteworthy are the nearly identical grains of some Gomphrena types, e.g., *Gomphrena elegans* and *Pfaffia iresinoides* to *Siphonychia diffusa* and *S. rugelii*. Eliasson (1988, Figs. 17 and 18) illustrates subcuboidal grains for *Pseudoplantago friesii* and there is at least one Caryophyllaceae that also has subcuboidal pollen, *Cardionema ramosissima* (Nowicke, unpubl. data).

Basellaceae (Plate 7.3,1–3). Most of the species examined in this study and in others (Nowicke 1975, Figs. 61–65, 77 and 83; Skvarla and Nowicke 1976, Fig. 19; Nowicke and Skvarla 1980, Figs. 22, 54 and 55), are pantocolpate, although the particular grain of *Anredera vesicaria* portrayed in Plate 7.3,1 does have short colpi. Even the large apertures on each of the six faces of the mostly cuboidal grains of *Basella* are colpi in an area of thin exine. *Anredera* (Plate 7.3.1 and 2) and *Ullucus* (Nowicke 1975, Fig. 62) are SP-PT/AP, whereas *Basella* ranges from reticulate (Nowicke 1975, Figs. 63–65) to deeply punctate (Plate 7.3,3; Nowicke and Skvarla 1980, Figs. 22 and 55). The pollen of the monotypic genus *Tournonia* is distinctive: around each of the seven to ten small pores, the foot layer is much thicker and has large, irregularly finger-like projections. In SEM, the pores appear only weakly crassimarginate – most of the increased wall is accommodated internally. The apertural endexine of *Anredera* has a solid inner component (Nowicke and Begle 1993).

Cactaceae (Plates 7.4–7.7). Because Cactaceae have a wide range of pollen variation (see Leuenberger 1976) but have not been the subject of a TEM investigation, 15 of the 20 species (Table 7.1) were examined in thin section. However, these species lack a distinctive exine structure, e.g., an endexine like many Didiereaceae, and four species examined in TEM are not illustrated. Cactaceae have the three common aperture types, but with a wide range of variation in the porate type, i.e., large pores and few in number to smaller and more numerous pores. There are species, mostly opuntias, with large thin-walled areas separated by ridges of reticulate exine (Plate 7.6,1 and 5), but not all thin areas function as apertures (exit for sperm nuclei): in *Opuntia erinacea* (Plate 7.6,1), the pattern of apertural areas versus thin areas and the orientation of the colpi indicate that this species is probably 12-pantocolpate, not pantoporate. SP-AP tecta are common, e.g., Plates 7.4,2 and 5, 7.5,2 and 7.7,1, 3 and 4; SP-PT are also present, e.g., Plates 7.5,7 and 7.6,4, although in the latter, *Opuntia macbridei*, there are only small vestigial spinules. *Opuntia triacantha* (Plate 7.4,6 and 7) has a finely punctate tectum and 15–20 pores, but the pores are irregular in size, shape and distribution; this was the only Cactaceae examined in TEM that had a massive foot layer.

The species of *Pereskia* and *Rhipsalidopsis* examined here (Table 7.1) have columellae with irregular surfaces (Plate 7.7,2 and 5).

One species, *Disocactus himatocladus*, has pollen in tetrads (Plate 7.4,1–4), mostly rhomboidal (for other examples of tetrad Cactaceae, see Leuenberger 1976). The exinous strands between adjacent tetrad members are visible in Plate 7.4,1 and 2. In thin section, where the tetrad members are in contact they appear to share, at least in some areas, a common tectum (Plate 7.4,3); the exterior or noncontact wall (Plate 7.4,4) has a foot layer, columellae and tectum as well as endexine, but while all appear somewhat irregular, the undulate tectum reflects the surface portrayed in Plate 7.4,2. A second species, *D. macranthus* (Plate 7.4,5), has monads with a somewhat similar tectum.

Caryophyllaceae (Plates 7.8–7.10). Although one of the larger families in the order with 70 genera and 1700 species (Pax and Hoffmann 1934b), only pantoporate and 3-colpate types have been reported. In an earlier publication (Nowicke 1975, Figs. 27 and 28) *Corrigola litoralis* appears 3-porate but a second species, *C. telephiifolia* is 3-colpate; in LM both appear to have thicker polar exines due mostly to elongation of the columellae (e.g., *Stegnosperma* has the same condition). To the best of the author's knowledge, thicker polar exines are associated with 3-colpate apertures, not 3-porate ones. The only variations from SP-PT/AP tecta are finely reticulate and microreticulate ones. *Acanthophyllum microcephalum* (Plate 7.8,1 and 2), *Agrostemma githago* (Plate 7.8,3 and 4), *Arenaria capillaris* (Plate 7.10,7 and 8), *A. macradenia* (Plate 7.8,5 and 6), *Lychnis flos-cuculi* (Plate 7.9,6 and 7), *Malachium aquaticum* (Plate 7.9,4 and 5) and *Saponaria officinalis* (Plate 7.9,1–3) have sunken pores separated by convex exines. These species have a distinct ultrastructure: threadlike endexines adherent to a thin foot layer, a thick arched tectum, and columellae that taper proximally, accounting for at least some of their stalactitic appearance. Moreover, even in the finely reticulate pollen of *Silene aperta* (Plate 7.10,1 and 2), tangential sections and a fracture in SEM (unpublished data) show a clear reduction in columellae diameter nearer the foot layer; a similar case could be made for tapering columellae in *Melandrium apricum* (Plate 7.10,3 and 4). Similar grains, i.e., sunken pores, thick arched tecta and thin endexine or foot layers, but not tapered columellae, occur throughout Chenopodiaceae, in many Amaranthaceae, and in certain other genera, e.g., *Microtea* (Plate 7.16,1–6; Skvarla and Nowicke 1976, Fig. 37). *Spergularia macrotheca* (Plate 7.10,5 and 6) is the only 3-colpate species examined and, in contrast to the pantoporate ones, it has a well-developed foot layer, continuous and uniform columellae, and a tectum equal in thickness to the foot layer.

Agrostemma githago (Plate 7.8,3 and 4) has echinate, cuplike aperture membranes over each pore; *Lychnis flos-cuculi* (Plate 7.9,6) has solid disks; and *Melandrium apricum* (Plate 7.10,3) and *Silene aperta* (unpublished data) have membranes consisting of fragments of muri over each pore.

Didiereaceae (Plate 7.3,4–6). This small family, four genera and 11 species endemic to Madagascar, can be distinguished from all other Caryophyllales by the 5–7-zonocolpate apertures, most of which have a distinctive aperture cover consisting of rows of broadly based spines. Many Didieraceae have pronounced annular perforations (Plate 7.3,4–6).

In a study of Didiereaceae pollen, which included LM and some TEM, Straka (1965a,b) published line drawings showing "bacula" beneath the colpi. Recently, pollen samples of three species, *Alluaudia comosa*, *Alluaudiopsis fiherenensis*, and *A. marnieriana*, were obtained from Missouri Botanical Garden and prepared for TEM. In *Alluaudiopsis* the prominent and uniformly present endexine consists of thin, densely spaced, vertical strands and is unique not only in the Caryophyllales but also probably in the angiosperms (Nowicke and Begle 1993). In *Alluaudia procera* (Skvarla and Nowicke 1976, Fig. 28) and *A. comosa* the strands are shorter, thicker and only under the aperture. The columellae in all three species have irregular or granular surfaces and are wider near the tectum.

Halophytaceae. The subcuboidal grains of this monotypic family were illustrated in SEM and TEM in an earlier publication (Skvarla and Nowicke 1976, Figs. 3 and 29). The pollen ultrastructure is similar to many Chenopodiaceae (thin endexine or foot layer and thick tectum), to which it had been assigned by Cronquist (1981, 1988), but the existence of subcuboidal grains in Amaranthaceae (Eliasson 1988) could make that family a better choice.

Molluginaceae. Ten species were examined in an earlier study (Nowicke 1975), all of which are variously SP-PT/AP and all but one are 3-colpate, *Mollugo verticillata* being 12-pantocolpate. In thin section, *Limeum viscosum* (Nowicke and Skvarla 1977, Figs. 2, 7 and 26) has a typical exine structure in the mesocolpus, but an apertural endexine similar to some Phytolaccaceae (Plate 7.14,8).

Nyctaginaceae (Plates 7.11 and 7.12). All three aperture types are present in Nyctaginaceae and there is a wide range of tectal variation: SP-AP, SP-PT, coarsely reticulate (*Abronia*, *Bougainvillea*, *Tripterocalyx*, Nowicke 1970, Figs. 13–18), and finely reticulate (*Neea*, Nowicke 1975, Fig. 9; *Reichenbachia*, Nowicke and Skvarla 1980, Figs. 20 and 56). Of the eight species examined, all are SP-PT/AP, except *Tripterocalyx micranthus* (Plate 7.11,5–7) which is reticulate with free columellae in the lumina. The pollen of subtribe Boerhaviinae, illustrated here by *Acleisanthes greggii* (Plate

7.11,4), *Anulocaulis leiosolenus* (Plate 7.11,1 and 2), and *Mirabilis tenuiloba* (Plate 7.11,3), can be easily distinguished from all remaining centrosperms: very large (60–200+ μm in diameter) grains, polypantoporate (>17 and usually much higher) apertures, with very thick exines (ca. 6–10 μm) due mostly to a massive foot layer and thick tectum. *Guapira costaricana* (Plate 7.12,1 and 2) has a better developed endexine than most Caryophyllales, and *Pisonia ambigua* (Plate 7.12,3 and 4) and *P. grandis* (Plate 7.12,5–7) have endexines "slightly similar" to certain Phytolaccaceae (Plates 7.13,7, 7.14,5 and 7.15,2 and 3): fibrous or lamellate under the colpi and separating from the ektexine (see discussion of Phytolaccaceae).

Phytolaccaceae (Plates 7.13–7.16). Phytolaccaceae s.l. (Heimerl 1934b; Nowicke 1969; all traditional genera except *Achatocarpus*, *Phaulothamnus*, and *Stegnosperma*) have all three common aperture types and are SP-PT/AP (*Seguieria langsdorffii* excepted). *Microtea maypurensis* (Plate 7.16,4–6) and *M. scabrida* (Plate 7.16,1–3) are 18–35 porate with the pores mostly regular, and *Petiveria alliacea* (Plate 7.13,6 and 7) is irregularly 8–10 porate (see discussion below); the remaining species (Table 7.1) are 3-colpate. *Seguieria langsdorffii* is granular or scabrate with small blunt spinules (Plate 7.14,4 and 5). Some Phytolaccaceae have an unusual endexine: directly under the aperture it consists of very thin, separated lamellae or even fibers, but on the inner boundary it is more dense. This endexine is present in *Barbeuia madagascariensis* (Plate 7.15,2 and 3), *Petiveria alliacea* (Plate 7.13,7), and two *Seguieria* species (Plate 7.14,5 and 8). To a lesser extent, this condition is found in *Ercilla spicata*, *Phytolacca dioica* and *P. pruinosa* (Nowicke, unpubl. data). Although the endexine is detached in *Lophiocarpus polystachyus* (Plate 7.13,3) it separates intact because it is a solid layer, not granular or lamellate. *Lophiocarpus*, like *Stegnosperma*, has elongate columellae at the poles.

The relationship of *Barbeuia madagascariensis* to the remaining Phytolaccaceae is problematical, mostly because of its two-carpellate capsule. But Phytolaccaceae have a wide range of fruiting structures, from fleshy to dry, including berry, drupe, achene and samara (Nowicke 1968). The pollen (Plate 7.15,1–4) is 3-colpate with an SP-PT tectum; in TEM the characteristics of the endexine (see above), fibrous or netlike beneath the colpus and denser on the inner surface, are similar to *Petiveria*, *Seguieria*, and a genus formerly assigned to Phytolaccaceae, *Stegnosperma* (see discussion below). Perhaps until more data are available *Barbeuia* should be retained in Phytolaccaceae.

Although a collection of *Petiveria* examined in an earlier study (Nowicke 1975: *Cutler 8291* from Brazil) was pantocolpate, and in this study (*Curtiss 5520* from Florida) was irregularly porate, their identifications were easily verified – *Petiveria* is a very distinctive plant. A comparison of the two collections in LM revealed that while not all pollen in *Cutler 8291* is regularly

12- or 15-pantocolpate (many grains are split), the apertures are narrow, even slitlike, whereas in the recent sample the grains are intact, thicker-walled and the apertures are clearly pores, only rarely elongated. The differences might reflect the varieties recognized by Nowicke (1968), but she noted that pollen in *P. alliacea* var. *tetrandra* appears acolpate, which is not the case here.

Another problematical genus is the vinelike and semiwoody *Agdestis* (Plate 7.16,7), which is further distinguished from other Phytolaccaceae by its ovary or fruit structure: it has a semi-inferior ovary which matures into an achene by abortion of three of the four ovules. However, the pollen, 3-colpate with an SP-PT tectum, does little to clarify its relationships.

For the most part, the transfer of *Microtea* to either Amaranthaceae or Chenopodiaceae is supported by pollen data. The grains are pantoporate (Plate 7.16,2 and 5), and in TEM (Plate 7.16,1, 3, 4 and 6) have thin endexines and foot layers but thick tecta, characters common to the Amaranthaceae and Chenopodiaceae. Moreover, the pollen of *Microtea* lacks the unusual endexine, fibrous or netlike under the aperture, found in many Phytolaccaceae.

Portulacaceae (Plate 7.17). Of the seven collections examined, only *Claytonia sarmentosa* (Plate 7.17,3 and 4) has 3-colpate grains (a few are irregularly pantocolpate); the remaining six were pantocolpate with 6, 12, or 15 colpi or, in the case of *Montiastrum lineare* (Plate 7.17,5 and 6), 28–35-pantocolpate and tholate. Pantoporate grains also occur in Portulacaceae (Nilsson 1966, 1967; Nowicke 1975). In addition to SP-PT/AP tecta, finely reticulate occurs in at least one species, *Mona meridensis* (Nowicke and Skvarla 1980). *Talinum patens* (Plate 7.17,1 and 2) has very small spinules and very small punctae. The exine structure in *Claytonia sarmentosa* (Plate 7.17,3 and 4) is perplexing: the thickened mesocolpus has large blocks of ektexine with one or two thin, parallel, vertical cracks, suggesting fusion of columellae.

Pollen samples representing three unidentified collections of *Talinella* have 3-3- or 3-3-3-pantocolpate grains with some variation in tectum morphology, e.g., fused spinules (Nowicke, unpubl. data).

Stegnospermataceae (Plate 7.18). The relationships and placement of *Stegnosperma* seem to have generated more interest than any other genus assigned to Phytolaccaceae, perhaps because its petaloid staminodia suggest a true corolla. The pollen, 3-colpate and SP-PT/AP, does have an endexine (Plate 7.18,2 and 5; Skvarla and Nowicke 1976, Fig. 35; Nowicke and Skvarla 1980, Fig. 47), similar to that found in *Barbeuia* and *Petiveria*. The sieve-element plastid data (Behnke 1976), specifically a polygonal central crystal not known to occur in other phytolaccs, support the separation of *Stegnosperma*.

7.4 Discussion

Overall, the results of the present pollen study do not change the traditional circumscription of the Centrospermae: Aizoaceae, Amaranthaceae, Basellaceae, Cactaceae, Caryophyllaceae, Chenopodiaceae (including *Dysphania*), Didiereaceae, Halophytaceae, Molluginaceae, Nyctaginaceae, Phytolaccaceae (and its segregates), Portulacaceae, and Stegnospermataceae. Although the pollen of Achatocarpaceae is scabrate and with four or five large, poorly defined pores that have no counterpart elsewhere in the Caryophyllales, the family is retained in the Caryophyllales mostly on the basis of the P-type sieve-element plastid.

Although pollen morphology and exine structure support a close relationship among Amaranthaceae, Chenopodiaceae and Caryophyllaceae, the differences in pigment and sieve-element plastids would be difficult to explain away.

Hershkovitz (1993) allied Portulacaceae, Basellaceae, Cactaceae, Didiereaceae, and possibly Hectorellaceae as a "portulacaceous alliance". Pollen morphology would not argue against a relationship among the first three, but the inclusion of Didiereaceae is problematical. The structure of the endexine or 5-8-zonocolpate apertures easily distinguishes Didiereaceae from all remaining centrosperms. However, *Alluaudiopsis* (Didiereaceae) and certain Cactaceae, *Pereskia* and *Rhipsalidopsis*, have similar columellae: granular surfaces and wider near the tectum.

The transfer of *Microtea* from Phytolaccaceae to either Chenopodiaceae or Amaranthaceae is supported by pollen data and the sieve-element plastid that lacks a central crystal (see Chap. 5). But in other characters, e.g., stamen number or absence of anomalous secondary thickening, *Microtea* would be out of place in either Amaranthaceae or Chenopodiaceae.

Other pollen characters (autapomorphies) are too restricted to suggest positive relationships, e.g., the very thick exine found in the Boerhaviinae (no other Caryophyllales have this), the unique endexine in some Didiereaceae, the tholate grains of *Montiastrum*, or the scabrate tectum in Achatocarpaceae.

The irregularity of apertures within a sample and the presence of more than one type within a collection (e.g., 3-colpate, 4-colpate and 6-pantocolpate occur in *Mesembryanthemum nodiflorum*), as well as in species and genera pose problems in using such characters to generate phylogenetic trees.

Undue emphasis should not be placed on the tectal openings, AP versus PT, because both types occur within families, e.g., Cactaceae, Nyctaginaceae, and Portulacaceae, and within genera. Admittedly, annular perforations have not been found in any supposedly related families, e.g., Ranunculaceae, Polygonaceae, or Plumbaginaceae.

More revelant, however, is whether all SP-PT/AP tecta are homologous; consider the extremes that occur within certain families. In Nyctaginaceae the 13 genera of Boerhaviinae (Plate 7.11,1–4) have large spinules and perforations that are conspicuously annular, as well as the thickest exines in the order. Also in Nyctaginaceae are several genera, e.g., *Guapira* and *Pisonia* (Plate 7.13,1–7), that have smaller spinules, punctae rather than annular perforations, and exine structures that are unremarkable. If we assume that Nyctaginaceae are monophyletic, then these two very divergent SP-PT/AP forms could be homologous. A similar argument could be made using Cactaceae pollen data.

Cronquist and Thorne (Chap. 1) and Leins and Erbar (Chap. 13) suggest that Caryophyllales are linked to paeonioid ancestors. Nowicke et al. (1986) examined the pollen of 20 collections representing 14 species of *Paeonia*. They described the pollen as 3-colpate/colporoidate/colporate with a punctate and finely rugulose tectum or with a perforate one; in thin section the exine consists of endexine, foot layer, columellae and tectum. Given the presence of some compound apertures, rugulose tecta, and ektexinous bridges over the endoaperture, the authors characterized the pollen as unspecialized rather than primitive. As to a relationship between Paeoniaceae and Caryophyllales, any common ancestor must have been 3-colpate when these two lineages diverged, since the order lacks compound apertures and three-aperturate pollen is the only type Paeoniaceae have (it seems unlikely that a pantoporate or pantocolpate ancestor would have evolved into the present 3-colpate/colporoidate *Paeonia*).

Of interest are the preliminary results of an analysis of nucleotide sequences from the plastid gene *rbc*L of 475 seed plants (Chase et al. 1993), in which Paeoniaceae are in the Asteridae II lineage with Gunneraceae, Viscaceae and Olacaceae. Of greater interest is that *Dillenia* (Dilleniaceae), *Drosera* (Droseraceae) and *Nepenthes* (Nepenthaceae) as well as *Plumbago* and *Rheum* were in the Rosidae II lineage with the Caryophyllales. However, these results should be treated with reserve until more taxa are examined.

7.4.1 Pollen Data and Molecular Results

Although similar to an earlier cladogram (Rettig et al. 1992), Fig. 9.2 of Manhart and Rettig (Chap. 9) has the traditionally circumscribed families associated into four clades. The following discussion of each clade considers only pollen morphology and exine structure.

Clade I, comprising two species of Amaranthaceae, *Amaranthus tricolor* and *A. hypochondriacus*, and three species of Chenopodiaceae, *Spinacia oleracea*, *Atriplex patula* and *A. rosea*, is supported by pollen morphology. *Amaranthus* (Nowicke 1975, Fig. 56; Skvarla and Nowicke Figs. 7 and 14) has pollen similar to that found in *Atriplex* (Lewis et al. 1983, Fig. 173) of numerous small pores with an SP-PT tectum. They also share thin endexines, thin foot layers and thick tecta. But composition of this clade is hardly surprising given the fact that Amaranthaceae and Chenopodiaceae have long been regarded as closely related, a relationship reinforced by their unique sieve-element plastids (Behnke 1976).

Clade II of four caryophyllaceous genera, *Dianthus*, *Silene*, *Arenaria*, and *Cerastium*, is supported by pollen data included here (Plates 7.8–7.10) and from previous studies (Nowicke 1975, Figs. 29, 30 and 32; Skvarla and Nowicke 1976, Figs. 43, 46 and 48). These four genera are pantoporate and, except for *Silene*, are SP-PT. These caryophylls also have a similar ultrastructure, very thin foot layer/endexine and thick tecta (although discontinuous, the tectum in *Silene* is thick). The 3-colpate caryophylls, illustrated here by *Spergularia macrotheca* (Plate 7.10,5 and 6) and in previous studies (Nowicke 1975, Figs. 25–27; Skvarla and Nowicke 1976, Figs. 41 and 42), have a different exine structure: a well-developed foot layer, more elongate columellae and a tectum about equal in thickness to the foot layer. The similarity of pantoporate pollen among certain Caryophyllaceae, Amaranthaceae and Chenopodiaceae has been documented earlier (Nowicke and Skvarla 1980, Figs. 25–36).

Clade III, composed of *Basella*, *Portulaca*, *Schlumbergera* (Cactaceae), *Pereskia* (Cactaceae) and *Alluaudia* (Didiereaceae), represents four families. To the best of the author's knowledge, there are SEM and TEM data for each genus except *Schlumbergera*. *Basella*, *Portulaca*, *Pereskia aculeata*, and *P. guamacho* are pantocolpate, and *Portulaca*, *Pereskia*, *Schlumbergera* (Leuenberger 1976), and *Alluaudia* are some form of SP-PT/AP. Although pantocolpate apertures and SP-PT/AP are common characters in the Caryophyllales, the pollen data would support this clade.

Clade IV, composed of *Phytolacca*, *Mirabilis*, *Bougainvillea*, *Rivina*, and *Gisekia*, is of more interest because diverse pollen types are present. *Phytolacca* and *Gisekia* are 3-colpate and SP-PT, *Bougainvillea* is 3-colpate and coarsely reticulate, *Mirabilis* is pantoporate with prominent spinules and prominent annular perforations, and *Rivina* is 12-15-pantocolpate and SP-PT. Even though the pollen of *Bougainvillea* is strikingly distinct from *Mirabilis*, they are the only two Nyctaginaceae examined and should be more closely related to each other than to any other genus in the tree.

The pollen data for the Caryophyllales need to be analyzed on a generic basis and may be treated in a future publication.

7.5 Summary

The pollen data presented here, simple apertures, SP-PT/AP tecta or sometimes reticulate, having mostly thin endexines, have not changed the composition of the order as including Achatocarpaceae, Aizoaceae, Amaranthaceae, Basellaceae, Cactaceae, Caryophyllaceae, Chenopodiaceae, Didiereaceae, Halophytaceae, Molluginaceae, Nyctaginaceae, Phytolaccaceae (including Petiveriaceae, *Agdestis*, and *Barbeuia*), Portulacaceae, and Stegnospermataceae. Furthermore, pollen data would: support the transfer of *Microtea* to Chenopodiaceae; support a close relationship between pantoporate Caryophyllaceae and certain Amaranthaceae or Chenopodiaceae; reinforce the distinction of Didiereaceae; suggest that *Barbeuia*, *Ercilla*, *Petiveria*, *Phytolacca*, and *Seguieria* are more closely related than currently thought; and support, although not strongly, the distinction of *Sarcobatus*. The widespread occurrence of well-developed aperture membranes or covers may be an adaptation to compensate for the usually reduced or very thin endexines. Finally, the organization of columellae in some *Anredera* species suggests that SP-PT/AP tecta and reticulate ones are more closely related than was previously thought and that the former tecta could be derived from the latter.

Acknowledgements. In an extensive study such as this a number of individuals assisted in the gathering of data and photography. I thank in particular Christine Begle (Smithsonian Institution) who, although only recently introduced to palynology and electron microscopy, made a substantial contribution, and also Muriel Poston (Howard University) for her help with TEM preparations. Mark Hershkovitz critically reviewed the manuscript and made many valuable suggestions and comments (not all of which were incorporated). I also thank Susann Braden and Walter Brown of the Scanning Electron Microscopy Laboratory, James S. Miller (Missouri Botanical Garden) for the pollen sample of *Barbeuia madagascariensis*, and Stanley Yankowski for compiling a summary of species examined in past investigations.

References

Behnke H-D (1976) Ultrastructure of sieve-element plastids in Caryophyllales (Centrospermae), evidence for the delimitation and classification of the order. Plant Syst Evol 126:31–54

Bittrich V (1986) Untersuchungen zu Merkmalsbestand, Gliederung und Abgrenzung der Unterfamilie Mesembryanthemoideae (Mesembryanthemaceae Fenzl). Mitt Inst Allg Bot Hamb 21:5–116

Bittrich V (1990) Systematic studies in Aizoaceae. Mitt Inst Allg Bot Hamb 23b:491–507

Bortenschlager S (1973) Morphologie pollinique des Phytolaccaceae. Pollen Spores 15: 227–253

Bortenschlager S, Auinger A, Blaha J, Simonsberger P (1972) Pollen morphology of the Achatocarpaceae (Centrospermae). Ber Naturwiss Med Ver Innsb 59:7–13

Chase MW, Soltis DE, Olmstead RG et al. (1993) Phylogenetics of seed plants: An analysis of nucleotide sequences from the plastid gene rbcL. Ann Mo Bot Gard 80:528–580

Cronquist A (1981) An integrated system of classification of flowering plants. Columbia University Press, New York

Cronquist A (1988) The evolution and classification of flowering plants, 2nd edn. New York Botanical Garden, Bronx

Eliasson U (1988) Floral morphology and taxonomic relations among the genera of Amaranthaceae in the New World and Hawaiian islands. Bot J Linn Soc 96:235–283

Erdtman G (1966) Pollen morphology and plant taxonomy: Angiosperms. Almquist and Wiksell, Stockholm xii + 533 pages
Gilbert MG (1987) The taxonomic position of the genera *Telephium* and *Corrigiola*. Taxon 36:47–49
Heimerl A (1934a) Nyctaginaceae. In: Engler A (ed) Die natürlichen Pflanzenfamilien 2nd edn, vol 16c. Engelmann, Leipzig, pp 86–134
Heimerl A (1934b) Phytolaccaceae. In: Engler A (ed) Die natürlichen Pflanzenfamilien 2nd edn, vol 16c. Engelmann, Leipzig pp 135–164
Hershkovitz MA Revised circumscriptions and subgeneric taxonomies of *Calandrinia* and *Montiopsis* (Portulacaceae) with notes on phylogeny of the portulacaceous alliance. Ann Mo Bot Gard 80:333–365
Leuenberger BE (1976) Die Pollenmorphologie der Cactaceae. Dissertationes Botanicae, Band 31. Cramer, Vaduz
Lewis WH, Vinay P, Zenger VE (1983) Airborne and allergenic pollen of North America. Johns Hopkins University Press, Baltimore
Mabry TJ (1976) Pigment dichotomy and DNA-RNA hybridization data for centrospermous families. Plant Syst Evol 126:79–94
Nilsson O (1966) Studies in *Montia* L. and *Claytonia* L. and allied genera. Two new genera *Mona* and *Paxia*. Bot Not 119:265–285
Nilsson O (1967) Studies in *Montia* L. and *Claytonia* L. and allied genera. III. Pollen morphology. Grana Palynol 7:279–364
Nowicke JW (1968) Palynotaxonomic study of the Phytolaccaceae. Ann Mo Bot Gard 55:294–364 (appeared 1969)
Nowicke JW (1970) Pollen morphology in the Nyctaginaceae. I. Nyctagineae (Mirabileae). Grana 10:79–88
Nowicke JW (1975) Pollen morphology in the order Centrospermae. Grana 15:51–77
Nowicke JW, Begle CM (1993) Exine Structure in the Caryophyllales: Basellaceae and Didiereaceae. *XV*th Int Bot Congr Tokyo, Abstr (in press)
Nowicke JW, Luikart TJ (1971) Pollen morphology of the Nyctaginaceae. II. Colignonieae, Boldoeae, and Leucastereae. Grana 11:145–150
Nowicke JW, Skvarla JJ (1977) Pollen morphology and the relationship of the Plumbaginaceae, Polygonaceae, and Primulaceae to the order Centrospermae. Smithson Contrib Bot 37:1–64
Nowicke JW, Skvarla JJ (1980) Pollen morphology: The potential influence in higher order systematics. Ann Mo Bot Gard 66:633–700
Nowicke JW, Skvarla JJ (1981) Pollen morphology and phylogenetic relationships of the Berberidaceae. Smithson Contrib Bot 50:1–83
Nowicke JW, Skvarla JJ (in press) Ranunculaceae. In: Hiepko P (ed) Engler and Prantl's "Die natürlichen Pflanzenfamilien" 2nd edn. Duncker and Humblot, Berlin
Nowicke JW, Bittner JL, Skvarla JJ (1986) *Paeonia*, exine substructure and plasma ashing. In: Blackmore S, Ferguson IK (eds) Pollen and spores: form and function. Academic Press, Orlando, pp 81–95
Pax F, Hoffmann K (1934a) Aizoaceae. In: Engler A (ed) Die natürlichen Pflanzenfamilien 2nd edn, vol 16c. Engelmann, Leipzig pp 179–233
Pax F, Hoffmann K (1934b) Caryophyllaceae. In: Engler A (ed) Die natürlichen Pflanzenfamilien 2nd edn, vol 16c. Engelmann, Leipzig, pp 275–364
Rettig JH, Wilson HD, Manhart JR (1992) Phylogeny of the Caryophyllales – gene sequence data. Taxon 41:201–209
Skvarla JJ, Nowicke JW (1976) Ultrastructure of pollen exine in centrospermous families. Plant Syst Evol 126:55–78
Skvarla JJ, Nowicke JW (1982) Pollen fine structure and relationships of *Achatocarpus* Triana and *Phaulothamnus* A Gray Taxon 31:244–249
Sladkov AN (1990) On the palynomorphology of Centrospermae. VII. Family Nyctaginaceae Juss. Biol NAUKI (Moscow) (O) 3:106–124
Straka H (1965a) Palynologia Madagassica et Mascarenica. Fam. 121: Didiereaceae. Pollen Spores 7:27–34

Straka H (1965b) Die Pollenmorphologie der Didiereaceen. Anhang zu: Rauh W, Schölch H-F Weitere Untersuchungen an Didiereaceen, 2. Teil. Sitzungsb Heidelberger Akad Wiss, Math-Naturwiss Kl 3. Abh., Springer, Berlin Heidelberg New York, pp 219–227

Takhtajan A (1987) Systema Magnoliophytorum. Nauka, Leningrad (in Russian)

8 Phylogenetic Relationships Using Restriction Site Variation of the Chloroplast DNA Inverted Repeat

STEPHEN R. DOWNIE and JEFFREY D. PALMER

8.1 Introduction

Despite intensive multidisciplinary studies using nonmolecular characters, relationships among families comprising the Caryophyllales remain unclear. Reconstruction of phylogenies from molecular data has become an increasingly common approach in systematics and has often provided valuable insight into historical relationships. Of the three genomes present in plants, the chloroplast genome is now the most widely used for phylogenetic inference.

The chloroplast genomes of photosynthetic land plants are circular DNA molecules ranging in size from 120 to 217 kilobase pairs (kb) (Palmer 1985). Complete restriction maps are available for several species of Caryophyllidae; their chloroplast genomes range between 147 and 158 kb (Palmer 1982, 1985; Downie and Palmer 1992a). Chloroplast genomes contain, with few exceptions, two duplicate regions in reverse orientation known as the inverted repeat (IR). In a typical angiosperm chloroplast genome of 150 kb, each of the two IR copies is about 25 kb. These repeated regions separate the remainder of the molecule into large single-copy and small single-copy regions (Fig. 8.1). The expansion or contraction of the IR into, or out of, adjacent single-copy regions, and changes in sequence complexity due to insertion or deletion of unique sequences are largely responsible for variation in size of the molecule.

Recent studies of chloroplast DNA (cpDNA) genome evolution have revealed a high degree of conservatism in size, structure, primary sequence, gene content, and linear order of genes among major lineages of land plants (Palmer 1985, 1991; Palmer and Stein 1986; Downie and Palmer 1992a). This conservative mode of cpDNA evolution suggests that any change in primary sequence, structure, arrangement, or content of the chloroplast genome may have significant phylogenetic implications.

Mutations in cpDNA are of two general kinds: point mutations (single nucleotide pair substitutions) and structural rearrangements. Point mutations can be detected either indirectly, through restriction site mapping (when mutations occur within restriction endonuclease recognition sites), or directly,

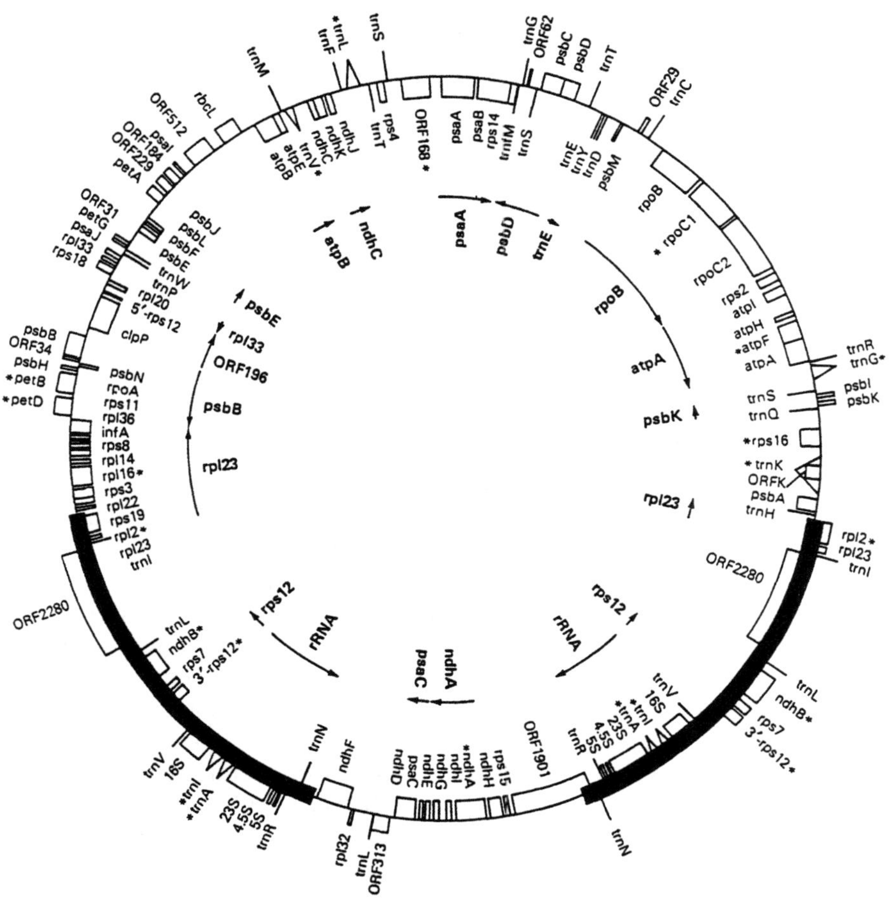

Fig. 8.1. Physical and gene map of the 156-kb *Nicotiana tabacum* chloroplast genome showing the densely packed arrangement of genes and open reading frames (ORFs). The genome is organized into large (87 kb) and small (18 kb) single-copy regions separated by two duplicate regions (about 25 kb each) in reverse orientation, known as the inverted repeat (indicated by the *thickened parts* of the circle). Genes transcribed clockwise are shown on the *inside* of the circle; those transcribed in the reverse direction are on the *outside*. *Arrows* on the inside of the circle indicate sets of genes thought to constitute operons; the operon names are indicated. *Asterisks* denote genes that contain introns

by DNA sequencing. Analyses of restriction site polymorphisms in cpDNA have almost invariably elucidated relationships among taxa at the rank of family or below (reviewed in Palmer et al. 1988). Recently, however, we have shown that by focusing on the highly conserved IR region, restriction site comparisons can be usefully extended to questions of relationships among families within a subclass (Downie and Palmer 1992b). Phylogenetic analyses of *rbc*L sequence data have been extensively used to examine relationships among several major lineages of angiosperms, including the

Caryophyllidae (Giannasi et al. 1992; Olmstead et al. 1992; Rettig et al. 1992; Chase et al. 1993), and will be discussed elsewhere (see Chap. 9). Structural rearrangements of the chloroplast genome (such as inversions and major deletions) are relatively infrequent events among photosynthetic land plants and can usually provide strong evidence of monophyly for a particular group (reviewed in Downie and Palmer 1992a).

In this chapter we present a preliminary analysis of the phylogenetic relationships of several caryophyllalean taxa based on restriction site variation in the highly conserved IR region of the chloroplast chromosome. This information ultimately will be synthesized with a second analysis, currently underway, with an emphasis on major genomic structural rearrangements, to identify the major lineages within the order.

8.2 Materials and Methods

Material of 24 species (Table 8.1), representing 13 families of Caryophyllales plus Polygonaceae and Plumbaginaceae, was field-collected or obtained from various sources as fresh leaf material. The isolation of cpDNA or total

Table 8.1. Species of Caryophyllidae examined for cpDNA IR restriction site and structural variation. A list of sources and voucher information for all taxa examined herein is available upon request

Caryophyllales	Nyctaginaceae
Aizoaceae	*Bougainvillea glabra*
Tetragonia tetragonioides	*Mirabilis nyctaginea*
Amaranthaceae	Petiveriaceae
Alternanthera dentata	*Rivina humilis*
Celosia plumosa	Phytolaccaceae
Basellaceae	*Phytolacca heterotepela*
Anredera cordifolia	Portulacaceae
Cactaceae	*Claytonia perfoliata*
Pereskia grandiflora	*Portulaca oleracea*
Caryophyllaceae	Stegnospermataceae
Agrostemma githago	*Stegnosperma halimifolium*
Corrigiola litoralis	
Silene schafta	Polygonales
Chenopodiaceae	Polygonaceae
Beta vulgaris	*Rheum rhaponticum*
Chenopodium murale	*Polygonum persicaria*
Spinacia oleracea	Plumbaginales
Didiereaceae	
Alluaudia montagnacii	Plumbaginaceae
Didierea madagascariensis	*Limonium gmelinii*
Molluginaceae	
Mollugo verticillata	

cellular DNA, restriction endonuclease digestion, agarose gel electrophoresis, bidirectional transfer of DNA fragments from agarose gels to nylon filters, labeling of recombinant plasmids with ^{32}P by nick-translation or random priming, filter hybridization, and autoradiography were performed according to Palmer (1986) and Downie and Palmer (1992a). All DNAs were digested singly with each of ten restriction endonucleases: *Ava*I, *Bam*HI, *Ban*II, *Bcl*I, *Bgl*II, *Cla*I, *Eco*RI, *Eco*RV, *Hinc*II, and *Hin*dIII. Nineteen subclones from the cpDNA IR region of *Nicotiana tabacum* were used as hybridization probes to survey for restriction site variation. These probes ranged in size from 0.2 to 3.3 kb, averaging approximately 1 kb. This list of probes is available upon request. The conservative nature of land plant chloroplast genome structure and primary sequence allows the use of these hybridization probes across divergent subclasses of angiosperms.

Unambiguous restriction site maps for each of the ten endonucleases were constructed for the *N. tabacum* IR by computer analysis of its completely known cpDNA sequence (Shinozaki et al. 1986). Because many restriction sites and fragment sizes among the taxa examined coincided with those known in *N. tabacum*, mapping efforts were greatly facilitated by scoring our data against these maps.

To assess the circumscription and possible monophyly of the Caryophyllales, three representatives of Polygonaceae and Plumbaginaceae were chosen as outgroups (Table 8.1). Among current classification systems, a consensus favors an association between the Caryophyllales and these two families. These two families are clearly excluded from the order, and several authors have even suggested that there is no strong evidence linking Polygonaceae and Plumbaginaceae to the Caryophyllales (e.g., Rodman et al. 1984; Giannasi et al. 1992). However, results from recent phylogenetic analyses of *rbc*L sequence data (Olmstead et al. 1992; Chase et al. 1993) strongly support the monophyly of the Caryophyllales and indicate that the Polygonaceae and Plumbaginaceae are their most appropriate outgroup.

8.3 Results and Discussion

The cpDNA IR sequences for each of 24 species (Table 8.1) are, with few exceptions, similar in structure and readily aligned with that of *N. tabacum* (and, thereby, also with the IRs of the majority of angiosperms so far examined). Differences in structure that were apparent included the previously documented loss of the *rpl*2 intron (Downie et al. 1991) and partial deletions of coding sequences within the gene ORF2280. We have greatly underestimated the actual extent of restriction fragment length variation in the species examined because: (1) most cpDNA length mutations (for the entire genome) are between 1 and 10 bp in size (Palmer 1985); (2) length variants less than 150 bp could not be detected on our gel systems; and (3) we could only deal with the IR region of the genome because, at the interfamilial level, little or no alignment of restriction sites was possible in single-copy regions. With the exception of the *rpl2* intron loss, which is discussed below, phylogenetic implications of the remaining deletions are discussed elsewhere (S. Downie and J. Palmer, unpubl. data).

8.3.1 *rpl2* Intron Loss

Introns (intervening noncoding sequences within gene coding regions) are highly stable components of land plant chloroplast genomes, with no cases of intron gains and few cases of intron losses known during land plant evolution (Palmer 1991). Plant cpDNAs contain approximately 20 introns (Fig. 8.1), most, if not all, of which were present in the common ancestor of land plants (Palmer 1991).

The chloroplast gene *rpl2*, encoding the ribosomal protein L2, is of systematic interest because it is known to be interrupted by a single intron in most, but not all, land plants. DNA sequencing first revealed that this chloroplast intron is present in *Nicotiana debneyi* (Solanaceae) but absent from *Spinacia oleracea* (Chenopodiaceae) (Fig. 8.2; Zurawski et al. 1984). Utilizing an *rpl2* intron-specific probe (Fig. 8.2), subsequent investigations revealed that this intron is absent from the chloroplast genomes of all examined taxa (ten families and 19 species) of Caryophyllales, yet present in cpDNAs of *Limonium gmelinii* (Plumbaginaceae) and three genera of Polygonaceae (*Polygonum*, *Rheum*, and *Rumex*) (Downie et al. 1991). Sequencing of the *rpl2* gene in five genera of the Caryophyllales and in *Rumex* not only confirmed the filter hybridization results, but also showed that for all taxa lacking the intron, the *rpl2* gene has undergone a precise

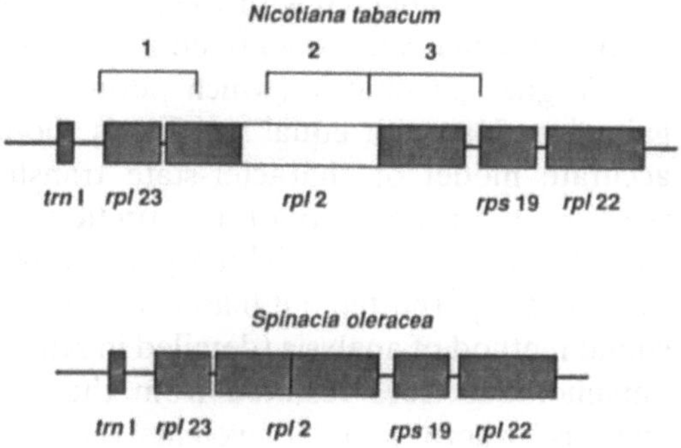

Fig. 8.2. Structural organization of the *rpl2* gene and flanking regions in *Nicotiana tabacum* and *Spinacia oleracea* cpDNAs. Coding regions are indicated by *shaded boxes*; the *rpl2* intron is indicated by an *open box*. The numbered *brackets* indicate the hybridization probes used to determine the presence or absence of coding regions (*1* and *3*) and the intron (*2*). Probe descriptions are presented in Downie et al. (1991)

deletion of the intron (Downie et al. 1991). This suggests that the intron was lost in the common ancestor of the order and supports the order as monophyletic, a concept in accordance with nonmolecular evidence (e.g., Eckardt 1976; Ehrendorfer 1976; Mabry 1977; Cronquist 1981; Rodman et al. 1984).

The relationships between Polygonaceae and Plumbaginaceae, and between these two families and the Caryophyllales, are not wholly clear (Nowicke and Skvarla 1977; Rodman et al. 1984; Giannasi et al. 1992). The presence of the chloroplast *rpl2* intron in Polygonaceae and Plumbaginaceae further distinguishes them from the Caryophyllales but provides no information with respect to the controversy concerning closeness of relationships among these three groups.

8.3.2 Phylogenetic Analysis of Inverted Repeat Restriction Site Mutations

A total of 161 different restriction sites was identified using ten endonucleases among the 24 taxa included in the phylogenetic analysis. Of these 62 (39%) were shared by two or more taxa and were informative for phylogenetic analysis; 60 (37%) of the remaining sites were unvarying, and 39 (24%) were unique to individual taxa and, therefore, provided no phylogenetic information. The occurrence of many invariant restriction sites and the ability to identify readily homologous sites across 14 families (including *N. tabacum*) is notable, and similar to what we observed across widely divergent families in a previous investigation of the Asteridae (Downie and Palmer 1992b).

Cladistic analysis using Wagner parsimony (Swofford 1990) resulted in 79 equally most-parsimonious topologies requiring 163 steps (consistency index including autapomorphies = 0.62; excluding autapomorphies = 0.50). From these, a strict consensus tree was derived (Fig. 8.3). A bootstrap analysis was conducted with 100 replications to provide a measure of internal support for the clades identified in the consensus tree (Fig. 8.3).

Wagner parsimony (which permits an individual restriction site to be gained or lost with equal weight) is thought not to provide a biologically accurate model of character-state transformation given that, at any one position, the probability of a restriction site loss is much greater than a site gain. Character-state weighted parsimony, where parallel gains of a restriction site are permitted but biased against, is believed to be a more biologically sound method of analysis (detailed in Albert et al. 1992a). Four equally most-parsimonious trees resulted from the character-state weighted parsimony analysis (where gain/loss weights ranged from 1.1:1.0 to 1.3:1.0). These trees did not, however, offer any greater resolution among the taxa than those obtained from the Wagner parsimony analysis and, thus, are not presented.

Results of the phylogenetic analysis show that the order is split into two major clades, one consisting of a basal Chenopodiaceae and Amaranthaceae,

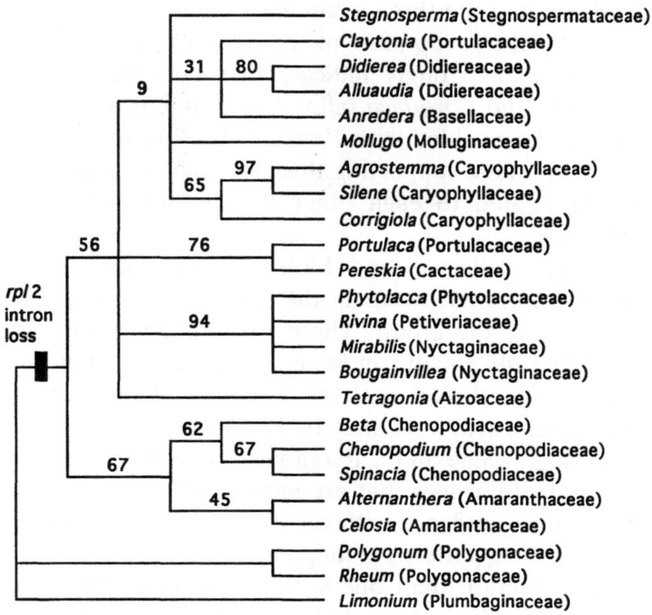

Fig. 8.3. Strict consensus of equally most-parsimonious Wagner trees based on cpDNA IR restriction site mutations. *Numbers* above nodes indicate the number of times that a monophyletic group occurred in 100 bootstrap replicates. This analysis produced 79 shortest trees of 163 steps and a consistency index of 0.62. The *rpl2* intron character was not used in the phylogenetic analysis, but was subsequently mapped onto the consensus tree

and the other consisting of all remaining families. The close relationship between Chenopodiaceae and Amaranthaceae has been expressed by many. On the basis of this limited sampling, both Chenopodiaceae and Amaranthaceae appear to be monophyletic.

Phytolacca (Phytolaccaceae), *Rivina* (Petiveriaceae), and *Mirabilis* and *Bougainvillea* (Nyctaginaceae) are strongly supported as a distinct clade. The consensus is that Nyctaginaceae and Phytolaccaceae sensu stricto are closely allied (Cronquist 1981; Rodman et al. 1984; Rettig et al. 1992; see also Chap. 9). *Stegnosperma*, long associated with Phytolaccaceae (e.g., Heimerl 1934; Cronquist 1981), is recognized by many as belonging to its own family (Stegnospermataceae; Hutchinson 1973; Dahlgren 1980; Bedell 1980; Takhtajan 1980; Brown and Varadarajan 1985). In our results, *Stegnosperma* is clearly excluded from the Phytolaccaceae–Nyctaginaceae clade. *Stegnosperma* shares numerous similarities with Caryophyllaceae (Behnke 1976; Bedell 1980; Narayana and Narayana 1986). Our results place, but with very weak bootstrap support, *Stegnosperma* in the same derived portion of the consensus tree as Caryophyllaceae (and several other taxa) but are equivocal in determining sister-group relationships. *Stegnosperma*,

Mollugo, and the three examined members of Caryophyllaceae constitute a monophyletic group in 11 of the 79 shortest Wagner trees.

The data strongly indicate that Portulacaceae are not monophyletic. *Claytonia* and *Portulaca* fall out in different portions of the consensus tree, with the latter allied with *Pereskia* (Cactaceae). The moderately high (76%) bootstrap value supporting this clade suggests that these two taxa are more closely related to each other than either is to any of the other genera examined.

Caryophyllaceae and Molluginaceae, the only anthocyanin-producing taxa in the order, occur in the same portion of the consensus tree and are not basal to the group. These two families form a monophyletic group in 28 of the 79 equally most-parsimonious Wagner trees. Consequently, as few as one coupled reversal (i.e., loss of betalain synthesis and regain of anthocyanin synthesis) may be necessary to generate anthocyanin production in Molluginaceae and Caryophyllaceae from a betalain-producing ancestor.

Several lineages that are indicated in the cladogram (Fig. 8.3) exhibit a close correspondence with currently recognized families, but, in many cases, the branching patterns among them remain unresolved. Of the seven families for which more than one species was examined, and on the basis of this limited sampling, five (Polygonaceae, Amaranthaceae, Chenopodiaceae, Caryophyllaceae, and Didiereaceae) constitute monophyletic groups; one (Portulacaceae) appears polyphyletic; and one (Nyctaginaceae) is unresolved with the data at hand.

8.3.3 *Nepenthes* and the Caryophyllales

Results from a recent cladistic analysis of *rbc*L sequence data (Albert et al. 1992b; Chase et al. 1993) indicate that *Nepenthes* (Nepenthaceae) may be surprisingly closely allied to the Caryophyllales. As part of an ongoing investigation to look for cpDNA structural changes in angiosperms (S. Downie and J. Palmer, unpubl. data), mapping data were available for *Nepenthes alata*. However, because its cpDNA was digested with only four restriction endonucleases (*Bam*HI, *Eco*RV, *Bgl*II, and *Hin*dIII), and not the ten used in the comparative IR restriction site survey, it was not used in the cladistic analysis.

Our results indicate that the cpDNA IR of *Nepenthes alata* is both colinear in gene arrangement and readily alignable with that of *Nicotiana tabacum* (and, thereby, also with the IR of all other Caryophyllales examined). Comparison of all maps revealed low levels of restriction site divergence. For example, of 41 restriction sites compared for four restriction endonucleases, *Nepenthes* and *Bougainvillea* (Nyctaginaceae) differed by only five sites. Pairwise nucleotide sequence divergence estimates of cpDNA IR sequences (expressed as $100 \times p$; Nei and Li 1979) between *Nepenthes* and *Bougainvillea*, *Nepenthes* and *Chenopodium*, and *Bougainvillea* and

Chenopodium were 1.1, 2.1, and 2.3%, respectively. The high sequence similarity between *Nepenthes* and members of the Caryophyllales is highly intriguing and generally supportive of the *rbc*L results (Chase et al. 1993).

*Rbc*L sequence data show both *Drosera* and *Nepenthes* to be closely allied to Polygonaceae, Plumbaginaceae, and the Caryophyllales (Albert et al. 1992b; Chase et al. 1993). Filter hybridizations revealed that the chloroplast *rpl2* intron is absent from *Drosera filiformis* and from all examined members of the Caryophyllales, but present in *Nepenthes alata* (Downie et al. 1991). Whether the loss of the *rpl2* intron occurred in parallel in *Drosera* and in the Caryophyllales or in a common ancestor shared by these two groups (but not *Nepenthes*) is an interesting problem that deserves further study.

8.4 Conclusions

The data presented here represent an initial analysis of Caryophyllales phylogeny using cpDNA IR restriction site characters. The lack of resolution in many portions of the consensus tree is due to an insufficient number of characters. Future analyses should, therefore, include additional restriction endonucleases to increase the number of informative restriction sites sampled. Moreover, because the current sampling is rather limited, further analyses would also benefit by the inclusion of additional taxa, particularly those representing problematic families such as Portulacaceae, Phytolaccaceae, Molluginaceae, and Caryophyllaceae. Nevertheless, our results provide some explicit hypotheses about relationships in the Caryophyllales that can be tested as more evidence becomes available.

Phylogenetic analyses of cpDNA IR restriction site variation and the distribution of structural rearrangements provide a means of reassessing the traditional classifications of the Caryophyllales. Phylogenetic relationships based on these molecular data should help to assess the relative importance of nonmolecular characters (e.g., morphological, anatomical, ultrastructural, phytochemical, and palynological) currently used in delimiting taxa within the order. The possible polyphyly of Portulacaceae, the putative sister-group relationship between *Portulaca* and *Pereskia*, and the close relationship of *Nepenthes* and *Drosera* to the Caryophyllales suggest that a reevaluation of these nonmolecular characters is in order. Similarly, the molecular data, including those derived from *rbc*L sequencing, need to be scrutinized carefully for various biological factors that may influence the assumptions of phylogenetic analysis (reviewed in Doyle 1992). One approach of phylogenetic estimation that should be considered in future analyses is the integration of molecular with nonmolecular data. This might offer the best hope in resolving evolutionary issues within the order.

Acknowledgements. Financial support for this study was provided by National Science Foundation grant BSR-8717600 to JDP. We thank the Missouri Botanical Garden, the W.J. Beal Botanical Garden (Michigan), Matthaei Botanical Garden (Michigan), and the Brooklyn Botanic Garden (New York) for generously providing us with leaf material used in this study.

References

Albert VA, Mishler BD, Chase MW (1992a) Character-state weighting for restriction site data in phylogenetic reconstruction, with an example from chloroplast DNA. In: Soltis PS, Soltis DE, Doyle JJ (eds) Molecular systematics of plants. Chapman and Hall, New York, pp 369–403

Albert VA, Williams, SE, Chase MW (1992b) Hierarchies of parallelism in carnivorous plants: Implications for structural evolution in angiosperms. Science 257:1491–1495

Bedell HG (1980) A taxonomic and morphological re-evaluation of Stegnospermaceae (Caryophyllales). Syst Bot 5:419–431

Behnke H-D (1976) Ultrastructure of sieve-element plastids in Caryophyllales (Centrospermae), evidence for the delimitation and classification of the order. Plant Syst Evol 126:31–54

Brown GK, Varadarajan GS (1985) Studies in Caryophyllales. I. Re-evaluation of classification of Phytolaccaceae s.l. Syst Bot 10:49–63

Chase MW, Soltis DE, Olmstead RG et al. (1993) Phylogenetics of seed plants: An analysis of nucleotide sequences from the plastid gene *rbc*L. Ann Mo Bot Gard 80:528–580

Cronquist A (1981) An integrated system of classification of flowering plants. Columbia University Press, New York

Dahlgren R (1980) A revised system of classification of angiosperms. Bot J Linn Soc 80:91–124

Downie SR, Palmer JD (1992a) Use of chloroplast DNA rearrangements in reconstructing plant phylogeny. In: Soltis PS, Soltis DE, Doyle JJ (eds) Molecular systematics of plants. Chapman and Hall, New York, pp 14–35

Downie SR, Palmer JD (1992b) Restriction site mapping of the chloroplast DNA inverted repeat: A molecular phylogeny of the Asteridae. Ann Mo Bot Gard 79:266–283

Downie SR, Olmstead RG, Zurawski G, Soltis DE, Soltis PS, Watson JC, Palmer JD (1991) Six independent losses of the chloroplast DNA *rpl2* intron in dicotyledons: Molecular and phylogenetic implications. Evolution 45:1245–1259

Doyle JJ (1992) Gene trees and species trees: Molecular systematics as one-character taxonomy. Syst Bot 17:144–163

Eckardt T (1976) Classical morphological features of centrospermous families. Plant Syst Evol 126:5–25

Ehrendorfer F (1976) Closing remarks: Systematics and evolution of centrospermous families. Plant Syst Evol 126:99–106

Giannasi DE, Zurawski G, Learn G, Clegg MT (1992) Evolutionary relationships of the Caryophyllidae based on comparative *rbc*L sequences. Syst Bot 17:1–15

Heimerl A (1934) Phytolaccaceae. In: Engler A (ed) Die natürlichen Pflanzenfamilien, 2nd edn, vol 16c. Engelmann Leipzig, pp 135–167

Hutchinson J (1973) The families of flowering plants, 3rd edn. Clarendon Press, Oxford

Mabry TJ (1977) The order Centrospermae. Ann Mo Bot Gard 64:210–220

Narayana PS, Narayana LL (1986) The embryology of Stegnospermataceae, with a discussion on its status, affinities and systematic position. Plant Syst Evol 154:137–146

Nei M, Li W (1979) Mathematical model for studying genetic variation in terms of restriction endonucleases. Proc Natl Acad Sci USA 76:5269–5273

Nowicke JW, Skvarla JJ (1977) Pollen morphology and the relationship of the Plumbaginaceae, Polygonaceae, and Primulaceae to the order Centrospermae. Smithson Contrib Bot 37:1–64

Olmstead RG, Michaels HJ, Scott KM, Palmer JD (1992) Monophyly of the Asteridae and identification of its major lineages inferred from DNA sequences of *rbc*L. Ann Mo Bot Gard 79:249–265

Palmer JD (1982) Physical and gene mapping of chloroplast DNA from *Atriplex triangularis* and *Cucumis sativa*. Nucleic Acids Res 10:1593–1605

Palmer JD (1985) Comparative organization of chloroplast genomes. Annu Rev Genet 19:325–354

Palmer JD (1986) Isolation and structural analysis of chloroplast DNA. Methods Enzymol 118:167–186

Palmer JD (1991) Plastid chromosomes: Structure and evolution. In: Bogoard L, Vasil IK (eds) The molecular biology of plastids, chap. 2. In: Vasil IK (ed-in-chief) Cell culture and somatic cell genetics of plants, vol 7A. Academic Press, San Diego, pp 5–53

Palmer JD, Stein DB (1986) Conservation of chloroplast genome structure among vascular plants. Curr Genet 10:823–833

Palmer JD, Jansen RK, Michaels HJ, Chase MW, Manhart JR (1988) Chloroplast DNA variation and plant phylogeny. Ann Mo Bot Gard 75:1180–1206

Rettig JH, Wilson HD, Manhart JR (1992) Phylogeny of the Caryophyllales – gene sequence data. Taxon 41:201–209

Rodman JE, Oliver MK, Nakamura RR, McClammer JU Jr, Bledsoe AH (1984) A taxonomic analysis and revised classification of Centrospermae. Syst Bot 9:297–323

Shinozaki K, Ohme M, Tanaka M et al. (1986) The complete nucleotide sequence of the tobacco chloroplast genome: Its gene organization and expression. EMBO J 5:2043–2047

Swofford DL (1990) PAUP: Phylogenetic analysis using parsimony, version 3.0. Illinois Natural History Survey, Champaign

Takhtajan A (1980) Outline of the classification of flowering plants (Magnoliophyta). Bot Rev 46:225–359

Zurawski G, Bottomley W, Whitfeld PR (1984) Junctions of the large single-copy region and the inverted repeats in *Spinacia oleracea* and *Nicotiana debneyi* chloroplast DNA: Sequence of the genes from $tRNA^{His}$ and the ribosomal proteins S19 and L2. Nucleic Acids Res 12:6547–6558

9 Gene Sequence Data

JAMES R. MANHART and JEFF H. RETTIG

9.1 Introduction

The use of chloroplast DNA comparisons to determine phylogenetic relationships in the angiosperms has become an active area of research. There are three basic types of chloroplast DNA data that have been used for phylogenetic reconstruction: (1) comparison of restriction endonuclease sites (Jansen and Palmer 1988); (2) major genome structural changes (Downie et al. 1991; Downie and Palmer 1992); and (3) comparative gene sequences (Clegg and Zurawski 1992). The first category is covered in Chapter 8. The goal of this chapter is to discuss the use of *rbc*L sequences to test phylogenetic hypotheses in the Caryophyllales.

Giannasi et al. (1992) constructed trees including *Dianthus*, *Amaranthus*, *Atriplex*, and *Spinacia* and a broad sampling of other angiosperm *rbc*L sequences using a number of tree-building methods. Their analyses indicate that *Rheum* (Polygonaceae) and *Plumbago* (Plumbaginaceae) comprise the sister group to the Caryophyllales. Within the Caryophyllales, *Spinacia* and *Atriplex* group with first each other, then *Amaranthus*, and, finally, *Dianthus* (Giannasi et al. 1992). Rettig et al. (1992) compared 19 members of the Caryophyllales using *Rheum* and *Plumbago* as the outgroup. Four major clades are supported by this analysis: (I) Amaranthaceae and Chenopodiaceae; (II) Caryophyllaceae; (III) Basellaceae, Portulacaceae, Cactaceae and Didiereaceae; and (IV) Phytolaccaceae, Petiveriaceae, Nyctaginaceae and *Gisekia*. These groupings generally correlate with those based on phenotypic characters (Cronquist 1981; Rodman et al. 1984). *Stegnosperma* (Phytolaccaceae s.l.) did not group with clade IV, which indicates that the Phytolaccaceae (s.l.) is not monophyletic (Rettig et al. 1992). *Mollugo* (Molluginaceae) did not group strongly with any taxa, including the Caryophyllaceae (Rettig et al. 1992). The Amaranthaceae and Chenopodiaceae were the basal element of the order but this was weakly supported (Rettig et al. 1992). It is clear from Giannasi et al. (1992) and Rettig et al. (1992) that *rbc*L sequence data are useful in testing phylogenetic hypotheses in the

Caryophyllales and that the inclusion of additional taxa will allow further resolution of these relationships.

9.2 Materials and Methods

9.2.1 Materials

Gene sequences from 24 caryophyllalean taxa, five of which were determined for this symposium (Table 9.1), in addition to *Rheum* and *Plumbago* were included in the analyses. The five new taxa are: *Gisekia pharnaceoides* L., *Phaulothamnus spinescens* Gray, *Pereskia aculeata* Mill., *Lithops* sp. and *Bougainvillea glabra* Choisy. Voucher specimens have been placed in TAMU.

Table 9.1. List of *rbc*L sequences used in this investigation

Species	Family	Sequence accession no.	Reference/voucher
Alluaudia procera Drake	Didiereaceae	M62563	Rettig et al. (1992)
Amaranthus hypochondriacus L.	Amaranthaceae	X51964	Michalowski et al. (1990)
Amaranthus tricolor L.	Amaranthaceae	X53980	Rettig et al. (1992)
Arenaria drummondii Shinners	Caryophyllaceae	M83541	Rettig et al. (1992)
Atriplex patula L.	Chenopodiaceae	X15925	Hudson et al. (1990)
Atriplex rosea L.	Chenopodiaceae	X15924	Hudson et al. (1990)
Basella alba L.	Basellaceae	M62564	Rettig et al. (1992)
*Bougainvillea glabra**	Nyctaginaceae	M88340	Dubrule #813
Cerastium glomeratum Thuill.	Caryophyllaceae	M83542	Rettig et al. (1992)
Dianthus caryophyllus L.	Caryophyllaceae	M77699	Giannasi et al. (1992)
Gisekia pharnaceoides L.*	Phytolaccaceae	M97890	Rettig #1662
Lithops sp.*	Aizoaceae	M97889	Manhart 1/2/91#1
Mirabilis jalapa L.	Nyctaginaceae	M62565	Rettig et al. (1992)
Mollugo verticillata L.	Molluginaceae	M62566	Rettig et al. (1992)
Pereskia aculeata Mill.*	Cactaceae	M97888	Dubrule #843
Phaulothamnus spinescens Gray*	Achatocarpaceae	M97887	*s.n.*
Phytolacca americana L.	Phytolaccaceae	M62567	Rettig et al. (1992)
Plumbago capensis Thumb.	Plumbaginaceae	M77701	Giannasi et al. (1992)
Portulaca grandiflora Hook.	Portulacaceae	M62568	Rettig et al. (1992)
Rheum x cultorum Hort.	Polygonaceae	M77702	Giannasi et al. (1992)
Rivina humilis L.	Petiveriaceae	M62569	Rettig et al. (1992)
Schlumbergera truncata (Haw.) Moran	Cactaceae	M83543	Rettig et al. (1992)
Silene gallica L.	Caryophyllaceae	M83544	Rettig et al. (1992)
Spinacia oleracea L.	Chenopodiaceae	J01443	Zurawski et al. (1981)
Stegnosperma halimifolium Benth.	Stegnospermataceae	M62570	Rettig et al. (1992)
Trianthema portulacastrum L.	Aizoaceae	M62572	Rettig et al. (1992)

X, EMBL accession number; M, GenBank number; *, new sequence

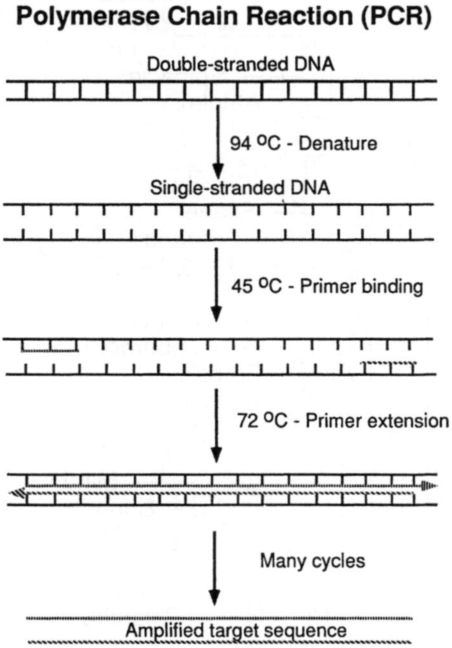

Fig. 9.1. Outline of polymerase chain reaction procedure

9.2.2 DNA Extractions, Cloning, Amplification, and Sequencing

Chloroplast (Palmer 1986; Milligan 1989) or total (Doyle and Doyle 1987) DNAs were extracted from living leaves. The *rbc*L genes in *Gisekia* and *Phaulothamnus* were cloned into the *Bam*HI site of pBS (Bluescript, Stratagene, Inc.). The *rbc*L gene was amplified from total DNAs in *Pereskia*, *Lithops*, and *Bougainvillea* using the polymerase chain reaction (Sambrook et al. 1989) (Fig. 9.1). Refer to Rettig et al. (1992) for a detailed description of methods. Sequences were determined from double-stranded cloned DNA and single-stranded amplified DNA by dideoxy-sequencing as described in the Sequenase manual (version 2.0) using a series of internal ca. 30-mer primers.

9.2.3 Analysis of Data

Sequence data have been deposited with GenBank (Table 9.1). Entire *rbc*L sequences are known for all the taxa with the exception of *Silene*, *Lithops*, and *Schlumbergera* in which the first 26 base pairs (bp) of sequence were not determined. This region is almost invariant and was excluded from the analysis. The extreme 3′ end of the gene is highly variable due to a number of insertion or deletion events involving whole codons after the first 1407 bp and is normally not included in analyses (Giannasi et al. 1992; Rettig et al. 1992). An additional 24 bp from the 3′ end were not included

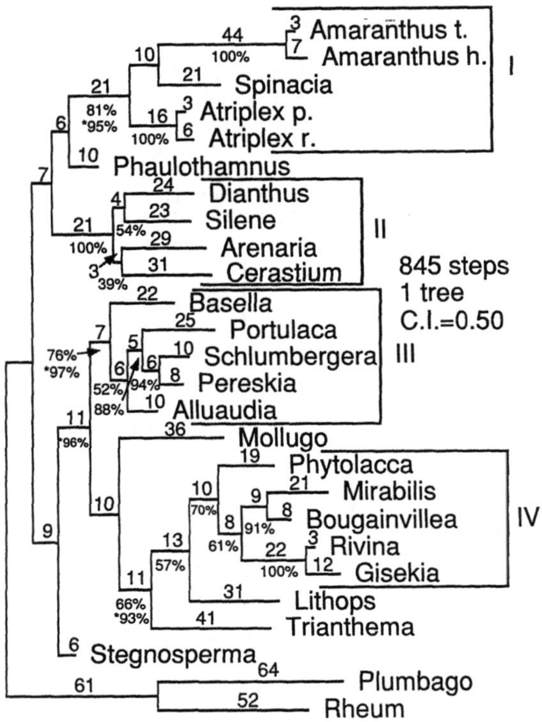

Fig. 9.2 Shortest tree found by PAUP (see Sect. 9.2.3). *Numbers* above branches indicate number of steps; *percentages* below branches indicate % of bootstrap replicates for each grouping; * *values* indicate % of bootstrap replicates obtained when *Phaulothamnus*, *Mollugo*, and *Stegnosperma* are excluded; *C.I.* is the consistency index; *Roman numerals* indicate major clades

due to technical problems in sequencing this region in *Lithops*, so the analysis utilized 1357 bp of sequence. *Rheum* and *Plumbago* were designated outgroup taxa based on the placement of these taxa as sister groups to the Caryophyllales in both phenotypic (Cronquist 1981) and molecular comparisons (Giannasi et al. 1992). An analysis of *rbc*L sequences from 450 taxa indicated that *Nepenthes* and *Drosera* are also members of the sister group to the Caryophyllales (Chase et al. 1993). The inclusion of *Nepenthes* and *Drosera* in preliminary analyses did not change tree topologies and they were not included in the final analyses due to computer runtime constraints. It has been our experience that outgroup selection has no affect on tree topologies within the Caryophyllales. This may be due to the very long branch length leading to the Caryophyllales (Chase et al. 1993). The TBR (tree-bisection-reconnection) branching option with random addition sequence (ten repetitions) of the PAUP 3.0i program (Swofford 1991) was employed to search for minimal trees (Fig. 9.2). Bootstrap analysis (Felsenstein 1985) of 100 replicates was employed to give confidence limits

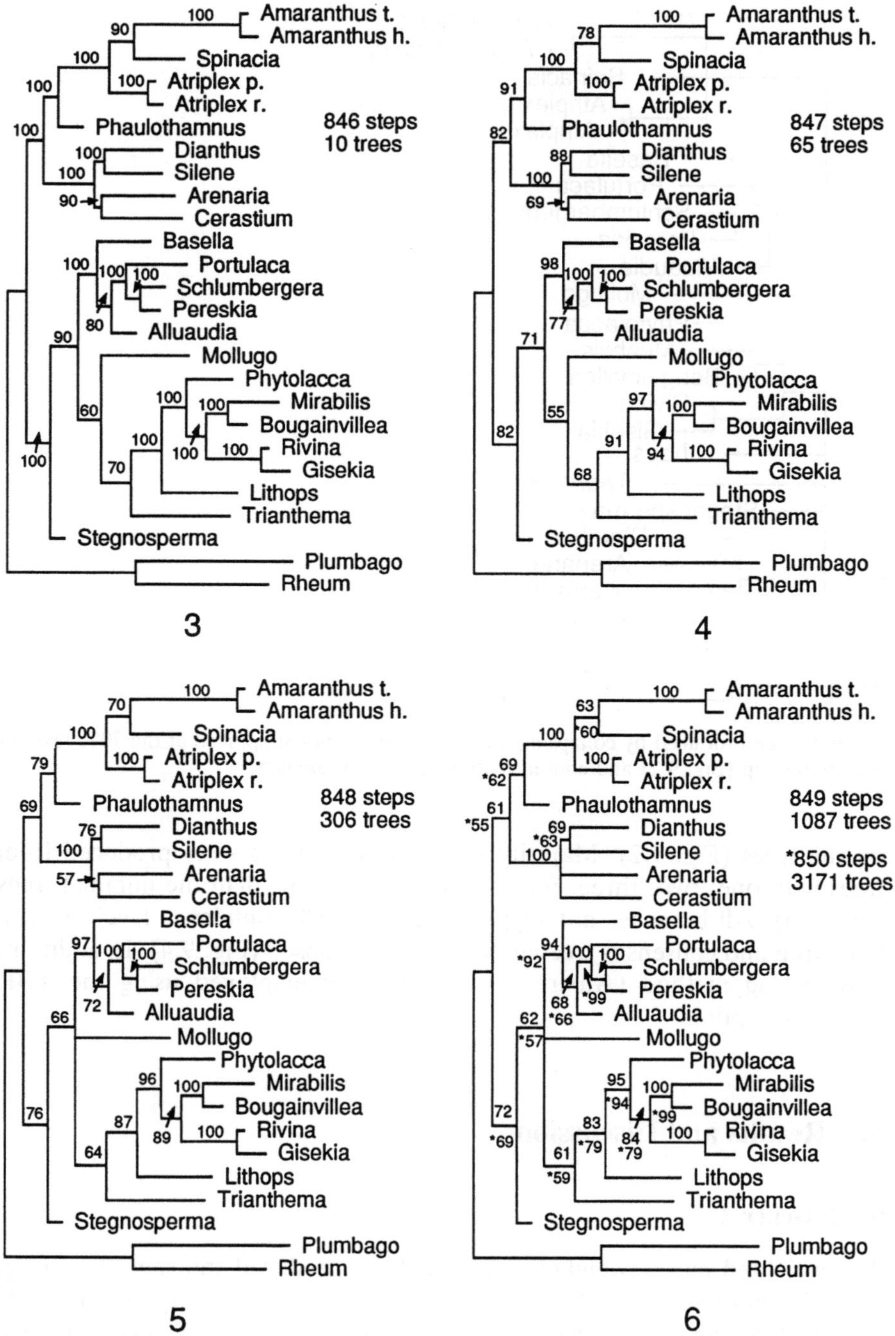

Fig. 9.3 Majority rule consensus trees for trees with one (*3*), two (*4*), three (*5*), and four and five (*6*) more steps than the minimal tree in Fig. 9.2. *Numbers* indicate % of trees in which groupings occurred

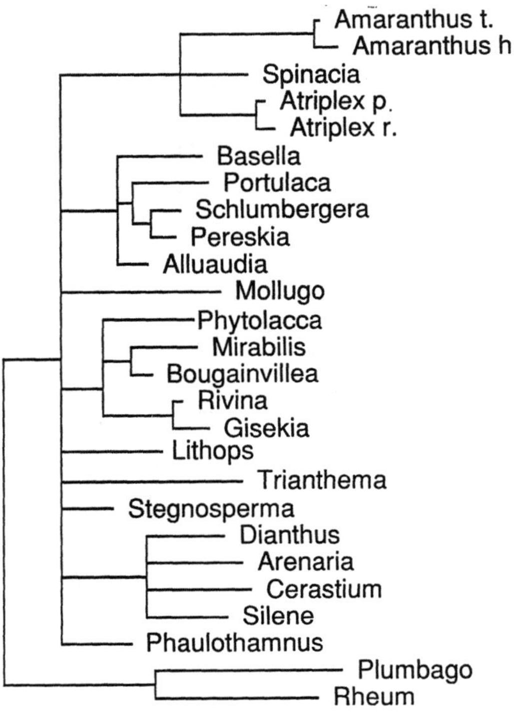

Fig. 9.4 Tree produced by collapsing branches that are not supported at the 70% level in both bootstrap (Fig. 9.2) and consensus tree (Figs. 9.3) analyses

on the trees (Fig. 9.2). Majority rule consensus trees were produced from trees with one, two, three, four and five more steps than the minimal trees (Fig. 9.3). All branches not supported at the 70% confidence level by both bootstrap and consensus tree analyses were collapsed (Fig. 9.4). In addition, a weighting method (Albert et al. 1993) was employed using the TBR branching option.

9.3 Results and Discussion

9.3.1 General

The weighted and unweighted analyses both produced the same tree (Fig. 9.2). The bootstrap (Fig. 9.2) and consensus tree analyses (Fig. 9.3) indicate that the following major clades are internally supported: (I) Chenopodiaceae and Amaranthaceae; (II) Caryophyllaceae; (III) Basellaceae, Portulacaceae, Cactaceae and Didiereaceae; and (IV) Phytolaccaceae, Petiveriaceae, Nyctaginaceae and *Gisekia* (Figs. 9.2–9.4). *Trianthema* and *Lithops* (Aizoaceae) group with IV but this alliance is not as strongly supported as the other major clades (Figs. 9.2 and 9.3). *Phaulothamnus* (Achatocarpaceae), *Mollugo* (Molluginaceae) and *Stegnosperma* (Stegnospermataceae) are not strongly allied with any other taxa (Figs. 9.2–9.4) and their presence in the analysis has a large effect on bootstrap values (Fig. 9.2). The bootstrap analysis done without these taxa indicates the following: stronger support for the major clades, placement of the Aizoaceae with clade IV, and the alliance of clades III and IV with each other and with the Aizoaceae. Therefore, the tree in Fig. 9.4, which was produced by collapsing all the branches that are not supported by bootstrap and consensus tree analyses, may represent an overly conservative approach to phylogenetic relationships in the Caryophyllales based on *rbc*L sequences.

9.3.2 Clade I: Chenopodiaceae and Amaranthaceae

The Chenopodiaceae and Amaranthaceae form one of the four major clades. *Spinacia* groups with *Amaranthus* rather than with *Atriplex*, which supports Rodman's (1990) hypothesis that the Chenopodiaceae are paraphyletic. However, the branch length leading to *Amaranthus* is the longest within the Caryophyllales (Fig. 9.2) and the grouping of *Amaranthus* with *Spinacia* may change when additional members of both families are included in the analysis. The long branch lengths indicate that there is adequate differentiation of *rbc*L sequences to allow further resolution of relationships within and between the Chenopodiaceae and Amaranthaceae. The inclusion of additional taxa should allow the testing of phylogenetic relationships involving these two families.

9.3.3 Clade II: Caryophyllaceae

The members of the Caryophyllaceae group together as one of the four major clades. In addition, the groupings within this clade are congruent with subfamilial designations; *Dianthus* and *Silene* are members of subfamily Caryophylloideae, while *Arenaria* and *Cerastium* are members of subfamily Alsinoideae. These subfamilial groupings are not supported by the bootstrap and consensus tree analyses. It is therefore unlikely that *rbc*L gene sequences will provide adequate resolution to test phylogenetic hypotheses of relationships between and within these two subfamilies.

Mollugo. Most current treatments place the Caryophyllaceae with the Molluginaceae, mainly on the basis of the lack of betalain production in these families (Mabry et al. 1963; Mabry 1977; Cronquist 1981; Rodman et al. 1984). Rodman (1990) goes further and "raises the possibility that genera of Molluginaceae are nested within Caryophyllaceae". The *rbc*L data do not support the placement of these families together (Figs. 9.2–9.4). Further more, *Mollugo* does not group strongly with any of the four major clades of the tree. The inclusion of additional genera of the Molluginaceae and members of subfamily Paronychioideae of the Caryophyllaceae will be required to further test the hypothesized relationship between the Molluginaceae and Caryophyllaceae. The Molluginaceae is a rather small family with 13 genera and 100 species (Cronquist 1981) and it would be feasible to include one representative of each of the genera in a broadened *rbc*L analysis.

Phaulothamnus. The inclusion of *Phaulothamnus* (Achatocarpaceae) has a major effect on the tree topology (Fig. 9.2) relative to an earlier analysis (Rettig et al. 1992). It "pulls" the Caryophyllaceae away from the succulents and Phytolaccaceae into a clade containing *Phaulothamnus*, Amaranthaceae, Chenopodiaceae, and Caryophyllaceae. The placement of the Chenopodiaceae and Caryophyllaceae as sister groups was suggested by Rodman et al. (1984) and the *rbc*L analysis provides support for this alignment. However, neither of the groupings produced from *rbc*L sequences is strongly supported by bootstrap and consensus tree analyses (Figs. 9.2–9.4, Rettig et al. 1992). The addition of *Achatocarpus*, the only other genus in the Achatocarpaceae, may help resolve the placement of the Achatocarpaceae and its relationship to Amaranthaceae, Chenopodiaceae, and Caryophyllaceae.

9.3.4 Clade III: Basellaceae, Portulacaceae, Cactaceae, and Didiereaceae

The succulent taxa, with the exception of the Aizoaceae, form a distinct clade. *Basella* (Basellaceae) is at the base of the succulent clade but bootstrap and consensus tree analyses (Figs. 9.2–9.4) indicate that *Alluaudia* (Didiereaceae) is also a basal element of this group. The most derived members of this group are *Schlumbergera*, *Pereskia* (Cactaceae), and *Portulaca* (Portulacaceae). Cronquist (1981) discusses the hypothesis that the Cactaceae have their origin "in or near the Portulacaceae". This analysis (Fig. 9.2) cannot distinguish between "in" or "near" since only one member of the Portulacaceae is included. The inclusion of additional taxa from both families should determine whether they are sister groups or whether the Cactaceae is derived from within the Portulacaceae.

9.3.5 Clade IV: Phytolaccaceae, Petiveriaceae, Nyctaginaceae, and Gisekia

The Phytolaccaceae, Petiveriaceae, Nyctaginaceae, and *Gisekia* form another major clade (Fig. 9.2). *Gisekia* was originally placed in the Molluginaceae but Mabry et al. (1976) demonstrated that two species of *Gisekia* produce betalains, which indicates that this genus may not belong in the Molluginaceae. *Gisekia* groups strongly with *Rivina* (Petiveriaceae), which is more congruent with Cronquist's (1981) placement of *Gisekia* the Phytolaccaceae than its inclusion in the Molluginaceae. The position of *Gisekia* also supports Bittrich and Hartmann's (1988) contention that "*Gisekia* cannot be considered a link between the Aizoaceae sensu stricto and the Phytolaccaceae." The other members of the Phytolaccaceae s.l., *Phaulothamnus* and *Stegnosperma*, are distant from this major clade. Brown and Varadarajan (1985) treat these two taxa as members of distinct families rather than as members of the Phytolaccaceae. A more restricted definition of the Phytolaccaceae that would include *Phytolacca*, *Rivina*, and *Gisekia* is still problematical since the Nyctaginaceae are nested within this group. There appears to be enough variability in *rbc*L sequences in this clade to allow further resolution of phylogenetic relationships in the group. The inclusion of all the genera in the Phytolaccaceae and Petiveriaceae and at least one representative genus in the subfamilies of the Nyctaginaceae should prove useful in understanding their relationships.

Aizoaceae. The placement of *Lithops* and *Trianthema* (Aizoaceae) with clade IV is supported by bootstrap analysis when *Mollugo*, *Stegnosperma*, and *Phaulothamnus* are removed from the data set (Fig. 9.2). Rodman et al. (1984) place the Aizoaceae close to the Cactaceae but Bittrich and Hartmann (1988) point out difficulties with this arrangement. Ehrendorfer (1976) concludes that the Aizoaceae s.s. may be close to the Phytolaccaceae, which is supported by the *rbc*L analyses. These taxa do not group with *Mollugo*, which supports Bittrich and Hartmann's (1988) observation that "no synapomorphies are known which could establish a sister-group relationship" between the Aizoaceae and Molluginaceae. *Lithops* and *Trianthema* form a grade rather than a clade and both have relatively long branch lengths, which indicates that they are not particularly closely related to each other. It may be that the Aizoaceae are a highly divergent family and that *Lithops* and *Trianthema* represent the extremes of variation in this family. The alternative explanation is that they are not closely related. The long branch lengths associated with *Lithops* and *Trianthema* indicate that further resolution of their relationships is possible with additional sampling of the Aizoaceae. This should also help determine whether the Aizoaceae belong in clade IV.

Stegnosperma. The placement of *Stegnosperma* as a member of the Phytolaccaceae is clearly not supported by the *rbc*L analyses (Figs. 9.2–9.4). It does not group strongly with any other taxon, which supports its treatment as a monogeneric family (Bedell 1980). *Stegnosperma* is on the first branch of a clade that contains clades III and IV, *Mollugo*, and the Aizoaceae (Fig. 9.2). It may therefore represent the most primitive taxon in this clade, and possibly in the Caryophyllales. However, the bootstrap and consensus tree analyses indicate its position is uncertain. In addition, *Stegnosperma* and *Mollugo* readily exchanged positions on trees with fewer taxa (Rettig and Manhart, unpublished results). Therefore, the position of *Stegnosperma* in Fig. 9.2 is tentative and this is reflected in Fig. 9.4.

9.4 Conclusions

The *rbc*L analyses indicate that four major lineages diverged relatively early in the evolution of the Caryophyllales (Figs. 9.2–9.4): (I) Chenopodiaceae and Amaranthaceae; (II) Caryophyllaceae; (III) Basellaceae, Portulacaceae, Cactaceae, and Didiereaceae; and (IV) Phytolaccaceae, Petiveriaceae, Nyctaginaceae, and *Gisekia*. The large clade containing IV, Aizoaceae, and *Mollugo* has support from phenotypic characters; all the members of this last grouping have form-P3cf or -P3cfs sieve-element plastids (see Chap. 5).

The *rbc*L analyses (Figs. 9.2–9.4) conflict with other treatments of the Caryophyllales in the composition of the Phytolaccaceae and its position in the order. Rodman (1990) comments, "If Polygonaceae are truly the sister group to the Centrospermae, then the alleged primitiveness of Phytolaccaceae (Cronquist 1981 and others) is seriously compromised". Giannasi et al.'s (1992) and Chase et al.'s (1993) analyses indicated that *Rheum* (Polygonaceae) is a member of the sister group to the Caryophyllales. It is not surprising, then, that the hypothesis that at least some elements of the Phytolaccaceae are derived is supported by the analyses reported here. The *rbc*L data indicate that the Phytolaccaceae, Nyctaginaceae, Petiveriaceae, *Gisekia*, and possibly the Aizoaceae are related to each other and relatively highly derived. However, *Stegnosperma* and *Phaulothamnus*, both members of the Phytolaccaceae (s.l.), are apparently remnants of less successful lineages and could be interpreted as primitive taxa. *Mollugo* may also fit into this category.

An earlier analysis of *rbc*L sequences (Rettig et al. 1992) indicated that the clade containing the Chenopodiaceae and Amaranthaceae was the basal element of the order. The trees produced by the inclusion of the additional taxa in these analyses indicate that this may not be the case (Figs. 9.2–9.4). It is not surprising that portions of the tree change with the inclusion of additional taxa, since the current sampling is rather limited. For this reason we are not prepared to suggest formal taxonomic changes based on these trees. These preliminary results do indicate that it will be possible to construct a well-resolved tree that resolves most interfamilial relationships within the Caryophyllales using *rbc*L sequences. The results obtained so far provide some indication of the number of taxa that will be required to accomplish this. Basically, all the genera in the small families should be included, while a broader sampling of the medium-sized families would include at least half of their genera. The Caryophyllaceae, Amaranthaceae, Chenopodiaceae, Portulacaceae, Cactaceae, and Nyctaginaceae present a problem due to their size but it is probably not efficient or necessary to include a majority of their genera. The Portulacaceae should be sampled rather extensively since chloroplast DNA restriction site data indicate that the family is not monophyletic (see Chap. 8). The additional sampling in the Caryophyllaceae would be limited to several members of the subfamily Paronychioideae since bootstrap and consensus tree analyses indicate that *rbc*L sequences may not be divergent enough to resolve phylogenetic relationships below the family level (Figs. 9.2–9.4). An adequate sampling of the other families could probably be obtained by including one or two genera in each of their subfamilies. In addition, taxa of uncertain affinity, such as *Barbeuia*, *Agdestis*, *Hectorella*, *Lyallia*, *Dysphania* and *Microtea*, would be included.

Acknowledgements. We wish to thank Hugh Wilson for assistance in obtaining plant materials and for text review, David Northcliffe for technical assistance and Monique Dubrule for plant materials and technical assistance.

References

Albert VA, Chase MW, Mishler BD (1993) Character-state weighting for cladistic analysis of protein-coding DNA sequences: Prospects, limitations, and an example from the plastid *rbc*L gene. Ann Mo Bot Gard 80:752–766

Bedell HG (1980) A taxonomic and morphological re-evaluation of Stegnospermaceae (Caryophyllales). Syst Bot 5:419–431

Bittrich V, Hartmann HEK (1988) The Aizoaceae – a new approach. Bot J Linn Soc 97:239–254

Brown GK, Varadarajan GS (1985) Studies in Caryophyllales. I. Re-evaluation of classification of Phytolaccaceae s.l. Syst Bot 10:49–63

Chase MW, Soltis DE, Olmstead RG et al. (1993) Phylogenetics of seed plants: An analysis of nucleotide sequences from the plastid gene *rbc*L. Ann Mo Bot Gard 80:528–580

Clegg MT, Zurawski G (1992) Chloroplast DNA and the study of plant phylogeny: Present status and future prospects. In: Soltis PS, Soltis DE, Doyle JJ (eds) Molecular systematics of plants. Chapman and Hall, New York, pp 1–13

Cronquist A (1981) An integrated system of classification of flowering plants. Columbia University Press, New York

Downie SR, Palmer JD (1992) Use of chloroplast DNA rearrangements in reconstructing plant phylogeny. In: Soltis PS, Soltis DE, Doyle JJ (eds) Molecular systematics of plants. Chapman and Hall, New York, pp 14–35

Downie SR, Olmstead RG, Zurawski G, Soltis DE, Soltis PS, Watson JC, Palmer JD (1991) Six independent losses of the chloroplast *rpl2* intron in dicotyledons: Molecular and phylogenetic implications. Evolution 45:1245–1259

Doyle JJ, Doyle JL (1987) A rapid DNA isolation procedure for small quantities of fresh leaf tissue. Phytochem Bull 19:11–15

Ehrendorfer F (1976) Closing remarks: systematics and evolution of centrospermous families. Plant Syst Evol 126:99–106

Felsenstein J (1985) Confidence limits on phylogenies with a molecular clock. Syst Zool 34:152–161

Giannasi DE, Zurawski G, Learn G, Clegg MT (1992) Evolutionary relationships of the Caryophyllidae based on comparative *rbc*L sequences. Syst Bot 17:1–15

Hudson GS, Mahon JD, Anderson PA, Gibbs MJ, Badger MR, Andrews TJ, Whitfeld PR (1990) Comparisons of *rbc*L genes for the large subunit of ribulose-bisphosphate carboxylase from closely related C3 and C4 plant species. J Biol Chem 265:808–814

Jansen RK, Palmer JD (1988) Phylogenetic implications of chloroplast DNA restriction site variation in the Mutisieae (Asteraceae). Am J Bot 75:753–766

Mabry TJ (1977) The order Centrospermae. Ann Mo Bot Gard 64:210–220

Mabry TJ, Taylor A, Turner BL (1963) The betacyanins and their distribution. Phytochemistry 2:61–64

Mabry TJ, Behnke H-D, Eifert IJ (1976) Betalains and P-type sieve-element plastids in *Gisekia* L. (Centrospermae). Taxon 25:112–114

Michalowski CB, Bohnert HJ, Klessig DF, Berry JO (1990) Nucleotide sequence of *rbc*L from *Amaranthus hypochondriacus* chloroplasts. Nucleic Acids Res 18:2187

Milligan BG (1989) Purification of chloroplast DNA using hexadecyltrimethylammonium bromide. Plant Mol Biol Rep 7:144–149

Palmer JD (1986) Isolation and structural analysis of chloroplast DNA. Methods Enzymol 118:167–186

Rettig JH, Wilson HD, Manhart JR (1992) Phylogeny of the Caryophyllidae – gene sequence data. Taxon 41:201–209

Rodman JE (1990) Centrospermae revisited, part I. Taxon 39:383–393

Rodman JE, Oliver MK, Nakamura RR, McClammer JU, Bledsoe AH (1984) A taxonomic analysis and revised classification of Centrospermae. Syst Bot 9:297–323

Sambrook J, Fritch EF, Maniatis T (1989) Molecular cloning: a laboratory manual. Cold Spring Harbor Laboratory, Cold Spring Harbor, NY

Swofford DL (1991) PAUP: Phylogenetic analysis using parsimony. Illinois Natural History Survey, Champaign

Zurawski G, Perrot B, Bottomley W, Whitfeld PR (1981) The structure of the gene for the large subunit of ribulose-1,5-bisphosphate carboxylase from spinach chloroplast DNA. Nucleic Acids Res 9:3251–3270

10 Chemical Review and Evolutionary Significance of the Betalains

John S. Clement, Tom J. Mabry, Hugo Wyler, and André S. Dreiding

10.1 Introduction

The betalains, which include the red-violet betacyanins and the yellow betaxanthins (Fig. 10.1), represent a structurally and biosynthetically distinct class of naturally-occurring pigments. These pigments are found in flowering plants only in members of certain families of the order Caryophyllales (Centrospermae) (Table 10.1), though betaxanthins have also been shown to occur in the fly agaric, *Amanita muscaria* (Döpp et al. 1982). Both betacyanins and betaxanthins possess a dihydropyridine moiety which is attached via a vinyl group to another nitrogenous group. In the betacyanins the latter is a glycosyl-hydroxylated dihydroindole ring; in the betaxanthins it is any one of several amino acids or amines. The basic chromophore of the betalain pigments is the 1,7-diazaheptamethinium system, as discussed by Mabry et al. (1967) and Mabry and Dreiding (1968). The different spectral properties of the betacyanins and betaxanthins are due to the extension of this conjugated system through the dihydroindole moiety in the betacyanins. The term "betalains" was introduced by Mabry and Dreiding (1968) to emphasize the common biogenetic and structural features of these two groups, which clearly distinguish this class of pigments from the more common group of plant pigments, the anthocyanins.

The crystallization of betanin from *Beta vulgaris* by two groups working independently (Wyler and Dreiding 1957; Schmidt and Schönleben 1957) permitted the first informative studies of a betacyanin, betanidin, the aglycone of betanin (Wyler and Dreiding 1959; Schmidt et al. 1960). Studies by Mabry et al. (1962, 1967) led to neobetanidin (the 14,15-dehydrogenated from of betanidin) and its derivatives. These compounds possessed the skeleton common to betacyanins and could be readily structurally elucidated on the basis of their characteristic degradation products: pyridine and indoline derivatives. This allowed rapid elucidation of many betacyanin structures, beginning with that of betanidin (Wyler et al. 1963).

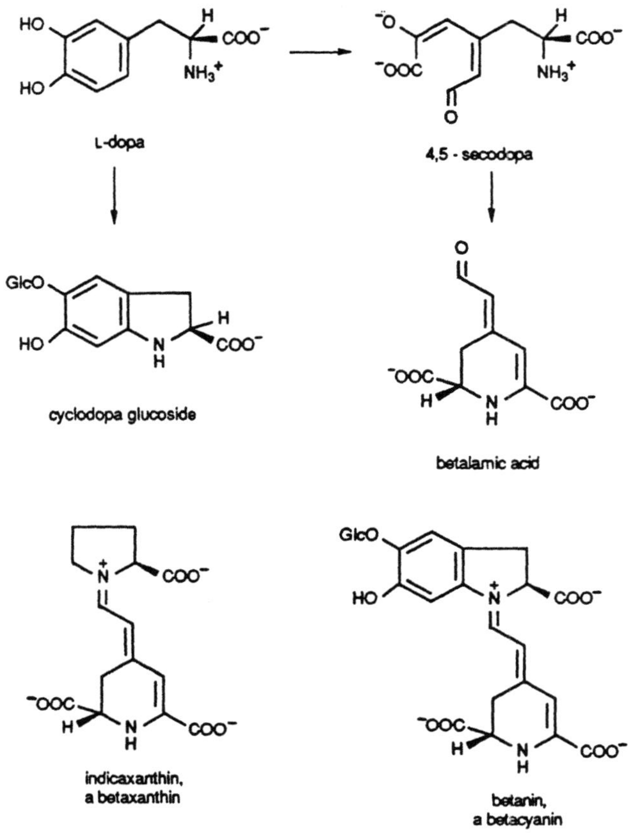

Fig. 10.1. Biosynthesis of betalains from L-DOPA

Structural similarities between the betacyanins and betaxanthins were noted by Piattelli et al. (1964) when the degradation of indicaxanthin from *Opuntia ficus-indica* yielded products typical of the dihydropyridine ring found in betacyanins. Conclusive evidence was afforded by the transformation of indicaxanthin into betanidin and the synthesis of indicaxanthin from betanin and L-proline (Wyler et al. 1965).

10.2 Biogenesis of Betalains

It was postulated in the early 1960s that betanin could possibly be synthesized in vivo from two molecules of L-DOPA (dihydroxyphenylalanine) (Wyler et al. 1963, 1965; Mabry 1964). Feeding experiments established that labelled L-DOPA is indeed incorporated into betanin in *B. vulgaris* (Hörhammer et al. 1964) and *Opuntia ficus-indica* (Minale et al. 1965). It is now known that

Table 10.1. Caryophyllalean genera known to contain betalains. Each reference is the earliest known report of a positive chemical test for presence of betalains in the genus. Synonymous/obsolete names (e.g., *Aloinopsis* = *Rabiaea* = *Nananthus*) are retained for reference. See Steglich and Strack (1990) for review of structure and distribution of individual betalains in the Caryophyllales

Genus	Reference
Agdestidaceae	
Agdestis	Behnke et al. (1974)
Aizoaceae	
Aloinopsis	Reznik et al. (1988)
Antimima	Strack et al. (1990)
Aptenia	Mabry (1966)
Argyroderma	Reznik (1978)
Astridida	Strack et al. (1990)
Bergeranthus	Reznik (1957)
Carpanthea	Reznik (1978)
Carpobrotus	Piattelli and Impellizzeri (1970)
Carruanthus	Reznik et al. (1988)
Cephalophyllum	Reznik (1978)
Chasmatophyllum	Reznik et al. (1988)
Conicosia	Reznik et al. (1988)
Conophytum	Lowrence et al. (1939)
Delosperma	Reznik et al. (1988)
Dinteranthus	Reznik (1978)
Dorotheanthus	Reznik (1957)
Dracophilus	Strack et al. (1990)
Drosanthemum	Impellizzeri et al. (1973)
Erepsia	Strack et al. (1990)
Faucaria	Mabry et al. (1990)
Fenestraria	Reznik (1957)
Frithia	Reznik et al. (1988)
Gibbaeum	Reznik (1957)
Glottiphyllum	Reznik (1957)
Hereroa	Reznik (1978)
Hymenocyclus	Reznik (unpubl. list)
Jensenobotrya	Strack et al. (1990)
Lamprathus	Reznik (1957)
Leipoldtia	Strack et al. (1990)
Lithops	Reznik (1957)
Malephora	Reznik (1957)
Mesembryanthemum	Gertz (1906)
Mestoklema	Strack et al. (1990)
Nananthus	Reznik (1978)
Ophthalmophyllum	Strack et al. (1990)
Oscularia	Strack et al. (1990)
Pleiospilos	Reznik (1957)
Psammophora	Strack et al. (1990)
Rabiaea	Reznik (1957)
Rhinephyllum	Reznik et al. (1988)
Rhombophyllum	Reznik (1957)
Ruschia	Strack et al. (1990)
Sesuvium	Taylor (1940)
Smicrostigma	Strack et al. (1990)
Tetragonia	Gertz (1906)
Tischleria	Reznik (1978)
Trianthema	Mabry et al. (1963)
Trichodiadema	Reznik (1957)
Amarathaceae	
Achyranthes	Gertz (1906)
Acnida	Kimler et al. (1970)
Aerva	Gertz (1906)
Altermanthera	Piatteli and Minale (1964b)
Amaranthus	Bischoff (1876)
Bosea	Kull and Kühn (1978)
Celosia	Bischoff (1876)
Chamissoa	Seeligman and Kraponikas (1969)
Froelichia	Mabry et al. (1963)
Gomphrena	Forsyth and Simmonds (1954)
Iresine	Weigert (1894)
Mogiphanes	Gertz (1906)
Tidestromia	Mabry et al. (1963)
Basellaceae	
Anredera	Mabry (1966)
Basella	Gertz (1906)

Table 10.1. *Continued*

Genus	Reference	Genus	Reference
Cactaceae		*Rhodocactus*	Piattelli and Imperato (1969)
Aporocactus	Piattelli and Imperato (1969)	*Selenicereus*	Lawrence et al. (1939)
Ariocarpus	Reznik (1957)	*Soehrensia*	Piattelli and Imperato (1969)
Aylostera	Reznik (1957)	*Stenocereus*	Piattelli and Imperato (1969)
Borzicactus	Piattelli and Imperato (1969)	*Submatucana*	Reznik (unpubl. list)
Cereus	Lawrence et al. (1939)	*Sulcorebutia*	Reznik (unpubl. list)
Chamaecereus	Reznik (1957)	*Thelocactus*	Dreiding (1961)
Cleistocactus	Reznik (1957)	*Weingartia*	Reznik (1978)
Coryphantha	Reznik (unpubl. list)	*Zygocactus*	Lawrence et al. (1939)
Echinocereus	Steglich and Strack (1990)		
Epicactus cultivars	Piattelli and Imperato (1969)	Chenopodiaceae	
Epiphyllum	Reznik (1955)	*Atriplex*	Gertz (1906)
Eriocereus	Piattelli and Imperato (1969)	*Beta*	Bischoff (1876)
Gymnocalycium	Reznik (1957)	*Chenopodium*	Bischoff (1876)
Haageocereus	Piattelli and Imperato (1969)	*Corispermum*	Gertz (1906)
Hariota	Reznik (1955)	*Cycloloma*	Mabry et al. (1963)
Heliocereus	Reznik (unpubl. list)	*Dysphania*	Mabry and Behnke (1976)
Hylocereus	Dreiding (1961)	*Enchylaena*	Wohlpart and Mabry (1968)
Lobivia	Reznik (1957)	*Grayia*	Kimler et al. (1970)
Mammillaria	Reznik (1957)	*Kochia*	Lawrence et al. (1939)
Mediolobivia	Reznik (1978)	*Monolepis*	Kimler et al. (1970)
Melocactus	Reznik (1957)	*Rhagodia*	Wohlpart and Mabry (1968)
Monvillea	Wyler and Dreiding (1961b)	*Salicornia*	Mabry et al. (1963)
Neoporteria	Reznik (1957)	*Salsola*	Piattelli and Minale (1964b)
Nopalea	Piattelli and Imperato (1969)	*Spinacia*	Piattelli and Minale (1964b)
Nopalxochia	Kryz (1926)	*Suaeda*	Lawrence et al. (1939)
Notocactus	Reznik (1957)		
Opuntia	Lawrence et al. (1939)	Didiereaceae	
Parodia	Reznik (1957)	*Alluaudia*	Rauh and Reznik (1961)
Pereskia	Reznik (1955)	*Alluaudiopsis*	Rauh and Reznik (1961)
Phyllocactus	Piattelli and Minale (1964a)	*Decaryia*	Rauh and Reznik (1961)
Pilosocereus	Piattelli and Imperato (1969)		
Psuedolobivia	Steglich and Strack (1990)		
Psuedorhipsalis	Reznik (unpubl. list)		
Rebutia	Reznik (1957)		
Rhipsalis	Reznik (unpubl. list)		

Table 10.1. *Continued*

Genus	Reference	Genus	Reference
Didierea	Rauh and Reznik (1961)	*Trichostigma*	Reznik (1955)
Halophytaceae		Phytolaccaceae	
Halophytum	Hunziker et al. (1974)	*Gisekia*	Mabry et al. (1976)
Hectorellaceae		*Phytolacca*	Bischoff (1876)
Hectorella	Mabry et al. (1978)	Portulacaceae	
Nyctaginaceae		*Anacampseros*	Reznik (1955)
Abronia	Mabry et al. (1963)	*Calandrinia*	Reznik (1957)
Allionia	Mabry et al. (1963)	*Ceraria*	Behnke et al. (1975)
Anulocaulis	Wohlpart and Mabry (1968)	*Claytonia*	Mabry et al. (1963)
Boerhavia	Taylor (1940)	*Lewisia*	Behnke et al. (1975)
Bougainvillea	Lawrence et al. (1939)	*Montia*	Mabry et al. (1963)
Cryptocarpus	Taylor (1940)	*Portulaca*	Bischoff (1876)
Cyphomeris	Mabry et al. (1963)	*Portulacaria*	Behnke et al. (1975)
Mirabilis	Reznik (1955)	*Spraguea*	Mabry et al. (1963)
Nyctaginia	Mabry et al. (1963)	*Talinella*	Behnke et al. (1975)
Oxybaphus	Gertz (1906)	*Talinopsis*	Behnke et al. (1975)
Petiveriaceae		*Talinum*	Forsyth and Simmonds (1954)
Petiveria	Forsyth and Simmonds (1954)	Stegnospermataceae	
Rivina	Reznik (1955)	*Stegnosperma*	Mabry et al. (1963)

betalamic acid, an intermediate in the formation of betalains, is produced by the cleavage of L-DOPA between the 4- and 5-positions and subsequent recyclization, forming the characteristic dihydropyridine ring (Fischer and Dreiding 1972; Impellizzeri and Piattelli 1972). The intermediate in the conversion of L-DOPA to betalamic acid, 4,5-secodopa, has been detected in an enzyme extract of *Amanita muscaria* (Terradas and Wyler 1991a,b), as has another cleavage product, 2,3-secodopa. The enzyme responsible for the 4,5-cleavage of L-DOPA (DOPA 4,5-dioxygenase) in *A. muscaria* has recently been isolated (Girod and Zryd 1991a). Ring-cleavage of tyrosine derivatives occurs in a number of species of flowering plants (Ellis 1973), generally by degradation of homogentisic acid (Durand and Zenk 1974; Mazelis 1980). The accumulation of recyclized product seems to occur less frequently. The only other enzymes in higher plants known to use L-DOPA as preferred substrate in cleavage and recyclization are stizolobic acid synthase and stizolobinic acid synthase from *Stizolobium* spp. (Leguminoseae) (Hattori and Komamine 1959; Saito and Komamine 1978). Interestingly, stizolobic acid and stizolobinic acid are also recyclized products of the 4,5-cleavage and 2,3-cleavage, respectively, of L-DOPA (Saito and Komamine 1975).

Early tracer studies on betacyanins (Sciuto et al. 1972, 1974) suggested that glycosylation could occur either before or after condensation of cyclodopa with betalamic acid. Wyler et al. (1984) have reported the accumulation of cyclodopa glucoside in *B. vulgaris*, while Heuer and Strack (1992) have synthesized betanin from betanidin and UDP-glucose by a protein preparation obtained from cell cultures of *Dorotheanthus bellidiformis*. This work suggests that, though both pathways (glycosylation of cyclodopa and glycosylation of betanidin) may be present, one or the other may be absent, depending on the plant (Steglich and Strack 1990). However, much of the earlier work may be confounded by the presence of decolorizing enzymes similar to those discussed by Elliot et al. (1983). These enzymes could lead to the accumulation of compounds assumed to be precursors.

Much of the diversity in betacyanin structures is due to glycosylation and acylation of the betanidin nucleus (Steglich and Strack 1990). Aside from the betanidin glucosyltransferase mentioned above, an enzyme responsible for acylation of amaranthin (betanidin 5-*O*-glucuronosylglucose) to form celosianin I (*p*-coumaroylamaranthin) and celosianin II (feruloylamaranthin) has been purified from *Chenopodium rubrum* (Bokern et al. 1992). Instead of addition of the hydroxycinnamic acids (HCAs) via the more common acyl donor, an HCA-CoA thioester, this enzyme utilizes HCA-glucose esters as donor molecules. This is presumably the only mechanism of acylation in betalain-producing plants (Bokern et al. 1992).

Interest in the hereditary basis of betalain production has only recently been renewed since Keller's (1936) original work on *B. vulgaris*. Wolyn and Gabelman (1989) describe a triallelic locus (R) which determines the ratio of betacyanin to betaxanthin in the roots of *B. vulgaris*, supposedly by regulating the synthesis of cyclodopa. Trezzini and Zryd (1990), using *Portulaca grandiflora* as a model, suppose two loci (R and C, respectively) for the control of synthesis of cyclodopa and betalamic acid, and introduce a third locus (I) for inhibition of betaxanthin production. However, there is still no agreement on whether the condensation reaction with betalamic acid leading to betalains in plants is spontaneous or enzymatically controlled. Workers have shown that the betaxanthins formed upon feeding of aminated precursors cannot be predicted either by the nature of the fed precursor or by the betaxanthins normally produced in the plant (Sciuto et al. 1974; Rink and Böhm 1985). Miller et al. (1968) and Wyler and Meuer (1979) have found that the feeding of labelled L-DOPA and dopaxanthin (labelled on the DOPA moiety) in *Opuntia* spp. gives much greater incorporation into the dihydropyridine ring than the dihydroindole moiety.

The accumulation of tyrosine immediately prior to the onset of pigmentation during floral development in *P. grandiflora* (Kishima et al. 1991) suggests that betalain biosynthesis (at least in this system) is regulated by availability of tyrosine and not L-DOPA.

Recent work on the physiology and enzymology of betalain biosynthesis has emphasized cell culture techniques (Bokern et al. 1991; Girod and Zryd 1991b; Böhm et al. 1991; Heuer and Strack 1992; see also Böhm and Rink 1988 for a review of betalain production in cell culture).

10.3 Evolutionary Significance of Betalains

It was recognized that members of a taxonomically distinct group of plants contained an unusual red-violet pigment several decades before the structural elucidation of betanidin, Bischoff being the first to equate the appearance of *Caryophyllinenroth* (betacyanin) with the Centrospermae (Bischoff 1876; see also Wyler and Dreiding 1961a; Dreiding 1961; and Mabry 1964 for historical reviews). The results of these early chemical tests have been confirmed by the more sophisticated techniques of chromatography and electrophoresis (Reznik 1955, 1957; Wyler and Dreiding 1961b). The term "betacyanin" was first coined by Schudel (1918), though the term "nitro-

Table 10.2. Relevant taxa known to contain anthocyanins

Genus	Reference
Caryophyllaceae	
Alsinidendron	Beck et al. (1962)
Arenaria	Beck et al. (1962)
Dianthus	Lawrence et al. (1939)
Geocarpon	Bogle et al. (1971)
Lychnis	Lawrence et al. (1939)
Melandrium	Lawrence et al. (1939)
Saponaria	Lawrence et al. (1939)
Silene	Lawrence et al. (1939)
Spergularia	Beck et al. (1962)
Viscaria	Reznik (1957)
Molluginaceae	
Hypertelis (questionable)	Beck et al. (1962)
Macarthuria	Behnke et al. (1983)
Mollugo[a]	Mears (1976)
Pharnaceum	Mabry (unpubl. data)
Non-caryophyllalean taxa	
Theligonaceae	Mabry et al. (1975)
Vivianiaceae	Behnke and Mabry (1977)

[a] Presence reported by Mears with no reference to experimental data.

genous anthocyanin" dominated much of the earlier literature (Robinson and Robinson 1932; Reznik 1955, 1957). Wyler and Dreiding (1961b), noting the co-occurrence of betacyanins and the yellow "flavocyanins" in Reznik's work, renamed the latter as betaxanthins.

The most striking aspect of betalain chemistry in relation to taxonomy is the mutual exclusion of this class of pigments and the otherwise ubiquitous anthocyanins. All centrospermous plants produce an array of flavonoids (Richardson 1978; Burret et al. 1981), including those yielding proanthocyanidin/leucoanthocyanidin hydrolysates (Bate-Smith 1962; Bittrich and Amaral 1991), but anthocyanins have been found only in members of the Caryophyllaceae and Molluginaceae (Table 10.2). Neither of these families has been shown to have species producing betalains, as found in all other pigmented members of the order (Table 10.1). It has been suggested both that the betalain- and anthocyanin-producing lineages diverged prior to the origin of floral pigmentation (Mabry 1973), and, alternatively, that the ability to produce betalains evolved subsequent to the loss of anthocyanin production in the betalain taxa (Ehrendorfer 1976). However, recent molecular treatments (see Chaps. 8 and 9) do not support the basal placement of the Caryophyllaceae and Molluginaceae. This being the case, the circumstances regarding the gain and loss of the two independent pathways remains a mystery.

10.4 Value of Chemotaxonomic Data in Studies of the Caryophyllales

The occurrence of bound ferulic acid in the unlignified cell walls (Harris and Hartley 1980; Hartley and Harris 1981) is occasionally regarded as a feature distinguishing the Caryophyllales from the rest of the dicots and linking it to the monocots (see Chap. 1). However, the application of more sensitive analytical techniques in a similar study on the Coniferae (Strack et al. 1988) shows that, while a particular hydroxycinnamic acid may predominate in the insoluble fraction of the cell wall, another HCA is often present in small quantities. A more careful analysis of the plants studied by Hartley and Harris may give similar results. Also, Strack and coworkers have shown that both *p*-coumaric and ferulic acids are incorporated into the insoluble portion of the cell war in significant amounts in cell-suspension cultures of *Chenopodium rubrum* (Bokern et al. 1991). It may be that the general ability to fix HCAs to the cell wall is of more phylogenetic significance than which cell wall conjugates are most readily detected.

Attempts have been made to utilize flavonoid data in the determination of within-family relationships (Cactaceae, Burret et al. 1982; Chenopodiaceae, Sanderson et al. 1988; Didiereaceae, Rabesa 1982; Portulacaceae, Nyananyo 1986). However, these investigations are largely anecdotal and provide little insight regarding phylogenetic relationships. Flavonoids have also shown to be of little use at higher taxonomic levels (M. Richardson pers. comm.), with the exception of *C*-glycosylflavonoids, which seem to be absent in the Aizoaceae and Cactaceae (Richardson 1978, 1981; Burret et al. 1981). Of minor chemotaxonomic interest, Paré et al. (1991) have reported the first known aurone phytoalexin (inducible antimicrobial compound), isolated from *Cephalocereus senilis* (Cactaceae). This is also the first report of an aurone appearing in the Caryophyllales. Flavonoids of the Caryophyllidae have been summarized by Giannasi (1988).

The distribution of Δ^7-, Δ^5- and Δ^0-sterols in the Caryophyllidae has recently been reviewed by Salt et al. (1991). There seems to be a general trend towards evolutionary reduction in accumulation of Δ^5- and Δ^0-sterols in the Caryophyllales, but there is no discernable pattern among families.

10.5 Current and Future Studies

With the increasing threat of global climate change and broad-scale desertification, interest is coming to focus on the use of drought-adapted plants, such as many members of the Caryophyllales, as possible food sources. Such drought-tolerant species as amaranth, quinoa and prickly pear are already in wide use. With increased reliance on these species, an understanding of their disease resistance must be sought. Current research in Mabry's laboratory has emphasized antiviral ribosome-inactivating proteins (RIPs) in *Phytolacca dodecandra* (Bonness and Mabry 1992) and phytoalexins in *Cephalocereus senilis* (Paré et al. 1991, 1992).

In the field of betalain research, Zryd and coworkers have raised polyclonal antibodies to DOPA 4,5-dioxygenase from *Amanita muscaria* and are currently screening corresponding enzymes from members of the Caryophyllales using both immunoprecipitation technique and Western blotting (Hinz et al. 1992). Wyler's laboratory is currently working on the characterization of DOPA 2,3-dioxygenase from *A. muscaria*. Additional enzymes involving the glycosylation and acylation of betacyanins are being sought by workers in Strack's laboratory, as is an understanding of the biochemical or physiological basis for their regulation (D. Strack, pers. comm.).

Also of interest is the recent cloning of the *Candi* gene from *Antirrhinum majus* (Martin et al. 1991) and the *A2* gene from *Zea mays* (Menssen et al. 1990). Both of these are believed to be structural genes responsible for the conversion of leucoanthocyanidin to anthocyanidin (Fig. 10.2). Using modern molecular techniques it is now possible to probe betalain-producing species for vestiges of anthocyanin synthesis. The taxonomic distribution of sequences homologous to the *Candi/A2* genes may help in resolving systematic relationships within Caryophyllales, especially regarding problematic unpigmented taxa (Table 10.3). Also, these sequences can be compared

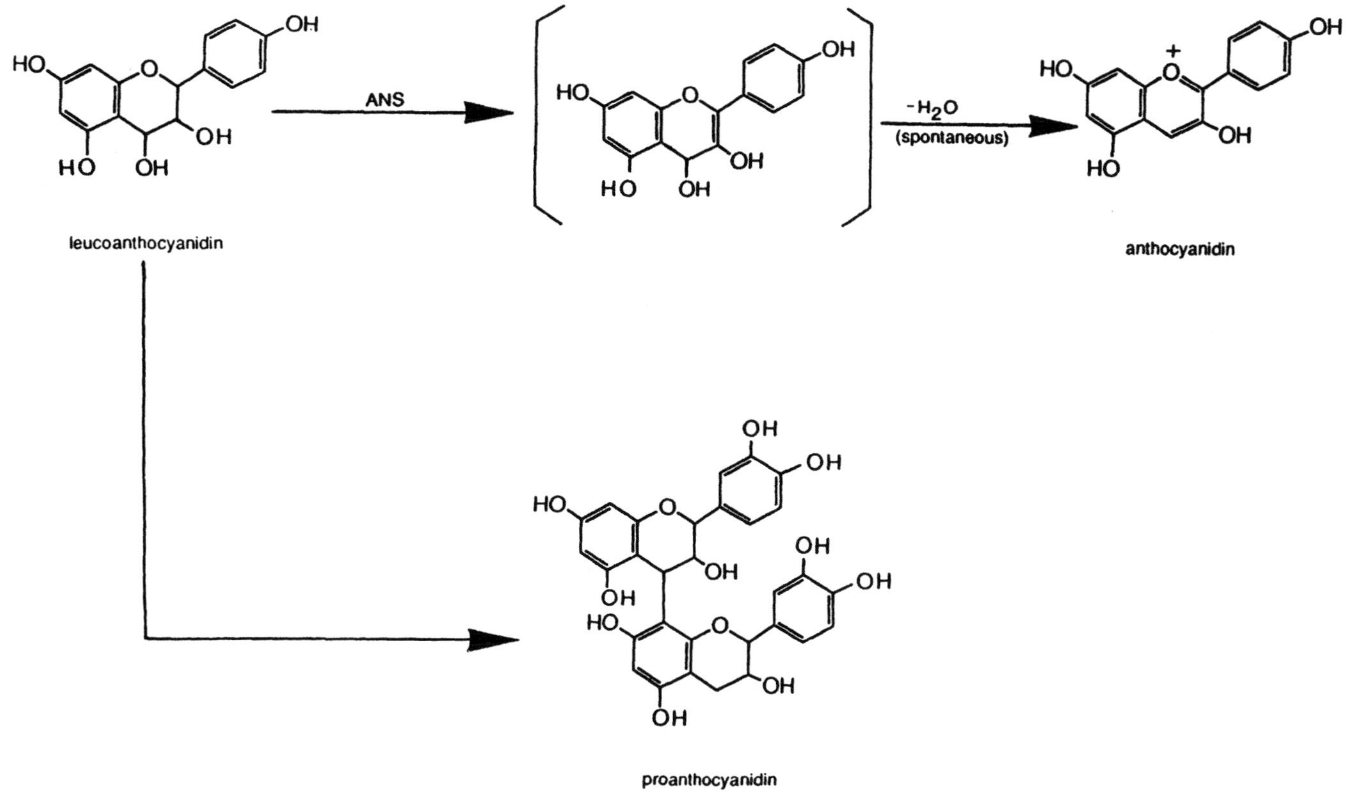

Fig. 10.2. Steps in anthocyanin biosynthesis showing the conversion of leucoanthocyanidin to anthocyanidin and proanthocyanidin. Although the enzymes involved in these steps have not yet been characterized, it has been proposed that the *Candi* gene in *Antirrhinum majus* and the *A2* gene in *Zea mays* code for an enzyme oxidizing the 2,3-bond in leucoanthocyanidin (tentatively named anthocyanidin synthase, ANS) and there is subsequently the spontaneous loss of water from the flavan 2-en-3,4-diol intermediate to form anthocyanidin (C. Martin pers. comm.; G. Forkmann pers. comm.). It is not yet known whether the condensation to proanthocyanidin is spontaneous or enzymatically controlled (Stafford 1989)

Table 10.3. Caryophyllalean taxa not known to produce pigment in detectable quantities

Achatocarpaceae
Barbeuia
Limeum
Lophiocarpus
Most members of Molluginaceae

with sequences from monocots and other dicots, potentially yielding information on the evolution of pigment biosynthesis. This presents a unique opportunity to combine molecular genetic information (presence or absence of *Candi/A2* homologues) with a phenotypic character of systematic and evolutionary relevance (presence or absence of anthocyanins) in an area of research that has already engendered nearly three decades of debate.

Acknowledgements. The authors wish to thank Cathie Martin and Gert Forkmann for their comments on anthocyanin biosynthesis, Hans Reznik for his list of betalain-containing species and Dieter Strack, Mick Richardson and W.H. Gabelman for their comments on the manuscript. Some of the work described in this report was supported by the National Institutes of Health (GM 35710) and the Robert A. Welch Foundation (Grant F-130).

References

Bate-Smith EC (1962) The phenolic constituents of plants and their taxonomic significance. J Linn Soc Bot 58:95–173

Beck E, Merxmüller H, Wagner H (1962) Über die Art der Anthocyane bei Plumbaginaceen, Alsinoideen, und Molluginaceen. Planta 58:220–224

Behnke H-D, Mabry TJ (1977) S-type sieve-element plastids and anthocyanins in Vivianiaceae: Evidence against its inclusion into Centrospermae. Plant Syst Evol 126:371–375

Behnke H-D, Chang C, Eifert IJ, Mabry TJ (1974) Betalains and P-type sieve-tube plastids in *Petiveria* and *Agdestis* (Phytolaccaceae). Taxon 23:541–542

Behnke H-D, Mabry TJ, Eifert IJ, Pop L (1975) P-type sieve-element plastids and betalains in Portulacaceae (including *Ceraria*, *Portulacaria*, and *Talinella*). Can J Bot 53:2103–2109

Behnke H-D, Mabry TJ, Neumann P, Barthlott W (1983) Ultrastructural, micromorphological and phytochemical evidence for a "central position" of *Macarthuria* (Molluginaceae) within the Caryophyllales. Plant Syst Evol 143:151–161

Bischoff H (1876) Das Caryophyllinenroth. Inaugural Dissertation. Universität Tübingen

Bittrich B, Amaral MCE (1991) Proanthocyanidins in the testa of centrospermous seeds. Biochem Syst Ecol 19:319–321

Bogle AL, Swain T, Thomas RD, Kohn ED (1971) *Geocarpon*: Aizoaceae or Caryophyllaceae? Taxon 20:473–477

Böhm H, Rink E (1988) Betalains. In: Constabel F, Vasil IK (eds) Cell culture and somatic cell genetics of plants, vol 5. Academic Press, New York, pp 449–463

Böhm H, Böhm L, Rink E (1991) Establishment and characterization of a betaxanthin-producing cell culture from *Portulaca grandiflora*. Plant Cell Tissue Organ Cult 26:75–82

Bokern M, Wray V, Strack D (1991) Accumulation of phenolic acid conjugates and betacyanins, and changes in the activities of enzymes involved in feruloylglucose metabolism in cell-suspension cultures of *Chenopodium rubrum* L. Planta 184: 261–270

Bokern M, Heuer S, Strack D (1992) Hydroxycinnamic acid transferases in the biosynthesis of acylated betacyanins: Purification and characterization from cell cultures of *Chenopodium rubrum* and occurrence in some other members of the Caryophyllales. Bot Acta 105:146–151

Bonness MS, Mabry TJ (1992) Tissue culture of endod (*Phytolacca dodecandra* L'Herit): Growth and production of ribosome-inactivating proteins. Plant Cell Rep 11:66–70

Burret F, Rabesa Z, Zandonella P, Viorin B (1981) Contribution biochimique à la systématique de l'ordre des Centrospermales. Biochem Syst Ecol 9:257–262

Burret F, Lebreton Ph, Voirin B (1982) Les aglycones flavoniques des cactées: Distribution, signification. J Nat Prod 45:687–693

Döpp H, Maurer S, Sasaki AN, Musso H (1982) Die Konstitution der Musca-aurine. Liebigs Ann Chem 1982:254–264

Dreiding AS (1961) The betacyanins, a class of red pigments in the Centrospermae. In: Ollis WE (ed) Recent developments in the chemistry of natural phenolic compounds. Pergamon Press. New York, pp 194–211

Durand R, Zenk MH (1974) Enzymes of the homogentisate ring-cleavage pathway in cell suspension cultures of higher plants. FEBS Lett 39:218–220

Elliot DC, Schultz CG, Cassar RA (1983) Betacyanin decolourizing enzyme in *Amaranthus tricolor* seedlings. Phytochemistry 22:383–387

Ellis BE (1973) Catabolic ring-cleavage of tyrosine in plant cell cultures. Planta 111: 113–118

Ehrendorfer (1976) Closing remarks: Systematics and evolution of centrospermous families. Plant Syst Evol 126:99–106

Fischer N, Dreiding AS (1972) Biosynthesis of betalaines. On the cleavage of the aromatic ring during the enzymatic transformation of dopa into betalamic acid. Helv Chim Acta 55:649–658

Forsyth WGC, Simmonds NW (1954) A survey of the anthocyanins of some tropical plants. Proc R Soc Lond Ser B 142:549–564

Gertz O (1906) Studier övfer Anthocyan. Dissertation, Lund

Giannasi DE (1988) Flavonoids and evolution in the dicotyledons. In: Harborne JB (ed) The flavonoids: Advances in research since 1980. Chapman and Hall, London, pp 479–504

Girod PA, Zryd JP (1991a) Biogenesis of betalains: Purification and partial characterization of dopa 4,5-dioxygenase from *Amanita muscaria*. Phytochemistry 30: 169–174

Girod PA, Zryd JP (1991b) Secondary metabolism in cultured red beet (*Beta vulgaris* L.) cells: Differential regulation of betaxanthin and betacyanin biosynthesis. Plant Cell Tissue Organ Cult 25:1–12

Harris PJ, Hartley RD (1980) Phenolic constituents of the cell walls of monocotyledons. Biochem Syst Ecol 8:153–160

Hartley RD, Harris PJ (1981) Phenolic constitutents of the cell walls of dicotyledons. Biochem Syst Ecol 9:189–203

Hattori S, Komamine A (1959) Stizolobic acid: A new amino-acid in *Stizolobium hassjoo*. Nature 183:1116–1117

Heuer S, Strack D (1992) Synthesis of betanin from betanidin and UDP-glucose by a protein preparation from cell suspension cultures of *Dorotheanthus bellidiformis* (Burm. f.) N.E.Br. Planta 186:626–628

Hinz U, Richard C, Bindschedler L, Zryd JP (1992) Betalain synthesis in *A. muscaria*. Abstract. Experientia 48:A10

Hörhammer L, Wagner H, Gritzsche W (1964) Zur Biosynthese der Betacyane I. Biochem Z 339:398–400

Hunziker JH, Behnke HD, Eifert IJ, Mabry TJ (1974) *Halophytum ameghinoi*: a betalain-containing and P-type sieve-tube plastid species. Taxon 23:537–539

Impellizzeri G, Piatelli M (1972) Biosynthesis of betalains: Formation of indicaxanthin in *Opuntia ficus-indica* fruits. Phytochemistry 11:2499–2502
Impellizzeri G, Piattelli M, Sciuto S (1973) A new betaxanthin from *Glottiphyllum longum*. Phytochemistry 12:2293–2294
Keller W (1936) Inheritance of some major color types in beets. J Agric Res 52:27–38
Kimler L, Mears J, Mabry TJ, Rösler H (1970) On the question of the mutual exclusiveness of betalains and anthocyanins. Taxon 19:875–878
Kishima Y, Suiko M, Adachi T (1991) Betalain pigmentation in petals of *Portulaca* is preceded by a dramatic tyrosine accumulation. J Plant Physiol 137:505–506
Kull J, Kühn B (1978) Vieraea 8:187 cited in: Hegnauer R Chemotaxonomie der Pflanzen, vol VIII. Birkhäuser, Basel, pp 28–32
Kryz F (1926) Ein Beitrag zur Kenntnis des Vorkommens und der Reaktionen der Anthocyanfarbstoffe der Rübenrotgruppe. Oester Chem Ztg 23:55–56
Lawrence WJC, Price JR, Robinson GM, Robinson R (1939) The distribution of anthocyanins in flowers, fruits and leaves. Philos Trans R Soc Lond Ser B 1939–1941(230):149–178
Mabry TJ (1964) The betacyanins, a new class of red violet pigments, and their phylogenetic significance. In: Leone CA (ed) Taxonomic biochemistry and serology. Roland Press, New York, pp 239–253
Mabry TJ (1966) The betacyanins and betaxanthins. In: Swain T (ed) Comparative phytochemistry. Academic Press, London, pp 231–244
Mabry TJ (1973) Is the order Centrospermae monophyletic? In: Bendz G, Santesson J (eds) Chemistry in botanical classification. Nobel Foundation, Stockholm and Academic Press, London, pp 275–285
Mabry TJ, Behnke H-D (1976) Betalains and P-type sieve-element plastids: The systematic position of *Dysphania* R. Br. (Centrospermae). Taxon 25:109–111
Mabry TJ, Dreiding AS (1968) The betalains. In: Mabry TJ, Alston RE, Runeckles VC (eds) Recent advances in phytochemistry, vol 1. Appleton-Century-Crofts, New York, pp 145–160
Mabry TJ, Wyler H, Sassu G, Mercier M, Parikh I, Dreiding AS (1962) Die Struktur des Neobetanidins: Über die Konstitution des Randenfarbstoffes Betanin. Helv Chim Acta 45:640–647
Mabry TJ, Taylor A, Turner BL (1963) The betacyanins and their distribution. Phytochemistry 2:61–64
Mabry TJ, Wyler H, Parikh I, Dreiding AS (1967) The conversion of betanidin and betanin to neobetanidin derivatives. Tetrahedron 23:3111–3127
Mabry TJ, Eifert IJ, Chang C, Mabry H, Kidd C, Behnke H-D (1975) Theligonaceae: Pigment and ultrastructural evidence with excludes it from the order Centrospermae. Biochem Syst 3:53–55
Mabry TJ, Behnke H-D, Eifert IJ (1976) Betalains and P-type sieve-element plastids in *Gisekia* L. (Centrospermae). Taxon 25:112–114
Mabry TJ, Neuman P, Philipson WR (1978) *Hectorella*: a member of the betalain-suborder Chenopodiineae of the order Centrospermae. Plant Syst Evol 130:163–165
Martin C, Prescott A, Mackay S, Bartlett J, Vrijlandy E (1991) Control of anthocyanin biosynthesis in flowers of *Antirrhinum majus*. Plant J 1:37–49
Mazelis M (1980) Amino acid catabolism. In: Stumpf PK, Conn EE (eds) The biochemistry of plants. Vol 5. Amino acids and derivatives. Academic Press, New York, pp 542–567
Mears JA (1976) Guide to research in plant taxonomy. Chem Plant Taxon Newslett 25:11–14
Menssen A, Höhmann S, Martin W, Schnable PS, Peterson PA, Saedler H, Gierl A (1990) The En/Spm transposable element of *Zea mays* contains splice sites at the termini generating a novel intron from a dSpm element in the *A2* gene. EMBO J 9:3051–3057
Miller HE, Rösler H, Wohlpart A, Wyler H, Wilcox ME, Frohofer H, Mabry TJ, Dreiding AS (1968) Biogenese der Betalaine: Biotransformation von Dopa und Tyrosin in den Betalaminsäureteil des Betanins. Helv Chim Acta 51:1470–1474

Minale L, Piattelli M, Nicolaus RA (1965) Pigments of Centrospermae. IV. On the biogenesis of indicaxanthin and betanin in *Opuntia ficus-indica* Mill. Phytochemistry 4:593–597

Nyananyo (1986) A survey of leaf flavonoids in the Portulacaceae. Biochem Syst Ecol 14:633–635

Paré PW, Dmitrieva N, Mabry TJ (1991) Phytoalexin aurone induced in *Cephalocereus senilis* liquid suspension culture. Phytochemistry 30:1133–1135

Paré PW, Mischke CF, Edwards R, Dixon RA, Norman HA, Mabry TJ (1992) Induction of phenylpropanoid pathway enzymes in elicitor-treated cultures of *Cephalocereus senilis*. Phytochemistry 31:149–153

Piattelli M, Impellizzeri (1970) 2-Descarboxybetanidin, a minor betacyanin from *Carpobrotus acinaciformis*. Phytochemistry 9:2553–2556

Piattelli M, Imperato F (1969) Betacyanins of the family Cactaceae. Phytochemistry 8:1503–1507

Piattelli M, Imperato F (1971) Betacyanins of some Chenopodiaceae. Phytochemistry 10:3133–3134

Piattelli M, Minale L (1964a) Pigments of Centrospermae. I. Betacyanins from *Phyllocactus hybridus* Hort. and *Opuntia ficus-indica* Mill. Phytochemistry 3:307–311

Piattelli M, Minale L (1964b) Pigments of Centrospermae. II. Distribution of betacyanins. Phytochemistry 3:547–557

Piattelli M, Minale L, Prota G (1964) Isolation, structure and absolute configuration of indicaxanthin. Tetrahedron 22:2325–2329

Rabesa ZA (1982) Weitere Untersuchungen an Didiereaceen. Teil 5. Recherche chimiosystématiques sur les flavonoides des Didiéréacées. Tropische und subtropische Pflanzenwelt 37:1–26. Steiner, Mainz

Rauh W, Reznik H (1961) Zur Frage der systematischen Stellung der Didiereaceen. Bot Jahrb 81:94–105

Reznik H (1955) Die Pigmente der Centrospermen als systematisches Element. Z Bot 43:499–530

Reznik H (1957) Die Pigmente der Centrospermen als systematisches Element. II. Untersuchungen über das ionophoretische Verhalten. Planta 49:406–434

Reznik H (1978) Das Vorkommen von Betalaminsäure bei Centrospermen. Z Pflanzenphysiol 87:95–102

Reznik H, Engel U, Wambach C (1988) Untersuchungen über die Betalainfarbstoffe der Aizoaceae. I. Die Betaxanthine der Ruschieae. Beitr Biol Pflanz 63:209

Richardson M (1978) Flavonols and C-glycosylflavonoids of the Caryophyllales. Biochem Syst Ecol 6:283–286

Richardson PM (1981) Flavonoids of some controversial members of the Caryophyllales (Centrospermae). Plant Syst Evol 138:227–233

Rink E, Böhm H (1985) Changed betaxanthin pattern in violet flowers of *Portulaca grandiflora* after the feeding of dopa. Phytochemistry 24:1475–1477

Robinson AM, Robinson R (1932) Synthetical experiments on the nature of betanin and related nitrogenous anthocyanins. J Chem Soc 1932:1439–1445

Saito K, Komamine A (1975) Biosynthesis of stizolobinic acid and stizolobic acid in the etiolated seedlings of *Stizolobium hassjoo*. Z Naturforsch 30c:659–662

Saito K, Komamine A (1978) Biosynthesis of stizolobinic acid and stizolobic acid in higher plants. Eur J Biochem 82:385–392

Salt TA, Xu S, Patterson GW, Adler JH (1991) Diversity of sterol biosynthetic capacity in the Caryophyllidae. Lipids 26:604–613

Sanderson SC, Ge-Ling C, McArthur ED, Stutz HC (1988) Evolutionary loss of flavonoids and other chemical characters in the Chenopodiaceae. Biochem Syst Ecol 16:143–149

Schmidt OT, Schönleben W (1957) Zur Kenntnis der Farbstoffe der Roten Rübe, II. Z Naturforsch 12b:262–263

Schmidt OT, Becher P, Hubner M (1960) Zur Kenntnis der Farbstoffe der Roten Rübe, III. Chem Ber 93:1296–1304

Schudel G (1918) Über die Anthocyane von *Beta vulgaris* L. (var. *rapacea* Koch, *forma rubra* L.) und *Raphanus sativus* L. (var. *radicula* Pers.). Promotionsarbeit, ETH, Zürich

Sciuto S, Oriente G, Piattelli M (1972) Betanidin glucosylation in *Opuntia dillenii*. Phytochemistry 11:2259–2262

Sciuto S, Oriente GP, Piatteli M, Impellizzeri G, Amico V (1974) Biosynthesis of amaranthin in *Celosia plumosa*. Phytochemistry 13:947–951

Seeligman P, Kraponikas A (1969) Report of current chemosystematic studies at the Department of Botany, Instituto Migvel Lillo, Universidad Nacional de Tucumán S.M. de Tucumán, Argentina. Chem Plant Taxon Newslett 13:8–11

Stafford HA (1989) The enzymology of proanthocyanidin biosynthesis. In: Hemingway RW, Karchesy JJ (eds) The chemistry and significance of condensed tannins. Plenum Press, New York, pp 47–70

Steglich W, Strack D (1990) Betalains. In: Brossi A (ed) The alkaloids, vol 39. Academic Press, New York, pp 1–62

Strack D, Heilemann J, Mömken M, Wray V (1988) Cell wall-conjugated phenolics from Coniferae leaves. Phytochemistry 27:3517–3521

Strack D, Marxen N, Reznik H, Ihlenfeld HD (1990) Distribution of betacyanins and hydroxycinnamic acid-glucose esters in flowers of the Ruschieae. Phytochemistry 29:2175–2178

Taylor TWJ (1940) Plant pigments in the Galápagos Islands. Proc R Soc Lond Ser B 129:230–237

Terradas F, Wyler H (1991a) 2,3- and 4,5-secodopa, the biosynthetic intermediates generated from L-DOPA by an enzyme system extracted from the fly agaric, *Amanita muscaria* L., and their spontaneous conversion to muscaflavin and betalamic acid, respectively, and betalains. Helv Chim Acta 74:124–140

Terradas F, Wyler H (1991b) The secodopas, natural pigments in *Hygrocybe conica* and *Amanita muscaria*. Phytochem 30:3251–3253

Trezzini GF, Zryd J-P (1990) *Portulaca grandiflora*: A model system for the study of the biochemistry and genetics of betalain synthesis. Acta Hortic 280:581–585

Weigert L (1894) Beiträge zur Chemie der rothen Pflanzenfarbstoffe. Jahresbericht und Programm der K.K. önologischen und pomologischen Lehranstalt in Klosterneuburg, Wien

Wohlpart A, Mabry TJ (1968) The distribution and phylogenetic significance of the betalains with respect to the Centrospermae. Taxon 17:148–152

Wolyn DJ, Gabelman WH (1989) Inheritance of root and petiole pigmentation in red table beet. J Hered 80:33–38

Wyler H, Dreiding AS (1957) Kristallisiertes Betanin. Helv Chim Acta 40:191–192

Wyler H, Dreiding AS (1959) Über die Konstitution des Randenfarbstoffes Betanin: Darstellung und Abbauprodukte des Betanidins. Helv Chim Acta 42:1699–1702

Wyler H, Dreiding AS (1961a) Phytolaccanin, der Farbstoff der Kermesbeere (*Phytolacca decandra* L.). 1. Mitteilung zur Kenntnis der Betacyane. Helv Chim Acta 44:249–257

Wyler H, Dreiding AS (1961b) Über Betacyane, die stickstoffhatigen Farbstoffe der Centrospermen. Vorläufige Mitteilung. Experientia 17:23–25

Wyler H, Meuer U (1979) Zur Biogenese der Betacyane: Versuche mit [2-^{14}C]-dopaxanthin. Helv Chim Acta 62:1330–1339

Wyler H, Mabry TJ, Dreiding AS (1963) Über die Konstitution des Randenfarbstoffes Betanin: Zur Struktur des Betanidins. Helv Chim Acta 46:1745–1748

Wyler H, Wilcox ME, Dreiding AS (1965) Umwandlung eines Betacyanes in ein Betaxanthin: Synthese von Indicaxanthin aus Betanin. Helv Chim Acta 48:361–366

Wyler H, Meuer U, Bauer J, Stravs-Mombelli L (1984) Cyclodopa glucoside (=(2S)-5-(β-D-glucopyranosyloxy)-6-hydroxyindoline-2-carboxylic acid) and its occurrence in red beet (*Beta vulgaris* var. *rubra* L.). Helv Chim Acta 67:1348–1355

11 Recent Advances in Betalain Analysis

DIETER STRACK and VICTOR WRAY

11.1 Introduction

Methodology for the analysis of betalains (betacyanins and betaxanthins) was, for many years, far behind research on the analogous water-soluble pigments, the anthocyanins (Strack and Wray 1989). The chemical nature of betalains was elucidated mainly by chemical means in the middle of this century (Wyler and Dreiding 1961). There are a number of chemical methods, which have been used extensively for structure elucidation (Steglich and Strack 1990), as well as a series of classical techniques, such as electrophoresis and column chromatography, which are still of importance.

Most of the betalains described in the 1960s have not been characterized by nuclear magnetic resonance spectroscopy (NMR) and mass spectrometry (MS), nor was it, until recently, feasible to gather exact quantitative data. With new, more sensitive techniques at hand, complete structural assignments and rapid quantifications of even small samples are possible.

The present article reviews advanced procedures for betalain analysis. Of late, research on natural products has received a significant impetus from recent, rapid developments in spectroscopic instrumentation. In particular, advances in the hardware and software for NMR and MS are such that it is often no longer necessary to resort to either derivatization or degradation in order to obtain structural information. In addition, methods of high performance liquid chromatographic (HPLC) separation have led in the last few years to a renaissance in the study of these compounds.

11.2 General Procedures

Methods for the isolation and characterization of betalains, such as layer chromatography, electrophoresis, and open column chromatography (CC), are reviewed elsewhere (Strack et al. 1993). Here we will only briefly consider CC which is, in most cases, a prerequisite to preparative work on betalains, e.g. in HPLC. The adsorbents commonly used are polyamide, strong acid ion-exchange resins and Sephadex ion-exchangers (citations of original papers are in Steglich and Strack 1990).

In the methodological development of betalain analysis the use of CC was a major step forward in the isolation of both betacyanins and betaxanthins. Selected examples of betalain CC are listed by Strack et al. (1993). A standard procedure utilizing this technique was developed by Piattelli and Minale (1964a,b). The aqueous plant extract is stirred with ion-exchange resin that adsorbs the betalains non-ionically. After washing the resin with 0.1% aqueous HCl, the pigments are eluted with water followed by a final separation and isolation of individual pigments on a polyamide column.

A standard CC procedure (prior to preparative HPLC) with crude extracts includes two successive CC steps: (1) ion exchange on Dowex 1X8 (stepwise gradient elution with water and increasing amounts of formic acid in water); and (2) gel filtration on Sephadex LH-20 with water and water–methanol mixtures (e.g. 1:1). This procedure was successfully applied in our work on betacyanins in flowers of *Lampranthus sociorum*, cell cultures of *Chenopodium rubrum* (Strack et al. 1988) and flowers of *Gomphrena globosa* (Heuer et al. 1992).

11.3 High Performance Liquid Chromatography

HPLC has become the method of choice in both analytical/quantitative and (semi-) preparative work on betalains. The most useful column support in HPLC is silica, the hydroxyl groups of which are derivatized with octyl- or octadecylsilyle groups (C_8- or C_{18}-reversed phase; particle size 3 to 10 μm). Useful solvents are water–methanol or water–acetonitrile mixtures, containing acetic, formic or phosphoric acid (Strack et al. 1993).

The key factors for separation are the overall polarity and stereochemistry of the betalains. In general, betaxanthins elute earlier than betacyanins. For the latter, not only is the nature and number of attached sugars as well as sugar acylation of particular importance, but also the site of attachment. Thus, betanin (betanidin 5-*O*-glucoside) exhibits a shorter retention time than gomphrenin I (betanidin 6-*O*-glucoside) (Heuer et al. 1992). The C-15 epimers (*S* and *R*) of betacyanins are easily separated from each other with the *S*-form eluting first. Possible neobetacyanins, such as neobetanin in fruits of *Opuntia ficus-indica* (Strack et al. 1987b), are clearly distinguishable by the lack of the typical second *R*-isomer peak and by sensitive detection at a lower wavelength (465 nm). Betacyanins are usually detected between 535 and 550 nm, the betaxanthins between 475 and 480 nm.

Table 11.1. Selected application data for (semi-) preparative reversed-phase HPLC of betalains

Application	Column[a]	Solvent[b]	Reference
Betanin[c]	μBondapak/Porasil B (35–75 μm; 610 × 7.8 mm i.d.)	MeOH-HOAc-0.05 *M* KH_2PO_4 (17.8 : 1 : 81.2; pH 2.75); 8 ml/min	Schwartz and von Elbe (1980)
Betanin, different betaxanthins	OPTI-UP C12 (65 × 25 mm i.d.)	Aqueous HOAc (1 m*M*, 100 m*M*); 4.5 ml/min	Trezzini and Zryd (1991a,b)
Humilixanthin	Multosorb-C_{18} (10 μm; 250 × 20 mm i.d.)	Gradient with 0.4% HCOOH and 50% MeCN; 5 ml/min	Strack et al. (1987a)
Betanin[d], gomphrenins	Silica C-18 (10 μm; 300 × 40 mm i.d.)	Gradient with 1% HCOOH and 80% MeOH; 20 ml/min	Heuer et al. (1992)

[a] Trade names of commercially available packing material.
[b] Abbreviations: HOAc, acetic acid; MeCN, acetonitrile; MeOH, methanol; i.d. = internal diameter
[c] In addition, isolation of cyclodopa 5-*O*-glucoside and betalamic acid was achieved; these were obtained after heating betanin (Schwartz and von Elbe 1983).
[d] See Fig. 11.1.

Acylation with hydroxycinnamic acids increases elution time in order of decreasing polarity of the ester moiety. By changing the detection wavelength to favour the detection of the acyl moiety, their number in the betacyanin molecule may be determined. In the case of monoacylation, peak ratios of absorbance at 535–540 to 315–325 nm should be about 2:1.

With regard to the separation of betaxanthin epimers, Terradas and Wyler (1991) were able for the first time to separate the C-11 epimers of a betaxanthin (*R*- and *S*-indicaxanthin). In contrast to the isomeric betacyanins, *R*-indicaxanthin exhibits a shorter retention time than *S*-indicaxanthin. The 11*R*-indicaxanthin has so far not been observed in nature.

There are several examples in the literature on semi-preparative and preparative reversed-phase HPLC of betalains, some of which are summarized in Table 11.1. Two isocratic systems are described by Schwartz and von Elbe (1980) and Trezzini and Zryd (1991a,b). The latter varied the amounts of acetic acid to improve separation. Betanin was isolated by using 1 m*M* acetic acid, whereas betaxanthins (indicaxanthin, vulgaxanthin I and portulacaxanthin III) were separated with 100 m*M* acetic acid. Taking advantage of this phenomenon, different amounts of acetic acid added to the solvents can be applied to optimize separations or even to design gradients (Strack and Reznik 1979). Two further examples of preparative HPLC in Table 11.1 describe gradient elutions with either aqueous acetonitrile (Strack et al. 1987a) or aqueous methanol (Heuer et al. 1992) against water

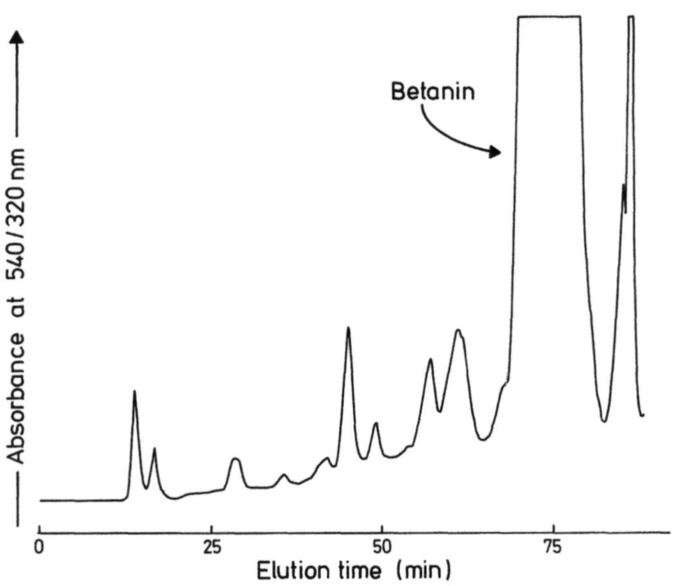

Fig. 11.1. HPLC purification of betanin (ca. 10 mg). Gradient elution: in 200 min from solvent A (1% aqueous formic acid) to solvent B (80% aqueous methanol); further details in Table 11.1; flow rate 20 ml min^{-1}

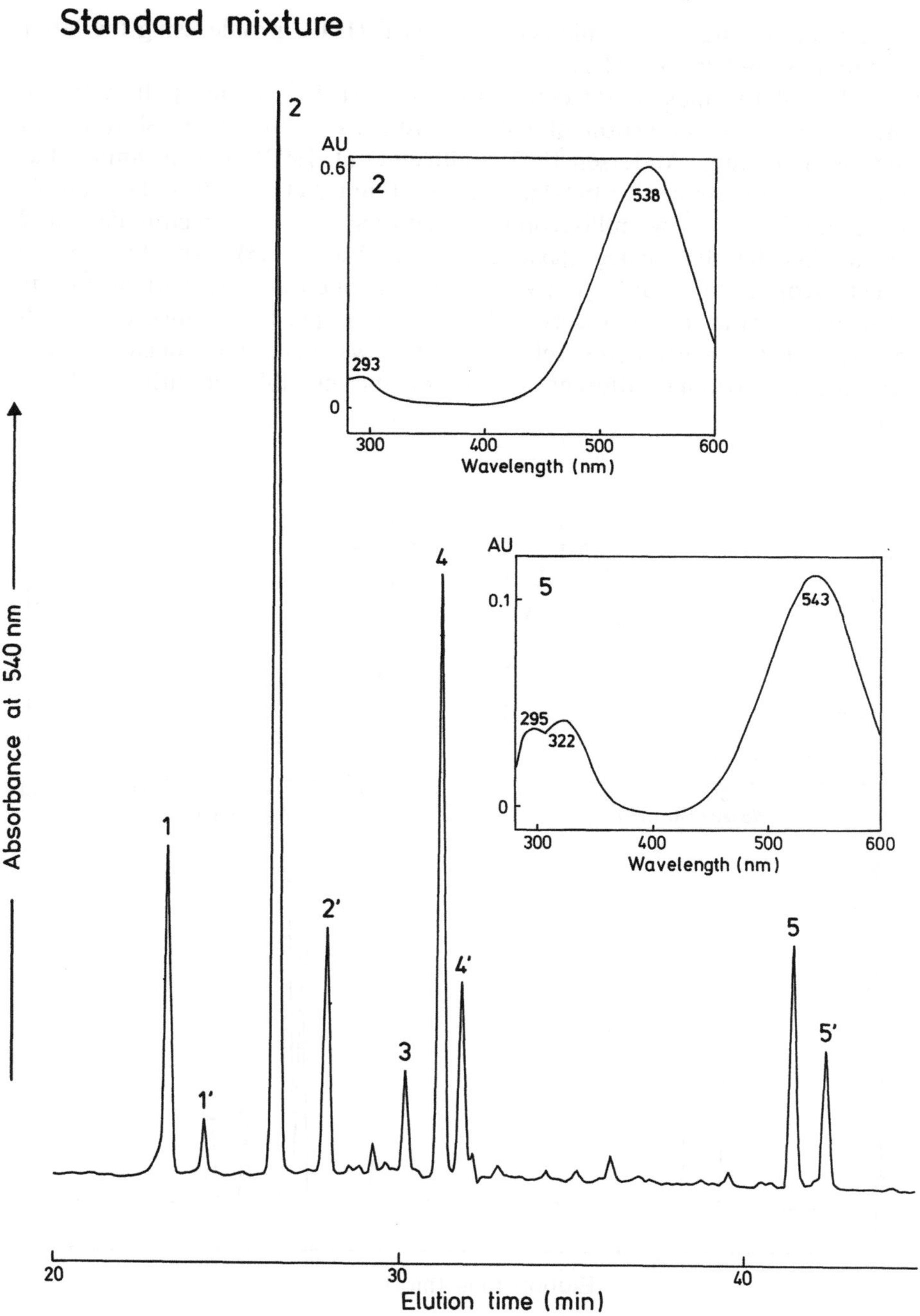

Fig. 11.2. HPLC separation of a betacyanin standard mixture. *Insets* On-line UV/vis. spectra of betanin (peak *2*) and lampranthin II (peak *5*). Peak identification: *1* and *1'*, amaranthin and isoamaranthin, respectively; (*2'*)*2*, (iso)betanin; *3*, celosianin I; (*4'*)*4*, (iso)celosianin II; (*5'*)*5*, (iso)lampranthin II. HPLC column: Nucleosil C_{18} (5 μm; 250 × 4 mm i.d.). Gradient elution: within 50 min from solvent A (1.5% H_3PO_4 in water) to solvent B (1.5% H_3PO_4, 20% acetic acid, 25% acetonitrile in water); flow rate 1 ml min^{-1}

with small amounts of formic acid. A typical HPLC profile using the latter system is shown in Fig. 11.1.

One of the most important advances in HPLC is the applications of photodiode-array detection of natural products, as has been shown with anthocyanins (e.g. Andersen 1985; Gläßgen et al. 1992). This technique has now also been applied to betalain analysis (Heuer et al. 1992). The sample is scanned every few milliseconds, generating UV/vis. spectral data and calculating the absorbance maxima (Figs. 11.2 and 11.3). With this on-line spectroscopy, the peak's purity can be examined by calculations of the absorbance ratio at two wavelengths or by multiple absorbance ratios calculated at all wavelengths relative to the wavelength of maximum absorbance, providing different spectra of the possible impurity and the analyte.

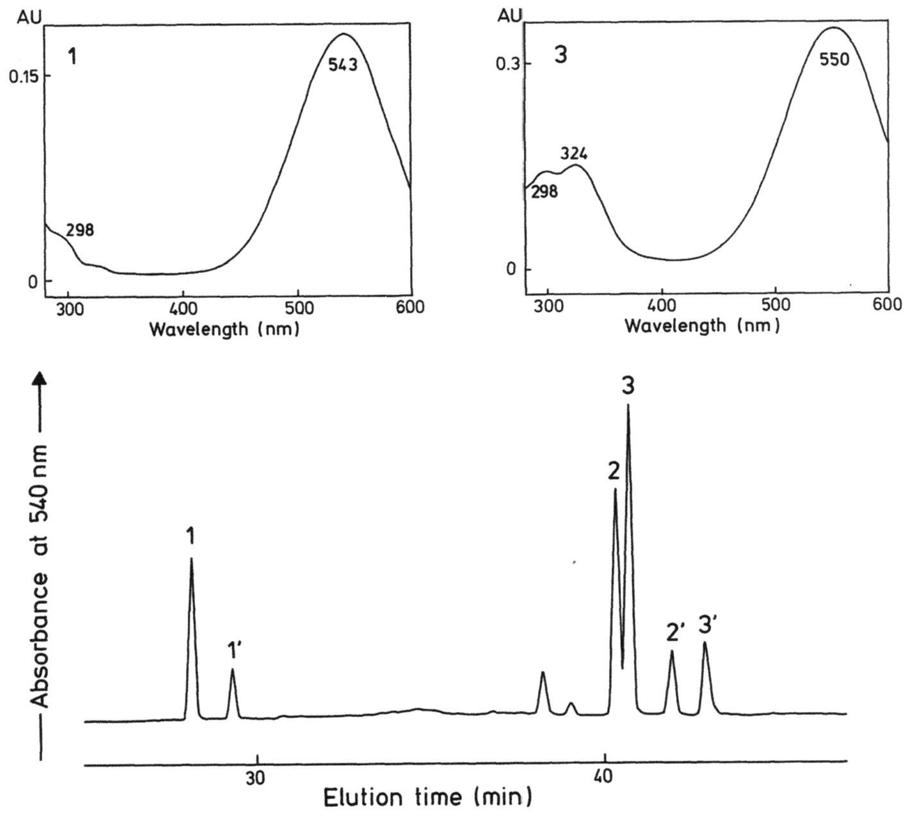

Fig. 11.3. HPLC separation of betacyanins (crude extract) from *Gomphrena globosa*. *Insets* On-line UV/vis. spectra of gomphrenin I (peak *1*) and gomphrenin III (peak *3*). Peak identification: *1* and *1'*, gomphrenin I and isogomphrenin I, respectively; (*2'*)*2*, (iso)gomphrenin II; (*3'*)*3*, (iso)gomphrenin III. Chromatographic conditions as in Fig. 11.2

With the on-line spectra obtained it is possible to easily detect betacyanins acylated with hydroxycinnamic acids. Comparison of the spectra of betanin and lampranthin II in Fig. 11.2 and gomphrenin I and gomphrenin III in Fig. 11.3 shows the additional absorbance bands at 322 nm (lampranthin II) and at 324 nm (gomphrenin III), in both compounds due to the esterification of the glucose moiety at C-6 with ferulic acid. The visible

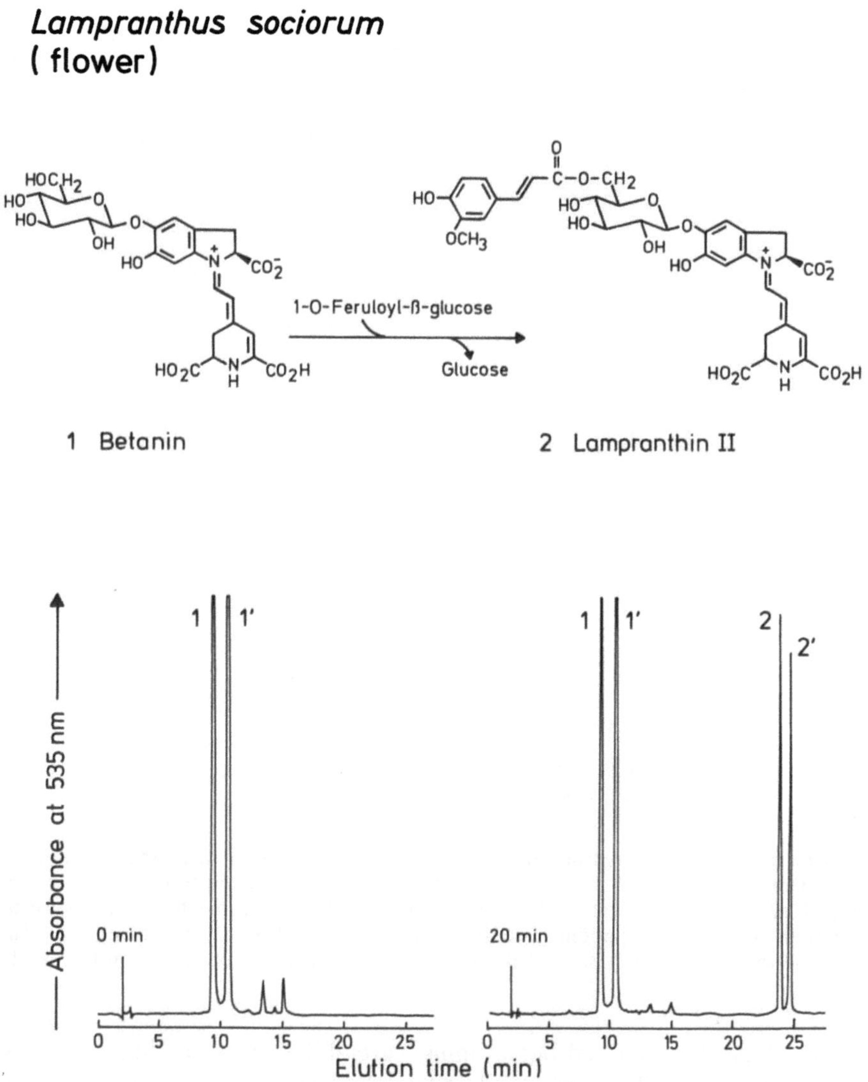

Fig. 11.4. Betanin hydroxycinnamoyltransferase HPLC assay immediately after introduction of protein (0 min; *left graph*; *1'*, isobetanin) and after 20 min reaction time at 30 °C (*right graph*; *2'* isolampranthin II) (Bokern and Strack 1988). Gradient elution: within 25 min from 10 to 60% solvent B in A; column as in Fig. 11.2

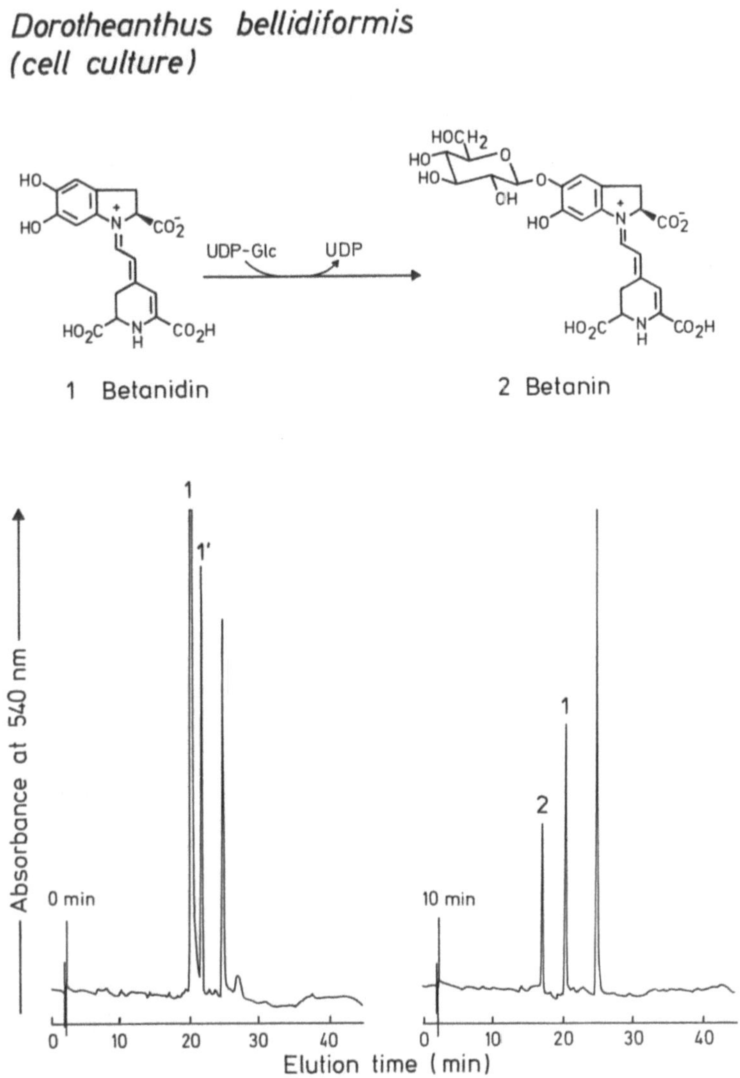

Fig. 11.5. Betanidin glucosyltransferase HPLC assay immediately after introduction of protein (0 min; *left graph*; *1'*, isobetanidin) and after 10 min reaction time at 30 °C (*right graph*) (Heuer and Strack 1992). The unidentified major peak came from an unknown degradation product of betanidin. Gradient elution: within 35 min from solvent A (1% formic acid in water) to 70% solvent B (80% aqueous methanol); column as in Fig. 11.2

maxima of the feruloylated betacyanins were shifted towards higher wavelengths compared with the non-acylated compounds, i.e. 5 nm with lampranthin II and 8 nm with gomphrenin III. This has been interpreted as being due to the interaction between the acyl moiety and the betanidin nucleus (Heuer et al. 1992).

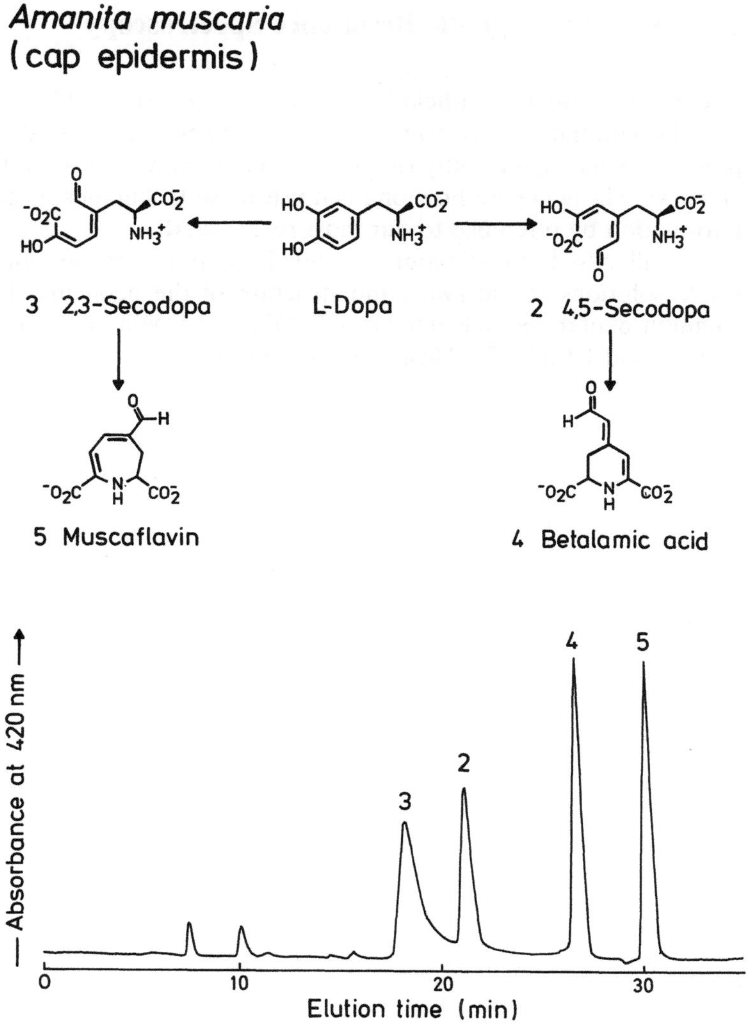

Fig. 11.6. Extradiol-cleavage HPLC assay with L-DOPA after a few days reaction time at 4 °C (Terradas and Wyler 1991). HPLC column: ODS Hypersil (5 μm; 25 × 5 mm i.d.). Gradient elution: 0.05 *M* Na_2HPO_4 buffer (pH 6) containing 0.03 *M* $(Bu_4N)H_2PO_4$ with 0–40% methanol within 25 min, then isocratic for 10 min; flow rate 0.9 ml min^{-1}

In addition, HPLC is the most powerful tool for the determination of enzyme activities in studies of betalain biosynthesis. For example, easy product quantification coupled with product identification has been demonstrated in studies on betacyanin acylation (Fig. 11.4) (Bokern and Strack 1988; Bokern et al. 1992), on betanidin glucosyltransferase (Fig. 11.5) (Heuer and Strack 1992) and on extradiol cleavage of L-dihydroxyphenylalanine (L-DOPA) in the formation of betalamic acid and muscaflavin via 4,5-secodopa and 2,3-secodopa, respectively (Fig. 11.6) (Terradas and Wyler 1991).

11.4 Nuclear Magnetic Resonance Spectroscopy

The introduction of high-field NMR spectrometers (up to 14 T, ^1H : 600 MHz) and the continued development of two-dimensional (2D) shift-correlation procedures has significantly simplified structural elucidation of the betalains. Here we will delineate how one can use these techniques to give structural information by reference to our most recent work.

In all NMR investigations acidified solutions are necessary to obtain stable solutions of the zwitterion structure of the betalains. The two most common deuterated solvents are CD_3OD and DMSO-d_6, containing traces of DCl and CF_3CO_2D. However, care must be taken as under these con-

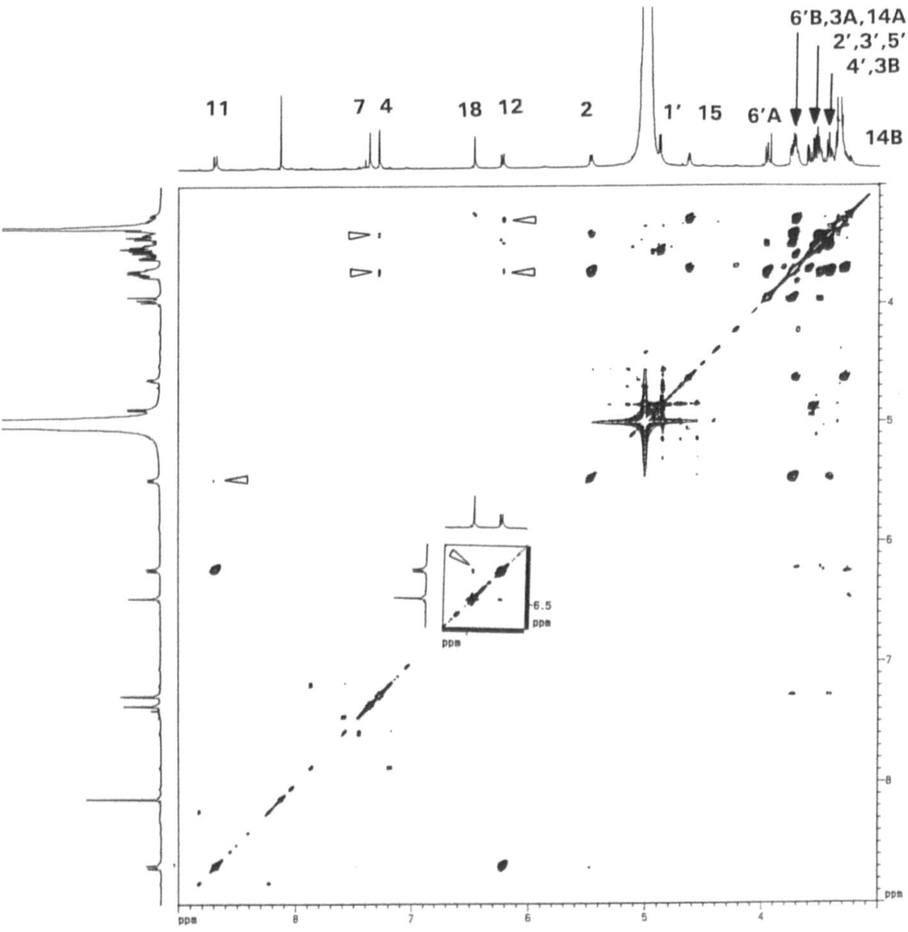

Fig. 11.7. A 600-MHz 2D ^1H COSY-45 NMR spectrum of betanin in CD_3OD/DCl. The cross peaks arising from long-range couplings are shown with *open arrowheads*. The *inset* is plotted at a lower contour level

ditions sensitive acyl ester linkages may be hydrolysed, product degradation may occur and deuterium exchange occurs not only of the expected OH and NH groups but also of several carbon-bound hydrogens. Hence, in the betacyanin system, H-12 and H-18 undergo exchange, and on prolonged exposure H-7 and H-15 show partial exchange. (Numbering is according to the 15*S*-betanin C-numbering system in the structure scheme.)

Characterization of the nature of the aglycone, together with the number and nature of sugar and acyl components, is usually ascertained by high-field 1D and 2D shift-correlated (COSY) ^{1}H NMR spectroscopy and is illustrated here by the COSY-45 spectrum of betanin (Heuer et al. 1992) (Fig. 11.7). In all cases the doublet signals of H-11 and H-12 are readily distinguishable by their low- and high-field shifts, respectively, and ca. 12 Hz coupling constant. The two three-spin systems (H-2/H-3AB and H-15/H-14AB) show easily distinguishable cross peaks in the COSY spectrum, from which H-2 is to low field of H-15. H-4, H-7 and H-18 appear as singlets which are assigned from their shifts, nuclear Overhauser enhancements (NOEs) and long-range couplings. Thus, H-7 is distinguished from the other singlets by the observation of a NOE at H-11 upon irradiation of H-7. Careful inspection of the 2D COSY spectrum often allows observation of characteristic long-range couplings between H-2 and H-11, H-3A and H-4, H-3B and H-4, H-12 and H-14A, and H-12 and H-14A.

In the most recent high-field data small differences in the ^{1}H chemical shifts of H-11, H-12, H-14A, H-14B and H-15 allow distinctions to be made between the normal (2*S*, 15*S*) and the isoforms (2*S*, 15*R*). In all the data to date the shifts of H-11 and H-12 in the isoform are to higher field.

Eight stereoisomers are, in principle, possible for isomerization of the 1,7-diazaheptamethinium system in the betalains (Fig. 11.8). However, Wyler and Dreiding (1984), from a series of careful, long-term experiments on betanidin, isobetanidin and indicaxanthin, were able to show that only the C-12/C-13 partial double bond showed detectable *E*/*Z*-diastereoisomerism, with both compounds consisting of an approximately 75:25 mix-

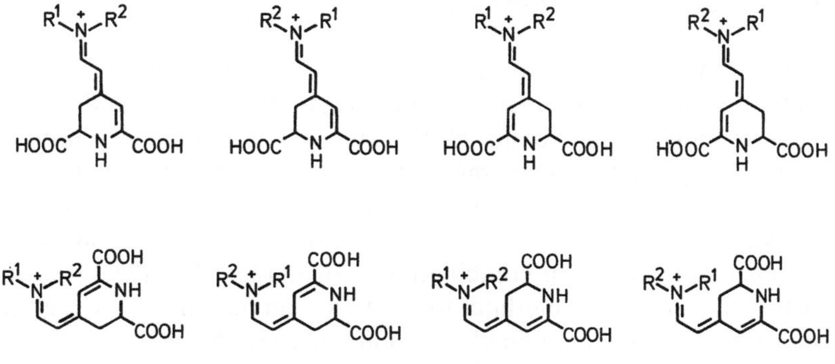

Fig. 11.8. Possible stereoisomers in the betacyanin and betaxanthin systems (Strack et al. 1993)

ture of the 12*E* and 12*Z* isomers, respectively. In all the cases we have investigated the same phenomenon is observed with characteristic signals of the minor isomer being found at about 9.0 and 5.9 ppm for H-11 and H-12, respectively. Proof that the two isomers are in a dynamic equilibrium is afforded by the observation of saturation transfer effects between the same proton in both isomers (Heuer et al. 1992). For betaxanthins this phenomenon may be more general, as we found that in humilixanthin there is one major isomer (60%) and three minor ones (22, 15 and 3%) (Strack et al. 1987a).

Characterization of the number and type of sugar and acyl substituents by NMR is now well documented and is the same in all areas of natural product chemistry. The position of O-glycosylation of the aglycone, at C-5 and C-6, is detected by an approximately 2 ppm downfield shift of the adjacent *ortho* proton and can be confirmed by the observation of a NOE to this same proton upon irradiation of the anomeric proton of the sugar and vice versa. Although data for acylated betalains are scarce, our recent results for the 4-coumaroyl and feruloyl C-6′ esters of the two isomers of gomphrenin I, where considerable chemical shift changes were observed for protons remote from the site of substitution, may be the first evidence for intramolecular copigmentation in these molecules (Heuer et al. 1992); this was also shown for lampranthin II (feruloylbetanin). The phenomenon is well known for anthocyanins (Strack and Wray 1989, 1993).

To date, there is very little ^{13}C NMR data for the betalains in the literature, although it can be anticipated that with the availability of more sensitive spectrometers and the possibility of using both one-bond and multiple 2D ^{1}H-detected ^{13}C-^{1}H correlations such data will gain more prominence in future structure elucidations, as seen with anthocyanins (Strack and Wray 1993). In particular, the multiple correlation method affords direct interfragment sequence information (aglycone–sugar, sugar–sugar, and sugar–acyl group).

11.5 Mass Spectrometry

Fast atom bombardment mass spectrometry (FAB-MS) is currently the method of choice for mass determination. It provides additional complementary information to the NMR data and is particularly useful in those situations where the NMR method encounters difficulties with substituents. The recent introduction of the alternative ion spray soft-ionization technique (IS-MS) also produces ions $[M+H]^+$ in high abundance directly from solutions of betalains. In combination with tandem MS (MS/MS) it allows a fast and convenient method for the determination of the aglycone and, additionally, when combined with liquid chromatography (HPLC), affords a rapid and sensitive method for identification of a series of betalains in complex mixtures, as has been shown with anthocyanins (Gläßgen et al.

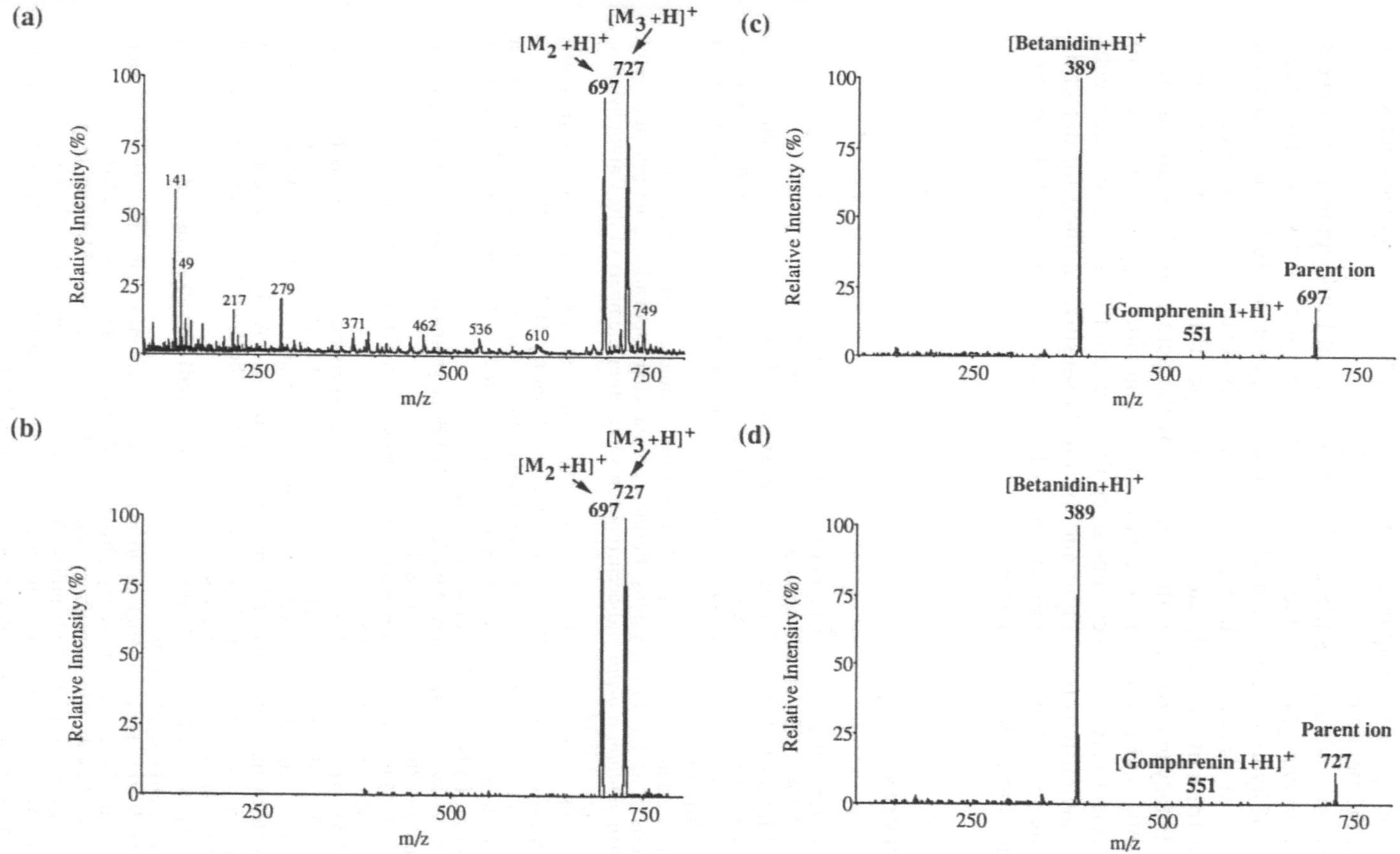

Fig. 11.9. **a** Ion spray mass spectrum of the mixture of gomphrenin II ($[M_2+H]^+$) and gomphrenin III ($[M_3+H]^+$). **b** Parent ion scan of m/z 389 $[betanidin+H]^+$. **c** Daughter ion scan of m/z 697 ($[M_2+H]^+$). **d** Daughter ion scan of m/z 727 ($[M_3+H]^+$). (Heuer et al. 1992)

1992). Ion spray techniques appear to be ideal for thermolabile compounds such as the betalains.

Recently we used IS-MS to identify a series of betacyanins from flowers of *Gomphrena globosa* (Heuer et al. 1992). Figure 11.9 shows ion spray spectra of hydroxycinnamic acid-acylated gomphrenins. In this investigation two types of tandem IS-MS experiments were particularly useful. A daughter ion spectrum of each individual peak in the spectrum was recorded to identify the betacyanin peaks in the spectrum by the detection of $[\text{betanidin}+\text{H}]^+$ fragments, and parent ion spectra of the $[\text{betanidin}+\text{H}]^+$ were recorded to detect selectively all betacyanins since they contain this aglycone. Such techniques obviously offer a rapid and extremely sensitive method for detecting betalains and will become the method of choice as advanced instrumentation becomes increasingly available.

References

Andersen OM (1985) Chromatographic separation of anthocyanins in cowberry (lingonberry) *Vaccinium vites-idaea* L. J Food Sci 50:1230–1232

Bokern M, Strack D (1988) Synthesis of hydroxycinnamic acid esters of betacyanins via 1-O-acylglucosides of hydroxycinnamic acids by protein preparations from cell syspension cultures of *Chenopodium rubrum* and petals of *Lampranthus sociorum*. Planta 174:101–105

Bokern M, Heuer S, Strack D (1992) Hydroxycinnamic acid transferases in the biosynthesis of acylated betacyanins: Purification and characterization from cell cultures of *Chenopodium rubrum* and occurrence in some other members of the Caryophyllales. Bot Acta 105:146–151

Gläßgen WE, Wray V, Strack D, Metzger JW, Seitz HU (1992) Anthocyanins from cell suspension cultures of *Daucus carota*. Phytochemistry 31:1593–1601

Heuer S, Strack D (1992) Synthesis of betanin from betanidin and UDP-glucose by a protein preparation from cell suspension cultures of *Dorotheanthus bellidiformis* (Burm. f.) N.E.Br. Planta 186:626–628

Heuer S, Wray V, Metzger JW, Strack D (1992) Betacyanins from flowers of *Gomphrena globosa*. Phytochemistry 31:1801–1807

Piattelli M, Minale L (1964a) Pigments of Centrospermae. II. Distribution of betacyanins. Phytochemistry 3:547–557

Piattelli M, Minale L (1964b) Pigments of Centrospermae. I. Betacyanins from *Phyllocactus hybridus* Hort. and *Opuntia ficus-indica* Mill. Phytochemistry 3:307–311

Schwartz SJ, von Elbe JH (1980) Quantitative determination of individual betacyanin pigments by high-performance liquid chromatography. J Agric Food Chem 28:540–543

Schwartz SJ, von Elbe JH (1983) Identification of betanin degradation products. Z Lebensm Unters Forsch 176:448–453

Steglich W, Strack D (1990) Betalains. In: Brossi A (ed) The alkaloids, vol 39. Academic Press, Orlando, pp 1–62

Strack D, Reznik H (1979) High-performance liquid chromatographic analysis of betaxanthins in Centrospermae (Caryophyllales). Z Pflanzenphysiol 94:163–167

Strack D, Wray V (1989) Anthocyanins. In: Harborne JB (ed) Methods in plant biochemistry, vol 1. Plant phenolics. Academic Press, London, pp 325–356

Strack D, Wray V (1993) The anthocyanins. In: Harborne JB (ed) The flavonoids: recent advances since 1986. Chapman and Hall, London (in press)

Strack D, Schmitt D, Reznik H, Boland W, Grotjahn L, Wray V (1987a) Humilixanthin, a new betaxanthin from *Rivina humilis*. Phytochemistry 26:2285–2287
Strack D, Engel U, Wray V (1987b) Neobetanin: a new natural plant constituent. Phytochemistry 26:2399–2400
Strack D, Bokern M, Marxen N, Wray V (1988) Feruloylbetanin from petals of *Lampranthus* and feruloylamaranthin from cell suspension cultures of *Chenopodium rubrum*. Phytochemistry 27:3529–3531
Strack D, Steglich W, Wray V (1993) Betalains. In: Waterman PG (ed) Methods in plant biochemistry, vol 8. Alkaloids and sulphur compounds. Academic Press, London, pp 421–450
Terradas F, Wyler H (1991) 2,3- and 4,5-secodopa, the biosynthetic inter-mediates generated from L-dopa by an enzyme system extracted from the fly agaric, *Amanita muscaria* L., and their spontaneous conversion to muscaflavin and betalamic acid, respectively, and betalains. Helv Chim Acta 74:124–140
Trezzini GF, Zryd J-P (1991a) Two betalains from *Portulaca grandiflora*. Phytochemistry 30:1897–1899
Trezzini GF, Zryd J-P (1991b) Characterization of some natural and semi-synthetic betaxanthins. Phytochemistry 30:1901–1903
Wyler H, Dreiding AS (1961) Über Betacyane, die stickstoffhaltigen Farbstoffe der Centrospermen. Vorläufige Mitteilung. Experientia 17:23–25
Wyler H, Dreiding AS (1984) Deuterierung von Betanidin und Indicaxanthin. (E/Z)-Stereoisomerie in Betalainen. Helv Chim Acta 67:1793–1800

12 Cladistic and Phenetic Studies

James E. Rodman

12.1 Summary

To improve cladistic and phenetic studies of the Centrospermae (Caryophyllales or Chenopodiales), character analyses are extended to chromosomal, morphological (vegetative), phytochemical, and ultrastructural features, augmenting floral morphological and palynological characters discussed earlier (Rodman 1990). A dataset of 76 characters was analyzed with the computer programs PAUP (phylogenetic analysis using parsimony; for cladistics) and NTSYS (numerical taxonomy system of computer programs; for phenetics), and the results are compared with previous findings and with recent macromolecular data. *Stegnosperma* is collateral to all remaining centrosperms, which constitute four major assemblages: Amaranthaceae-Chenopodiaceae; Caryophyllaceae-Molluginaceae; Nyctaginaceae and Phytolaccaceae s.s. with *Agdestis* and Rivinaceae; and the succulents Aizoaceae, Cactaceae, and Portulacaceae (including Basellaceae, Didiereaceae, and Hectorellaceae). Paraphyly afflicts all four groups, and several additional taxa remain problematic: Achatocarpaceae, *Barbeuia*, *Halophytum*, *Lophiocarpus*, and *Microtea*. New macromolecular data from restriction site analyses and nucleotide sequencing promise to invigorate discussion of centrosperm phylogeny but will create debate over morphological versus molecular approaches in plant systematics.

12.2 Introduction

Somewhere between the optimism of Constance's (1964) unending synthesis and the pessimism of Heywood's (1974) stone lies the impetus that drives our search for taxonomic order among plants. Unending also seems to be the criticism of early efforts (Rodman et al. 1984) to bring an explicit taxometric focus to the long-standing problem of centrosperm relationships (e.g. Levin 1985; Gibson and Nobel 1986; Carolin 1987; Bittrich and

Hartmann 1988; Hershkovitz 1989). Fortunately, criticism's stones turn to building blocks when they stimulate improved character definitions and foster new evidence, especially "primary semantide" or macromolecular data (Zuckerkandl and Pauling 1965). The saga of *Batis* and Gyrostemonaceae, once allied with Centrospermae (Eckardt 1964; Airy Shaw 1966; Cronquist 1968; Thorne 1968; Takhtajan 1969) but now confirmed as affiliates of Capparales (Rodman 1991a,b; Rodman et al. 1993), illustrates the productive interplay of morphological and molecular evidence when analyzed with explicit numerical methods. That success strengthens the resolve to continue with numerical studies of centrosperms.

Building upon the constructive criticism elicited by that first taxometric treatment of Centrospermae (or Caryophyllales, although Chenopodiales appears to have priority as a name at ordinal rank: Thorne 1968, 1992), I sought to create from the literature a dataset of characters appropriate for cladistic and phenetic analyses. The first revision (Rodman 1985, 1990) described anatomical or morphological features of the flower and seed as well as pollen for a total of 40 characters. Here, I revise and recode these characters and add new chromosomal, morphological (vegetative), phytochemical, and ultrastructural features, for a total of 76 characters (Appendix A). Using this dataset, I present cladistic and phenetic results derived from appropriate computer programs (see Rodman 1991a,b for rationale and references). The results should be considered heuristic because the dataset is not especially large and problems remain with the diagnosis of monophyletic taxa within the order.

12.3 Materials: Taxa

Problems in the diagnosis of both ingroup and outgroup taxa continue to beset the study of Centrospermae. The order itself (if such be the rank we choose to recognize) is stable in composition and characterized by an impressive concordance of phytochemical (Mabry 1977; Hartley and Harris 1981; Piatelli 1981; see also Chap. 10), ultrastructural (Chap. 5), and embryological features (Cronquist 1981; Dahlgren et al. 1981). However, several constituent taxa remain poorly delimited, and four of 16 recently diagnosed families may prove paraphyletic: Caryophyllaceae, Chenopodiaceae, Portulacaceae, and Rivinaceae (Rodman 1990). An intensive study of variation in multiple character systems is required to diagnose all the centrosperm ingroups (Hershkovitz 1989). It is reassuring that such work proceeds on a number of fronts: Bittrich and Hartmann (1988) on Aizoaceae; Bohlin (1988) on *Colignonia* in Nyctaginaceae; Brown and Varadarajan (1985) and Rogers (1985) on Phytolaccaceae s.l.; Carolin (1987) and Hershkovitz (1991a,b) on Portulacaceae; Eliasson (1988) on New World Amaranthaceae;

and Leuenberger (1986) on *Pereskia* in Cactaceae. Unfortunately, large families like the Caryophyllaceae, Chenopodiaceae, and Portulacaceae have received only partial attention; Nyctaginaceae, probably nested within a paraphyletic Rivinaceae (a name with priority over Petiveriaceae: see Rogers 1985), are also poorly known. Progress in understanding relationships is likely only when these are dissolved into homogeneous subunits (see Bittrich and Hartmann 1988 for an example with Aizoaceae), an effort requiring genus-by-genus, if not species-by-species, studies (Hershkovitz 1991a,b). The present exercise is a compromise that may serve to focus study on problematic character systems while providing a phyletic scaffold for ingroup analyses (see Carolin 1987 for an example with Portulacaceae). Although Hershkovitz (1989) has argued that no progress can be made until all constituent taxa are clearly and robustly diagnosed in a cladistic context, I am willing to accept a period of "groping".

The first taxometric study of centrosperms focused upon 11 core groups and ignored several taxa that were poorly known and hence could not be analyzed with then-current computer programs (Rodman et al. 1984; Levin 1985; Rodman 1985). Subsequent revision (Rodman 1990, Table 1) has brought the number of recognized groups to 16. That list of diagnosing characteristics can be supplemented with the following: uniseriate or biseriate wood rays in Achatocarpaceae (see Chap. 3); idioblastic cortical fibers in *Barbeuia* (see Chap. 3); fiber tracheids in wood of *Stegnosperma* (see Chap. 3); and for a presumptive outgroup, Plumbaginaceae, tetrasporic development of the embryo sac may be a family-wide feature (Haig 1990). Several additional taxa remain problematic, and for lack of so many characters, several have been excluded from my current analyses: *Halophytum*, *Lophiocarpus*, and *Microtea* for example. Gibson (1978) describes anatomical similarities of *Halophytum* with *Tetragonia* and *Trianthema* of Aizoaceae; *Lophiocarpus* is allied with Phytolaccaceae s.s. by Takhtajan (1986) and Thorne (1992); Cronquist (1981) suggests that *Microtea* belongs in the Amaranthaceae or Chenopodiaceae, an insight consonant with its Pfs-type plastids (see Chap. 5) and polyporate pollen (see Chap. 7). *Dysphania* is here accepted as part of Chenopodiaceae (Mabry and Behnke 1976); *Gisekia* is accepted as part of Phytolaccaceae s.s. (Brown and Varadarajan 1985); and *Hectorella* and *Lyallia* are considered to be derived members of a portulacaceous alliance and likewise are not treated separately (Cronquist 1981).

Outgroup taxa present their own set of problems. My choices of just three, Dilleniaceae, Plumbaginaceae, and Polygonaceae, are conservative and influenced by past results (Rodman et al. 1984) and recent DNA-based findings (Giannasi et al. 1992; Rettig et al. 1992; Chase et al. 1993). The work by Chase and colleagues focuses attention on taxa in superorder Theanae (Thorne 1992) as possible centrosperm sister groups but not on Theaceae themselves, contrary to Leins and Erbar's (see Chap. 13) speculation.

12.4 Materials: Characters

My rationale for selecting characters has been described before (Rodman et al. 1984; Rodman 1991a): to amass a large number of characters for which an argument of homology seems reasonable; for which character-state coding discriminates between taxa; and for which character-state transformations (polarities) can be assigned on the basis of outgroup analysis. To implement these goals, three main issues are confronted: homology, degree of taxon sampling, and variability within a taxon. For the centrosperms and their likely outgroups, character definitions, component states, and codings were recently discussed for two classes of phenotypic traits: floral anatomy or morphology (including ovule and seed features) and pollen (Rodman 1990, Table 2). These are revised and recoded here, and additional characters in four broad categories (chromosomes, phytochemistry, ultrastructure, and vegetative anatomy) are added. Appendix A lists the 76 characters and the codings; identifies major literature sources; discusses disputes, if present, about homology; and describes sampling intensity and variability. Appendix B presents the complete matrix of character-state codings for all 19 taxa. The goal of a large number of characters has led me to use information even if based on incomplete coverage for a family. In coding families that are variable for the character under consideration, an effort was made to choose the state present in putatively primitive members (Cantino 1982), for example, *Pereskia* in Cactaceae. In all, 76 characters have been compiled, 69 of which are coded as binary and seven as multistate characters, for a total of 161 character states.

12.5 Methods

For phenetic analysis of overall or aggregate similarity (Ridley 1986), Rohlf's (1987) NTSYS-pc computer programs were used. With the discrete, qualitative character data assembled here, association coefficients are the appropriate measures of phenetic resemblance (Sneath and Sokal 1973). The coefficient denoted "unnamed 3" by Rohlf (1987: matches in character states divided by nonmatches) was found empirically to yield the least distorted phenograms when checked with the cophenetic correlation coefficient, with UPGMA (unweighted pair-group method, arithmetic averages) clustering (Sokal 1986). A principal coordinates analysis (PCRDA) was also performed in the hope that taxa with large numbers of unknown characters would be associated with their nearest phenetic neighbors (Rohlf 1972).

For cladistic analysis of phylogenetic branching, Swofford's (1985) PAUP computer programs were used. All multistate characters except character #11 (carpel number) were treated as unordered transformations. Trees were generated by designating Dilleniaceae as outgroup and by invoking the SWAP = GLOBAL and MULPARS options. To choose among Wagner trees of equal length generated from PAUP runs, Farris's (1972) goodness-of-fit measure, his "F-value", was used as an optimizing criterion (see Swofford and Maddison 1987) with the MINF option.

12.6 Results and Discussion

Aggregate similarity was studied by two methods, clustering and ordination, and these produced complementary results. Figure 12.1 presents the UPGMA phenogram based on Rohlf's (1987) "unnamed 3" association measure, which yielded a high cophenetic correlation coefficient (0.95; Sokal 1986). Figure 12.2 plots the first two principal coordinates from PCRDA with third-axis values (positive or negative) recorded after each taxon name.

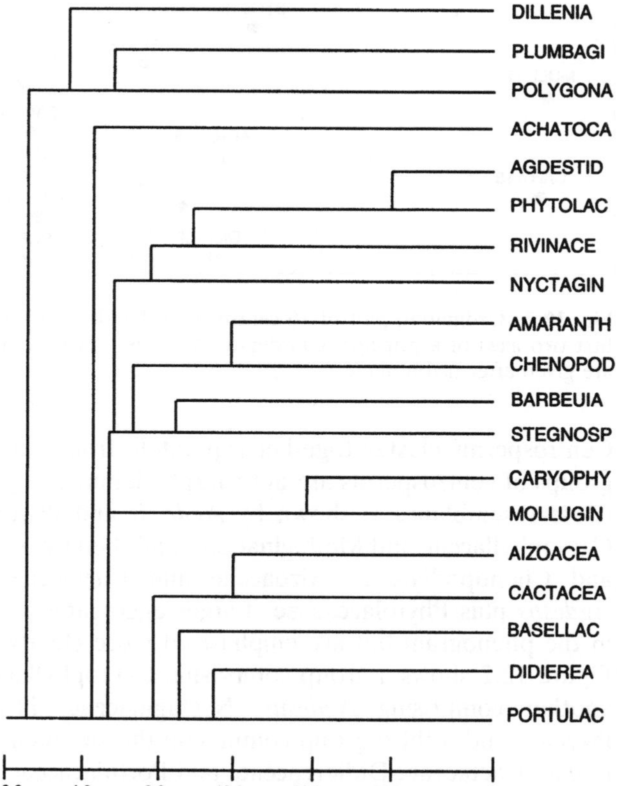

Fig. 12.1. UPGMA phenogram of 16 centrosperm families and three putative relatives (Dilleniaceae, Plumbaginaceae, and Polygonaceae) based on Rohlf's (1987) "unnamed 3" association coefficient; the cophenetic correlation coefficient is 0.95

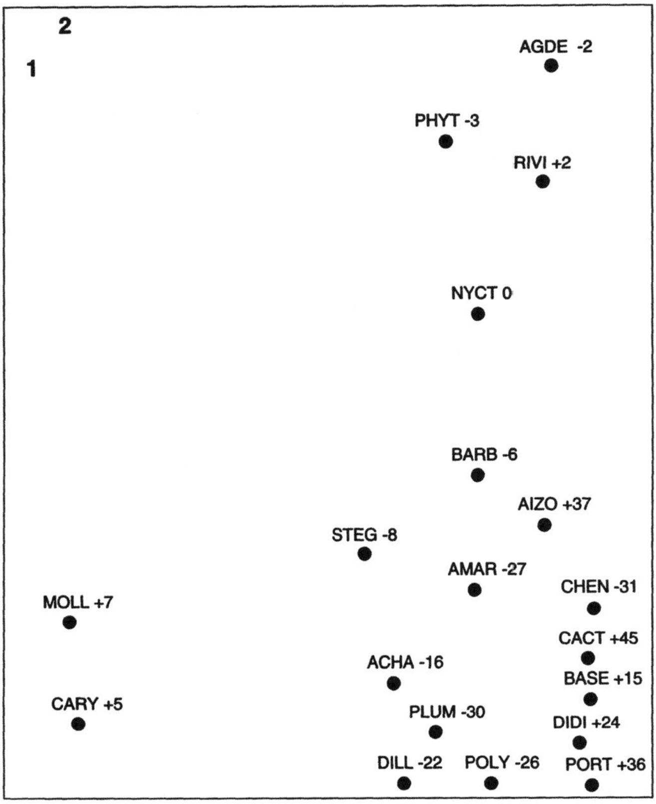

Fig. 12.2. Ordination plot of 16 centrosperm families and three putative relatives along first two axes of a principal coordinates analysis; third-axis values (positive or negative) are given after taxon name

Centrosperms cluster together separately from outgroup families, but subgroups of centrosperms are not sharply delimited by gaps in the phenogram. Close resemblance is shown by *Agdestis* and Phytolaccaceae s.s., then by Caryophyllaceae and Molluginaceae, and then by pairings of Amaranthaceae and Chenopodiaceae, Aizoaceae and Cactaceae, and Rivinaceae with *Agdestis* plus Phytolaccaceae. Larger aggregates are only weakly separated in the phenogram but are emphasized more clearly in the ordination plots. Figure 12.2 shows a group comprising Caryophyllaceae and Molluginaceae; another comprising *Agdestis*, Nyctaginaceae, Phytolaccaceae, and Rivinaceae; and a third group comprising the succulent taxa Aizoaceae, Basellaceae, Cactaceae, Didiereaceae, and Portulacaceae (these aggregating most clearly along the third-axis ordination). All these groups are consistent with the phenogram clusters. The PCRDA results for Amaranthaceae and Chenopodiaceae align them with Achatocarpaceae, *Barbeuia*, and *Stegnosperma* and near the outgroup families.

A few results remain problematic. Achatocarpaceae separate first from all other centrosperms in the phenogram (Fig. 12.1) and are intermediate between centrosperms, such as Amaranthaceae and Chenopodiaceae, and outgroup families in the ordination (Fig. 12.2). *Barbeuia* and *Stegnosperma* cluster together in the phenogram and lie near Achatocarpaceae, Amaranthaceae, and Chenopodiaceae in the ordination. These are intriguing results, especially in light of Manhart and Rettig's (Chap. 9) nucleotide sequencing data, but they are compromised by the large number of missing characters for Achatocarpaceae (20 characters) and *Barbeuia* (40). However, *Agdestis* associates tightly with the Nyctaginaceae plus Phytolaccaceae cluster in both phenogram and ordination, a result that seems robust despite the large number of missing characters (31) for this genus. Two general points emerge. First, the phenetic results emphasize the great heterogeneity of the old concept of "Phytolaccaceae." Second, when a taxonomist's judgment is strongly influenced by an assessment of resemblance or similarity, it becomes understandable that close taxonomic relationship will be advocated for such alliances as Aizoaceae and Cactaceae (see Buxbaum 1961), Amaranthaceae and Chenopodiaceae (Cronquist 1968), Basellaceae and Portulacaceae (Bogle 1969), Caryophyllaceae and Molluginaceae (Gilbert 1987), and Nyctaginaceae and Phytolaccaceae (Cronquist 1968).

These new results, based on revised character codings, are largely congruent with the earlier phenetic findings for the centrosperms (Rodman et al. 1984). The Molluginaceae present a conspicuous exception. The association of Molluginaceae with Nyctaginaceae and Phytolaccaceae in the previous study appears to have resulted from poor character codings (Levin 1985; Hershkovitz 1989). The present association with Caryophyllaceae reflects similarities remarked by several workers: morphology (Gilbert 1987), floral anatomy (Hofmann 1973), flavonoid chemistry (Valant-Vetschera 1985), and leaf wax features (Engel and Barthlott 1988). It remains disturbing that Molluginaceae may not constitute a monophyletic group (Bittrich and Hartmann 1988; Rodman 1990); more convincing analyses will require detailed studies of the individual genera.

Cladistic affinities were inferred using a maximum parsimony approach implemented with Swofford's (1985) PAUP programs. With the dataset of 76 characters, 20 equally parsimonious trees of 165 steps were found, with a consistency index (Kluge and Farris 1969) of 0.515. The strict consensus (Swofford 1985) of these 20 is shown in Fig. 12.3. Centrosperms constitute a monophyletic group of seven unresolved lineages: three individual taxa (Achatocarpaceae, *Barbeuia*, and *Stegnosperma*) and four major subgroups (*Agdestis*, Nyctaginaceae, Phytolaccaceae, and Rivinaceae; Amaranthaceae plus Chenopodiaceae; Caryophyllaceae plus Molluginaceae; and a clade of succulents, Aizoaceae, Basellaceae, Cactaceae, Didiereaceae, and Portulacaceae). These same four groups are found also in the phenetic results. Of the 20 equally parsimonious trees, the one with the lowest F-ratio (0.217; see Swofford and Maddison 1987) is selected for description as the least

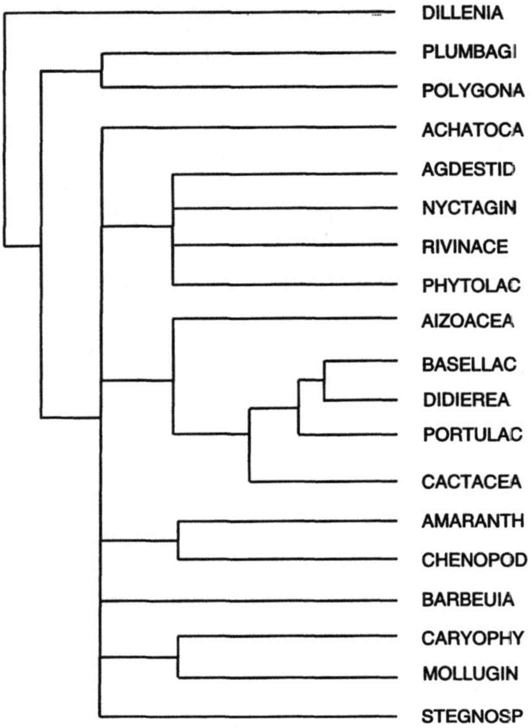

Fig. 12.3. Strict consensus of the 20 equally parsimonious Wagner trees of 165 steps and consistency index 0.515 generated by PAUP (Swofford 1985) from the 76-character dataset

homoplastic tree and is shown in Fig. 12.4. For this tree, 28 characters (26 binary, two multistate) provide unique synapomorphies and are marked on the appropriate branches. Characters from all six phenotypic categories are represented among these synapomorphies: floral anatomy, vegetative anatomy, chromosomes, pollen, phytochemistry, and ultrastructure. Eight of the 28 characters serve to distinguish Dilleniaceae from the other taxa and could be construed as autapomorphies; alternatively, they diagnose a clade comprising centrosperms with Plumbaginaceae and Polygonaceae, derived from dicot stocks for which Dilleniaceae serve as placeholders. All the remaining characters show homoplasy on the tree, either parallelism or reversal or both. Table 12.1 lists all the supporting characters, whether uniquely synapomorphic or homoplastic, for several clades or subclades of this tree, along with a consistency score for that branch (see Rodman 1991b for explanation of the calculation of this score).

Is Fig. 12.4 the true phylogeny of centrosperms? Probably not, but it is an efficient summary of a large number of character changes within Centro-

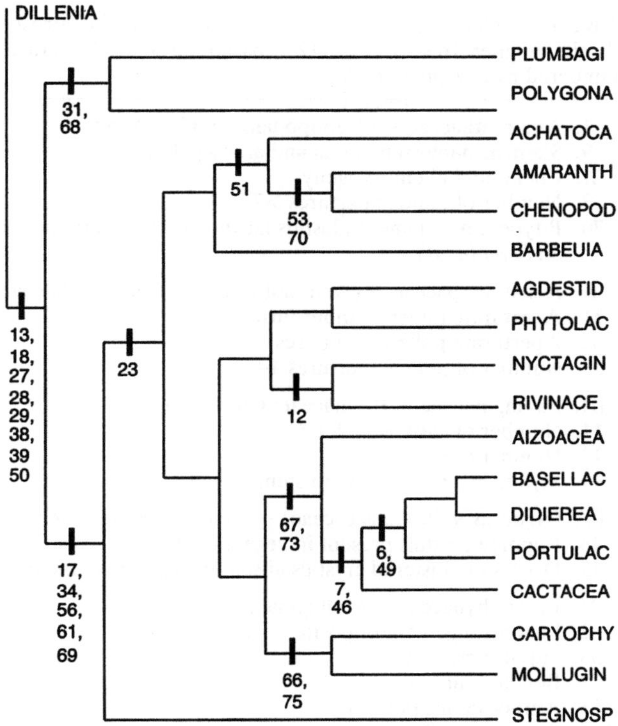

Fig. 12.4. Least homoplastic of the 20 equally parsimonious trees generated by PAUP, with F-value 0.217 (MINF option) and with unique synapomorphies listed on branches (characters numbered as in Appendix A)

spermae and between these and some likely phylogenetic relatives. Several aspects of this tree invite comment:

1. *Stegnosperma* is collateral to all remaining centrosperms, a result found in nine of the 20 equally parsimonious trees. In four trees Achatocarpaceae occupied this basal position; in seven trees the clade of Caryophyllaceae plus Molluginaceae was collateral to all remaining centrosperms. That third topology is congruent with our earlier results (Rodman et al. 1984) that postulated a fundamental evolutionary split between anthocyanin and betalain lineages. It is interesting that Achatocarpaceae, Caryophyllaceae, and *Stegnosperma* are among the earliest diverging lineages in Manhart and Rettig's (see Chap. 9) analysis of *rbc*L sequence data. The fundamental position of *Stegnosperma* in Fig. 12.4 rests upon numerous generalized (plesiomorphic) features along with the absence of a nucellar beak or cap (Narayana and Narayana 1986) and the presence of endosperm (Bedell 1980); presence of a nucellar beak and loss of endosperm appear to characterize all remaining centrosperms.

Table 12.1. Character evidence for select clades of centrosperms from the least homoplastic Wagner tree of Fig. 12.4 based on the 76-character dataset. Characters are numbered as in Appendix A

I. Amaranthaceae + Chenopodiaceae; CS = 0.60
9. Stamens haplostemonous and antesepalous
21. Chenopodiad embryogeny
*53. Number of pollen apertures >7
*70. P-type sieve-element plastids lacking a central protein inclusion (but see Chap. 5, on *Sarcobatus*)

II. Achatocarpaceae + (Amaranthaceae + Chenopodiaceae); CS = 0.72
*51. Aperturate pollen without furrows
52. Aperturate pollen with pores
53. Number of pollen apertures 4–7

III. Nyctaginaceae + Rivinaceae; CS = 0.63
11. Number of carpels 3, 2, or 1
*12. Number of carpels 1
44. Styloid crystals present in stem or leaf

IV. (*Agdestis* + Phytolaccaceae) + (Nyctaginaceae + Rivinaceae); CS = 0.50
43. Raphide crystals present in vegetative tissues
45. Druses or clustered crystals absent in vegetative tissues

V. Caryophyllaceae + Molluginaceae; CS = 0.61
32. Leaf arrangement opposite
33. Stipules present
57. Betalains absent
58. Anthocyanins present
*66. C-arabinosylflavones present
*75. Wax crystalloids present as lobed platelets and erect tubular rodlets

VI. (Basellaceae + Didiereaceae) + Portulacaceae; CS = 0.68
*6. Two transverse involucral bracts present
11. Number of carpels 3, 2, or 1
22. Funicular obturator absent
*49. Chromosome base $x = 12$

VII. Cactaceae + [Portulacaceae + (Basellaceae + Didiereaceae)]; CS = 0.75
*7. Styles fused (Portulacaceae coded as unknown)
40. Successive cambia absent
*46. Calcium oxalate crystals present in stem epidermal cells

VIII. Aizoaceae + {Cactaceae + [Portulacaceae + (Basellaceae + Didiereaceae)]}; CS = 0.70
36. Leaf mucilage cells present
48. Chromosome base other than $x = 9$
*67. CAM present
*73. Phytoferritin present in phloem parenchyma cells
76. Wax crystalloids present as ribbonlike platelets and/or rodlets

* Unique synapomorphies; CS, consistency score
For each clade a consistency score (see Rodman 1991b) was calculated as the average of the individual consistency-index values for the characters on that branch.

2. Amaranthaceae and Chenopodiaceae are sister groups and affiliate with Achatocarpaceae and *Barbeuia*. The cladistic findings echo the phenetic results, especially the PCRDA plots, and find partial corroboration in the *rbc*L sequence results on *Phaulothamnus* (see Chap. 9). The consistent association of Amaranthaceae and Chenopodiaceae in taxonomic treatments obscures the troublesome fact that the two are only weakly diagnosed in a cladistic sense: haplostemonous, antesepalous androecium (a homoplasy); chenopodiad embryogeny (another homoplasy); pollen primitively with more than seven apertures (a unique synapomorphy, but the condition is achieved within other centrosperm families); and P-type sieve-element plastids without a central inclusion (another unique synapomorphy but now compromised by findings on *Sarcobatus*). Both families lack convincing subfamilial classifications, and one may prove paraphyletic (Rodman 1990). Carolin (1983) suggested that Amaranthaceae are nested within Chenopodiaceae, but such an arrangement appears to require a secondary reversion to multiovulate gynoecia in such putatively primitive amaranths as *Celosia* and *Deeringia*.

3. Phytolaccaceae s.s. affiliate with *Agdestis* and these with Nyctaginaceae plus Rivinaceae in a lineage only weakly diagnosed by homoplastic characters (presence of raphides in vegetative tissues, absence of druses or clustered crystals). Again, the cladistic findings echo the phenetic results. Also, the restriction endonuclease site data (see Chap. 8) and the *rbc*L sequence data (Rettig et al. 1992; see also Chap. 9) are consistent with this assemblage. Rivinaceae may prove to be paraphyletic, with Nyctaginaceae nested within a clade diagnosed by a unicarpellate gynoecium and styloid crystals in leaf and stem tissues (Rodman 1990).

4. Caryophyllaceae and Molluginaceae are sister groups, collateral to a major clade of succulents. This embedded position within the Centrospermae would imply the loss or repression of betalain biosynthesis and a reversion to anthocyanin biosynthesis, a scenario that had been avoided by our earlier results (Rodman et al. 1984). An embedded position is consistent with restriction site (see Chap. 8) and *rbc*L sequence results (see Chap. 9). However, the sequence data also challenge the affiliation of Caryophyllaceae with Molluginaceae; indeed, the macromolecular results require the inference of multiple betalain/anthocyanin switches among centrosperms. Such multiple biosynthetic switching may prove an ironic reprimand to those of us who sought neat taxonomic markers in secondary compound biochemistry (Mabry 1976; Rodman et al. 1984).

5. Highly derived within Centrospermae is a lineage of succulent taxa with Aizoaceae at the base (diagnosed by the presence of phytoferritin and occurrence of crassulacean acid metabolism, CAM), Cactaceae next (fused styles, calcium oxalate crystals in stem epidermis), and finally a portulacaceous alliance including Basellaceae and Didiereaceae and presumed to include *Hectorella* and *Lyallia* (presence of two transverse involucral bracts, chromosome base $x = 12$). The same alliance of succulent taxa is recovered in the phenetic results, and it is consistent in many respects with the earlier

cladistic findings (Rodman et al. 1984). Nucleotide sequence data reinforce an affiliation of Cactaceae with portulacaceous taxa including Basellaceae and Didiereaceae (Rettig et al. 1992; see also Chap. 9). Restriction site results, however, seem bizarre, with Downie and Palmer (see Chap. 8) arguing a diphyletic origin of Portulacaceae on the basis of shared restriction sites. The homology of these needs checking. Both nucleotide sequence and restriction site data agree, however, in associating Aizoaceae not with other succulent taxa but with Nyctaginaceae, Phytolaccaceae s.s., and Rivinaceae (a clade that might be diagnosed by the presence of raphides and/or styloids in vegetatives tissues). The CAM and phytoferritin characters employed here have not been surveyed extensively and may prove unreliable. In addition, taxon sampling needs to be increased, and more Molluginaceae and primitive Caryophyllaceae need analysis.

12.7 Conclusion

Since that first taxometric analysis of Centrospermae (Rodman et al. 1984), several studies have addressed problems of character coding and taxon diagnosis, but remarkably little new data have been added to the discussion of centrosperm phylogeny. Sources of traditional kinds of data that could be expected to provide taxonomic insight – like chromosome number, leaf venation pattern, betalain composition – have not been exploited. Instead, novel characters of restriction endonuclease sites (see Chap. 8) and of nucleotide sequences (see Chap. 9) have emerged, with interpretations of each that show agreement and disagreement over particular linkages among component centrosperms. Limited taxon sampling may account for some of the disagreement between the two macromolecular approaches. The determination of homologous restriction sites may also present difficulties. Nonetheless, these powerful new methods promise a wealth of comparable data, and we can look forward to an energetic exchange between morphology- and molecule-based viewpoints.

Appendix A

Characters, States, and Codings, with Notes on Literature Sources, Homology, Sampling, and Variability

For purposes of coding, the character-state present in Dilleniaceae (a presumptive outgroup) arbitrarily is coded as 0. The code 9 is used when the condition is unknown, inappropriate, or problematic. Except for character #11, no unidirectional morphocline is implied by the numerical order of coding for multistate characters (#4, 8, 10, 15, 17, and 53), and PAUP computer programs allow the use of unordered transformations.

Anatomical/Morphological (Reproductive) Characters

Literature sources are identified and issues of homology, sampling, and variability discussed by Rodman (1990). Heterostyly is sporadic in dicots (Lewis 1979), and homologies across families are difficult to assess; I suspect that dimorphic heterostyly in Plumbaginaceae and Polygonaceae may be an "apomorphic tendency" (Cantino 1985; Rodman 1985).

1. Inflorescence cymose (0) or racemose (1).
2. Flowers bisexual (0) or unisexual (1).
3. Perianth differentiated into sepals and petals (0) or not (1).
4. Perianth parts pentamerous (0), pleiomerous (1), trimerous (2) or tetramerous (3).
5. Calyx of fused sepals absent (0) or present (1).
6. Two transverse involucral bracts absent (0) or present (1).
7. Styles free (0) or fused (1).
8. Stamens pleiomerous (0), trimerous (1), tetramerous (2) or pentamerous (3).
9. Stamens spiral or diplostemonous (0), or haplostemonous and antesepalous (1).
10. Nectaries absent (0), stamen- (1) or perianth-associated (2).
11. Number of carpels indefinitely large (0), five or four (1), or three, two, or one (2).
12. Gynoecium pluricarpellate (0) or unicarpellate(1).
13. Gynoecium apocarpous (0) or syncarpous (including monomerous) (1).
14. Carpel position episepalous/epitepalous (0) or alternisepalous (1).
15. Ovule placentation marginal (0), basal (1) or axile/free central (2).
16. Number of ovules per carpel or locule: many (0) or one (1).
17. Ovule anatropous (0), campylotropous (1) or orthotropous (2).
18. Micropyle formed by both integuments (0) or inner only (1).
19. Nucellar beak or cap absent (0) or present (1).
20. Embryogeny of caryophyllad type absent (0) or present (1).
21. Embryogeny of chenopodiad type absent (0) or present (1).
22. Funicular obturator absent (0) or present (1).
23. Endosperm present (0) or absent (1) in seed.
24. Perisperm not persisting (0) or persisting (1) in seed.
25. Embryo without chlorophyll (0) or with chlorophyll (1).
26. Seeds arillate (0) or lacking funicular aril (1).
27. Exotestal layer of seed not thickened (0) or thickened (1).
28. Endotestal layer thickened (0) or not (1).
29. Exotegmic layer tracheidal/fibrous (0) or unspecialized (1).
30. Endotegmic layer not thickened (0) or thickened (1).
31. Dimorphic heterostyly absent (0) or present (1).

Anatomical/Morphological (Vegetative) Characters

Major sources of data include Metcalfe and Chalk (1950), Lawrence (1951), Cronquist (1981), Carlquist (1988), and Gregory and Baas (1989). Some characters are well surveyed (leaf arrangement), while others have received only sporadic attention. Bicollateral bundles in Basellaceae (Bogle 1969; Sperling 1987) may prove synapomorphic with those of selected Portulacaceae. Gibson (Chap. 3) describes the unique occurrence of calcium oxalate crystals in stem epidermal cells of Cactaceae, Didiereaceae, and Portulacaceae; he also describes pigmented mucilage cells in stem cortex of Didiereaceae and Portulacaceae, reinforcing similarity between these two. Occasional bordered pits in Caryophyllaceae (Metcalfe and Chalk 1950) may prove vestigial and are yet another intriguing similarity with *Stegnosperma*. Anomalous growth by successive cambia occurs in several Plumbaginaceae (Metcalfe and Chalk 1950) and could be primitive in the family, whereas it occurs rarely in Dilleniaceae and only in putatively derived lianas (*Doliocarpus*). Likewise, it may be primitive for Caryophyllaceae, albeit occasional, and also occurs in some Molluginaceae.

32. Leaf arangement alternate (0) or opposite (1).
33. Stipules present (0) or absent (1).
34. Nodal vasculature tri- or multilacunar (0) or unilacunar (1).
35. Idioblastic leaf sclereids present (0) or absent (1).
36. Leaf mucilage cells present (0) or absent (1).
37. Pitting of imperforate tracheary elements bordered (0) or simple (1).
38. Vessel end-plates scalariform (0) or simple (1).
39. Wood parenchyma apotracheal (0) or paratracheal (1).
40. Successive cambia absent (0) or present (1).
41. Bicollateral bundles absent (0) or present (1).
42. Wood fibers nonseptate (0) or septate (1).
43. Raphide crystals in stem and/or leaf commonly present (0) or absent (1).
44. Styloid crystals absent (0) or commonly present (1) in stem and/or leaf.
45. Druses or clustered crystals in stem and/or leaf present (0) or absent (1).
46. Calcium oxalate crystals absent (0) or present (1) in stem epidermal cells.
47. Pigmented mucilage cells absent (0) or present (1) in stem cortex.

Chromosomal Characters

Raven (1975) and Ehrendorfer (1976) are major sources of data. Several codings are little more than guesses, an obvious one being the guess that $x =$ 6 or 12 in basal Portulacaceae (see Carolin 1987), despite Bernardello's (1989) claim that $x = 9$ or Turner's (see Chap. 2) claim that $x = 8$. Likely base numbers of $n = 12$ in Basellaceae and Didiereaceae (Rodman et al. 1984) and in *Hectorella* would thus constitute a synapomorphy with Portulacaceae. Several taxa remain unknown for chromosome number, including many of the segregates from Phytolaccaceae s.l. (Achatocarpaceae, *Agdestis*, *Barbeuia*, *Lophiocarpus*, and *Stegnosperma*). In Bittrich and Hartmann's (1988) subfamilial classification of Aizoaceae, three of their five taxa are characterized by $n = 8$, the others by $n = 9$, and the former may include the more primitive taxa.

48. Chromosome base number other than 9 (0) or equal to 9 (1).
49. Chromosome base number other than 12 (0) or equal to 12 (1).

Pollen Characters

Literature sources are identified and issues of homology, sampling, and variability are discussed in Rodman (1990).

50. Pollen shed in binucleate (0) or trinucleate (1) condition.
51. Aperturate pollen with furrows (0) or not (1).
52. Aperturate pollen without pores (0) or with pores (1).
53. Number of apertures three (0), four–seven (1) or more than seven (2).
54. Pollen surface not spinulose (0) or spinulose (1).
55. Pollen surface punctate/perforate (0) or not (1).

Phytochemical Characters

Flavonoid and other phenolic characters are compiled mostly from Richardson (1978, 1981), Gornall et al. (1979), Burret et al. (1981), Gurni and Kubitzki (1981), Hartley and Harris (1981), Young (1981), Harborne and Mabry (1982), Valant-Vetschera (1985), and Barron et al. (1988). When references conflicted, positive occurrences were given preference over reports of absence. I have tried to emphasize biosynthetic classes or end-products thought to be characteristic at upper taxonomic ranks. Homology is assumed between families with the same identified constituent. Although phytochemical sampling is often sporadic, the occurrence of rare side-chains or other substituents is here treated as taxonomically significant, even if interpretable only as an apomorphic tendency. Betalain occurrence is compiled from Mabry (1974, 1977), Behnke (1976b), and Piatelli (1976, 1981). The CAM feature is derived from Kluge and Ting (1978) and Teeri (1982). The saponin character is compiled from Paris (1963), Rizk (1986), and Mahato et al. (1988).

56. Bound ferulic acid in unlignified cell walls absent (0) or present (1).
57. Betalains absent (0) or present (1).
58. Anthocyanins present (0) or absent (1).
59. The flavonol myricetin present (0) or absent (1).

60. 6-Hydroxy flavonols absent (0) or present (1).
61. 5-Methoxy flavonols present (0) or absent (1).
62. Flavonol sulfates present (0) or absent (1).
63. Apigenin and/or luteolin present (0) or absent (1).
64. 6-Hydroxy flavones absent (0) or present (1).
65. C-glycosylflavones present (0) or absent/lost (1).
66. C-arabinosylflavones absent (0) or present (1).
67. CAM (malic acid storage) absent (0) or present (1).
68. Triterpenoid saponins present (0) or absent (1).

Ultrastructural Characters

Major sources of data are Behnke (1976a,b, 1981, 1982; Behnke et al. 1983a) for sieve-element features and Engel and Barthlott (1988) for leaf cuticles. Behnke et al. (1983b) report globular and cuboid protein inclusions in Molluginaceae, including "even polygonal" outlines in *Glinus*, and thus they raise a suspicion of variability in the geometry of these inclusions and of transition from polygonal shapes to cuboid or globular shapes within families. Such may be the case with "Molluginaceae" nested within a paraphyletic Caryophyllaceae (Rodman 1990). Achatocarpaceae were first reported as possessing globular inclusions (Behnke 1976a,b; repeated by Skvarla and Nowicke 1982) but were soon corrected to polygonal (Behnke 1981, 1982). *Sarcobatus*, traditionally allied with Chenopodiaceae, is now aberrant in that family in possessing Pcf-type plastids (see Chap. 5); this finding challenges the conclusion that Amaranthaceae-Chenopodiaceae are sharply delimited by plastid morphology. Behnke (Chap. 5) notes the occasional presence of phytoferritin in some Caryophyllaceae and doubts the phyletic significance of this feature, suggesting instead an ecophysiological correlation with succulence; I code four families for the feature (Behnke 1978). Taxonomic coverage for leaf wax characters is sporadic. Engel and Barthlott (1988) remark on similarities between Aizoaceae-Cactaceae and Portulacaceae, but they do not report the former's characteristic ribbonlike forms in any Portulacaceae they surveyed. Although irregular wax platelets seem to be rare in Nyctaginaceae (*Oxybaphus* and *Pisonia*) and Portulacaceae (*Ceraria*, *Hectorella*, and *Portulacaria*), I code them as present for these families, in line with their occurrence in many other centrosperms, and suggest secondary loss for the noncrystalloid members.

69. Sieve-element plastid type starch (0) or protein (1).
70. Protein crystals in sieve elements present (0) or absent/missing (1).
71. Protein crystals in sieve elements not globular (0) or globular (1) if present.
72. Protein crystals in sieve elements not polygonal (0) or polygonal (1) if present.
73. Phytoferritin in phloem parenchyma cells absent (0) or present (1).

74. Epicuticular leaf wax layer continuous (0) or crystalloid (1).
75. Wax crystalloids absent (0) or present as lobed platelets and erect (tubular) rodlets (1).
76. Wax crystalloids absent (0) or present as ribbonlike platelets and/or rodlets (1).

Appendix B

Matrix of Coding Assignments

Matrix for the 19 study taxa, named by their first eight letters, for the 76 characters listed in Appendix A.

DILLENIA PLUMBAGI POLYGONA ACHATOCA AGDESTID
AIZOACEA AMARANTH BARBEUIA BASELLAC CACTACEA
CARYOPHY CHENOPOD DIDIEREA MOLLUGIN NYCTAGIN
RIVINACE PHYTOLAC PORTULAC STEGNOSP

Anatomical/morphological (reproductive)

0 1 0 1 0 0 0 0 0 0 0 0 0 0 0 0 0 0 1
0 0 0 1 0 0 0 0 0 0 0 0 1 0 0 0 0 0 0
0 0 1 1 1 1 1 1 1 1 1 1 1 1 1 1 1 1 1
0 0 2 0 3 0 0 0 0 1 0 0 3 0 0 3 0 0 0
0 1 1 0 0 0 0 0 0 0 0 0 0 0 1 0 0 0 0
0 0 0 0 0 0 0 0 1 0 0 0 1 0 0 0 0 1 0
0 0 0 0 0 9 0 0 1 1 0 0 1 0 9 9 0 9 0
0 3 1 3 2 3 3 0 3 3 3 3 2 3 3 2 3 3 3
0 9 0 0 0 0 1 0 1 0 0 1 0 0 0 0 0 0 0
0 2 1 9 9 1 1 9 1 1 1 1 1 1 1 0 1 1 1
0 1 2 2 1 1 2 2 2 1 1 2 2 1 2 2 1 2 1
0 0 0 0 0 0 0 0 0 0 0 0 0 0 1 1 0 0 0
0 1 1 1 1 1 1 1 1 1 1 1 1 1 1 1 1 1 1
0 0 0 0 0 0 9 0 9 9 1 9 9 0 9 9 0 9 1
0 1 1 1 1 2 1 1 1 2 2 1 1 2 1 1 1 2 1
0 1 1 1 1 0 9 1 1 0 0 1 1 0 1 1 1 0 1
0 0 2 1 1 1 1 1 1 1 1 1 1 1 1 1 1 1 1
0 1 1 1 1 1 1 9 1 1 1 1 1 1 1 1 1 1 1
0 0 1 9 9 1 1 9 1 1 1 1 9 1 1 1 1 1 0
0 0 0 9 9 0 0 9 0 9 1 0 0 9 0 0 0 9 1
0 0 0 9 9 0 1 9 1 9 0 1 1 9 0 1 1 9 0
0 9 1 9 9 1 0 9 0 1 1 0 0 1 9 1 1 0 0
0 0 0 1 9 1 1 9 1 1 1 1 9 1 1 1 1 1 0

DILLENIA PLUMBAGI POLYGONA ACHATOCA AGDESTID
AIZOACEA AMARANTH BARBEUIA BASELLAC CACTACEA
CARYOPHY CHENOPOD DIDIEREA MOLLUGIN NYCTAGIN
RIVINACE PHYTOLAC PORTULAC STEGNOSP

0 0 0 1 9 1 1 9 1 1 1 1 0 1 1 1 1 1 1
0 1 0 9 9 0 0 9 1 0 0 1 0 0 1 0 0 0 9
0 1 1 1 1 0 0 0 1 9 1 1 0 0 1 0 1 0 0
0 1 1 9 9 1 1 9 1 1 1 1 1 1 1 1 1 1 9
0 1 1 9 9 1 1 9 1 1 1 1 1 1 1 1 1 1 9
0 1 1 9 9 1 1 9 1 1 1 1 1 1 1 1 1 1 9
0 0 0 9 9 1 1 9 1 1 1 1 1 9 1 1 0 1 9
0 1 1 0 0 0 0 0 0 0 0 0 0 0 0 0 0 0 0

Anatomical/morphological (vegetative)
0 0 0 0 0 9 0 0 0 0 1 0 0 1 1 0 0 9 0
0 1 0 1 1 1 1 1 1 1 0 1 1 0 1 1 1 1 1
0 0 0 9 9 1 1 9 1 1 1 1 1 1 1 1 1 1 1
0 0 1 9 9 1 9 9 1 1 1 0 0 1 9 1 1 0 0
0 0 0 9 9 0 1 9 0 0 1 1 0 1 1 1 1 0 1
0 1 1 1 1 1 1 1 1 1 0 1 1 1 1 1 1 1 0
0 1 1 1 1 1 1 1 1 1 1 1 1 1 1 1 1 1 1
0 1 1 1 9 1 1 1 9 1 1 1 1 1 1 1 1 1 1
0 1 0 0 1 1 1 1 0 0 1 1 0 1 1 1 1 0 1
0 0 0 0 0 0 0 0 1 0 0 0 0 0 0 0 0 1 0
0 1 1 0 0 0 0 0 1 0 0 0 0 0 0 0 0 0 0
0 1 1 1 0 9 1 1 1 1 1 1 1 9 0 0 0 1 1
0 0 0 0 0 1 0 0 0 0 0 0 0 0 1 1 0 0 0
0 1 0 0 1 0 0 0 0 0 0 0 0 0 1 1 1 0 0
0 0 0 0 0 0 0 0 9 1 0 0 1 0 0 0 0 1 0
0 0 0 0 0 0 0 0 0 0 0 0 1 0 0 0 0 1 0

Chromosomal
0 0 0 9 9 0 1 9 0 0 1 1 0 1 9 1 1 0 9
0 0 0 9 9 0 0 9 1 0 0 0 1 0 0 0 0 1 9

Palynological
0 1 1 9 9 1 1 9 1 1 1 1 1 1 1 1 1 1 1
0 0 0 1 0 0 1 0 0 0 0 1 0 0 0 0 0 0 0
0 0 1 1 0 0 1 0 0 0 0 1 0 0 0 0 0 0 0
0 0 0 1 0 0 2 0 1 0 0 2 1 0 0 0 0 0 0
0 0 0 0 1 1 1 1 1 1 1 1 1 1 1 1 1 1 1
0 1 0 1 0 0 0 0 0 0 0 0 0 0 0 0 0 0 0

DILLENIA PLUMBAGI POLYGONA ACHATOCA AGDESTID
AIZOACEA AMARANTH BARBEUIA BASELLAC CACTACEA
CARYOPHY CHENOPOD DIDIEREA MOLLUGIN NYCTAGIN
RIVINACE PHYTOLAC PORTULAC STEGNOSP

Phytochemical

9 0 0 9 9 1 1 9 1 1 1 1 1 9 1 1 1 1 9
0 0 0 0 1 1 1 9 1 1 0 1 1 0 1 1 1 1 1
0 0 0 1 1 1 1 9 1 1 0 1 1 0 1 1 1 1 1
0 0 0 1 9 0 1 9 1 1 1 0 0 1 1 1 1 1 1
0 0 1 0 9 1 0 9 0 0 0 1 0 0 0 0 0 0 0
0 0 0 1 9 1 1 9 1 1 1 1 1 1 1 1 1 1 1
0 0 0 1 9 1 1 9 1 1 1 0 1 1 1 9 1 1 1
0 0 0 1 9 1 1 9 1 1 0 1 1 1 1 9 1 0 1
0 0 1 0 9 0 1 9 0 0 0 0 0 0 0 0 0 0 0
0 0 0 0 9 1 0 9 0 1 0 0 0 0 0 9 0 0 1
0 0 0 0 9 0 0 9 0 0 1 0 0 1 0 9 0 0 0
0 0 0 9 9 1 0 9 9 1 0 0 1 0 0 0 0 1 0
0 1 1 9 9 0 0 9 9 0 0 0 9 0 0 0 0 0 0

Ultrastructural

0 0 0 1 1 1 1 9 1 1 1 1 1 1 1 1 1 1 1
9 9 9 0 0 0 1 9 0 0 0 1 0 0 0 0 0 0 0
9 9 9 0 1 1 0 9 1 1 0 0 1 9 1 1 1 1 0
9 9 9 1 0 0 0 9 0 0 1 0 0 9 0 0 0 0 1
0 0 0 9 9 1 0 9 9 1 0 0 1 0 0 9 0 1 9
0 0 0 1 1 1 0 9 0 1 1 1 1 1 1 9 1 1 1
0 0 0 0 0 0 0 9 0 0 1 0 0 1 0 9 0 0 0
0 0 0 0 0 1 0 9 0 1 0 0 1 0 0 9 0 9 0

Acknowledgements. I am especially grateful to Arthur Gibson and Ursula Hofmann for their generous comments on characters and codings.

References

Airy Shaw HK (ed) (1966) J.C. Willis' A dictionary of the flowering plants and ferns, 7th edn. Cambridge University Press, Cambridge

Barron D, Varin L, Ibrahim RK, Harborne JB, Williams CA (1988) Sulphated flavonoids – an update. Phytochemistry 27:2375–2395

Bedell HG (1980) A taxonomic and morphological re-evaluation of Stegnospermaceae (Caryophyllales). Syst Bot 5:419–431

Behnke HD (1976a) Ultrastructure of sieve-element plastids in Caryophyllales (Centrospermae), evidence for the delimitation and classification of the order. Plant Syst Evol 126:31–54
Behnke HD (1976b) A tabulated survey of some characters of systematic importance in centrospermous families. Plant Syst Evol 126:95–98
Behnke HD (1978) Elektronenoptische Untersuchungen am Phloem sukkulenter Centrospermen (including Didiereaceen). Bot Jahrb Syst 99:341–352
Behnke HD (1981) Sieve-element characters. Nord J Bot 1:381–400
Behnke HD (1982) *Geocarpon minimum*: Sieve-element plastids as additional evidence for its inclusion in the Caryophyllaceae Taxon 31:45–47
Behnke HD, Mabry TJ, Neumann P, Barthlott W (1983a) Ultrastructural, micromorphological and phytochemical evidence for a "central position" of *Macarthuria* (Molluginaceae) within the Caryophyllales. Plant Syst Evol 143:151–161
Behnke HD, Pop L, Sivarajan VV (1983b) Sieve-element plastids of Caryophyllales: Additional investigations with special reference to the Caryophyllaceae and Molluginaceae. Plant Syst Evol 142:109–115
Bernardello LM (1989) The chromosomes of *Grahamia* (Portulacaceae). Plant Syst Evol 163:127–131
Bittrich V, Hartmann HEK (1988) The Aizoaceae – a new approach. Bot J Linn Soc 97:239–254
Bogle AL (1969) The genera of Portulacaceae and Basellaceae in the southeastern United States. J Arnold Arbor 50:566–598
Bohlin JE (1988) A monograph of the genus *Colignonia* (Nyctaginaceae). Nord J Bot 8:231–252
Brown GK, Varadarajan GS (1985) Studies in Caryophyllales. I. Re-evaluation of classification of Phytolaccaceae s.l. Syst Bot 10:49–63
Burret F, Rabesa Z, Zandonella P, Voirin B (1981) Contribution biochimique à la systematique de l'ordre des Centrospermales. Biochem Syst Ecol 9:257–262
Buxbaum F (1961) Vorläufige Untersuchungen über Umfang, systematische Stellung und Gliederung der Caryophyllales (Centrospermae). Beitr Biol Pflanz 36:1–56
Cantino PD (1982) Affinities of the Lamiales: A cladistic analysis. Syst Bot 7:237–248
Cantino PD (1985) Phylogenetic inference from nonuniversal derived character states. Syst Bot 10:119–122
Carlquist S (1988) Comparative wood anatomy. Systematic, ecological, and evolutionary aspects of dicotyledon wood. Springer, Berlin Heidelberg New York
Carolin RC (1983) The trichomes of the Chenopodiaceae and Amaranthaceae. Bot Jahrb Syst 103:451–466
Carolin RC (1987) A review of the family Portulacaceae. Aust J Bot 35:383–412
Chase MW, Soltis DE, Olmstead RG et al. (1993) Phylogenetics of seed plants: An analysis of nucleotide sequences from the plastid gene *rbc*L. Ann Mo Bot Gard 80:528–580
Constance L (1964) Systematic botany, an unending synthesis. Taxon 13:257–273
Cronquist A (1968) The evolution and classification of flowering plants. Houghton Mifflin, Boston
Cronquist A (1981) An integrated system of classification of flowering plants. Columbia University Press, New York
Dahlgren R, Rosendal-Jensen S, Nielsen BJ (1981) A revised classification of the angiosperms with comments on correlation between chemical and other characters. In: Young DA, Seigler DS (eds) Phytochemistry and angiosperm phylogeny. Praeger, New York, pp 149–204
Eckardt T (1964) Reihe Centrospermae. In: Melchior H (ed) A. Engler's Syllabus der Pflanzenfamilien, 12th edn, vol II. Borntraeger, Berlin, pp 79–101
Ehrendorfer F (1976) Chromosome numbers and differentiation of centrospermous families. Plant Syst Evol 126:27–30
Eliasson UH (1988) Floral morphology and taxonomic relations among the genera of Amaranthaceae in the New World and the Hawaiian Islands. Bot J Linn Soc 96: 235–283

Engel T, Barthlott W (1988) Micromorphology of epicuticular waxes in centrosperms. Plant Syst Evol 161:71–85
Farris JS (1972) Estimating phylogenetic trees from distance matrices. Am Nat 106: 645–668
Giannasi DE, Zurawski G, Learn G, Clegg MT (1992) Evolutionary relationships of the Caryophyllidae based on comparative *rbc*L sequences. Syst Bot 17:1–15
Gibson AC (1978) Rayless secondary xylem of *Halophytum*. Bull Torrey Bot Club 105:39–44
Gibson AC, Nobel PS (1986) The cactus primer. Harvard University Press, Cambridge
Gilbert MG (1987) The taxonomic position of the genera *Telephium* and *Corrigiola*. Taxon 36:47–49
Gornall RJ, Bohm BA, Dahlgren R (1979) The distribution of flavonoids in the angiosperms. Bot Not 132:1–30
Gregory M, Baas P (1989) A survey of mucilage cells in vegetative organs of the dicotyledons. Isr J Bot 38:125–174
Gurni AA, Kubitzki K (1981) Flavonoid chemistry and systematics of the Dilleniaceae. Biochem Syst Ecol 9:109–114
Haig D (1990) New perspectives on the angiosperm female gametophyte. Bot Rev 56:236–274
Harborne JB, Mabry TJ (eds) (1982) The flavonoids: advances in research. Chapman and Hall, London
Hartley RD, Harris PJ (1981) Phenolic consitutents of the cell walls of dicotyledons. Biochem Syst Ecol 9:189–203
Hershkovitz MA (1989) Phylogenetic studies in Centrospermae: a brief appraisal. Taxon 38:602–610
Hershkovitz MA (1991a) Phylogenetic assessment and revised circumscription of *Cistanthe* Spach (Portulacaceae). Ann Mo Bot Gard 78:1009–1021
Hershkovitz MA (1991b) Leaf morphology of *Cistanthe* Spach (Portulacaceae). Ann Mo Bot Gard 78:1022–1060
Heywood VH (1974) Systematics – the stone of Sisyphus. Biol J Linn Soc 6:169–178
Hofmann U (1973) Centrospermen-Studien. 6. Morphologische Untersuchungen zur Umgrenzung und Gliederung der Aizoaceen. Bot Jahrb Syst 93:247–324
Kluge AG, Farris JS (1969) Quantitative phyletics and the evolution of anurans. Syst Zool 18:1–32
Kluge M, Ting IP (1978) Crassulacean acid metabolism. Springer, Berlin Heidelberg New York
Lawrence GHM (1951) Taxonomy of vascular plants. Macmillan, New York
Leuenberger BE (1986) *Pereskia* (Cactaceae). Mem N Y Bot Gard 41:1–141
Levin GA (1985) Character analysis and cladistics: A response to Rodman et al. Syst Bot 10:496–500
Lewis D (1979) Sexual incompatibility in plants. Arnold, London
Mabry TJ (1974) Is the order Centrospermae monophyletic? In: Bendz G, Santesson J (eds) Chemistry in botanical classification. Nobel Foundation, Stockholm and Academic Press, New York, pp 275–285
Mabry TJ (1976) Pigment dichotomy and DNA-RNA hybridization data for centrospermous families. Plant Syst Evol 126:79–94
Mabry TJ (1977) The order Centrospermae. Ann Mo Bot Gard 64:210–220
Mabry TJ, Behnke HD (1976) Betalains and P-type sieve-element plastids: The systematic position of *Dysphania* R.Br. (Centrospermae). Taxon 25:109–111
Mahato SB, Sarkar SK, Poddar G (1988) Triterpenoid saponins. Phytochemistry 27: 3037–3067
Metcalfe CR, Chalk L (1950) Anatomy of the dicotyledons, I and II. Clarendon Press, Oxford
Narayana PS, Narayana LL (1986) The embryology of Stegnospermataceae with a discussion on its status, affinities and systematic position. Plant Syst Evol 154:137–146
Paris R (1963) The distribution of plant glycosides. In: Swain T (ed) Chemical plant taxonomy. Academic Press, London

Piatelli M (1976) Betalains. In: Goodwin TW (ed) Chemistry and biochemistry of plant pigments, 2nd edn. Academic Press, New York, pp 560–596
Piatelli M (1981) The betalains: Structure, biosynthesis, and chemical taxonomy. In: Conn EE (ed) The biochemistry of plants. A comprehensive treatise, vol 7. Secondary plant products. Academic Press, New York, pp 557–575
Raven PH (1975) The bases of angiosperm phylogeny: Cytology. Ann Mo Bot Gard 62:724–764
Rettig JH, Wilson HD, Manhart JR (1992) Phylogeny of the Caryophyllales – gene sequence data. Taxon 41:201–209
Richardson M (1978) Flavonols and C-glycosylflavonoids of the Caryophyllales. Biochem Syst Ecol 6:283–286
Richardson PM (1981) Flavonoids of some controversial members of the Caryophyllales (Centrospermae). Plant Syst Evol 138:227–233
Ridley M (1986) Evolution and classification. The reformation of cladism. Longman, London
Rizk AM (1986) The phytochemistry of the flora of Qatar. University of Qatar, Doha
Rodman JE (1985) Reply to Levin: "Honest effort was better than none at all". Syst Bot 10:500–503
Rodman JE (1990) Centrospermae revisited, part I. Taxon 39:383–393
Rodman JE (1991a) A taxonomic analysis of glucosinolate-producing plants, part 1: Phenetics. Syst Bot 16:598–618
Rodman JE (1991b) A taxonomic analysis of glucosinolate-producing plants, part 2: Cladistics. Syst Bot 16:619–629
Rodman JE, Oliver MK, Nakamura RR, McClammer JU, Bledsoe, AH (1984) A taxonomic analysis and revised classification of Centrospermae. Syst Bot 9:297–323
Rodman J, Price RA, Karol K, Conti E, Sytsma KJ, Palmer JD (1993) Nucleotide sequences of the *rbc*L gene indicate monophyly of mustard oil plants. Ann Mo Bot Gard 80:686–699
Rogers GK (1985) The genera of Phytolaccaceae in the southeastern United States. J Arnold Arbor 66:1–37
Rohlf FJ (1972) An empirical comparison of three ordination techniques in numerical taxonomy. Syst Zool 21:271–280
Rohlf FJ (1987) NTSYS-pc. Numerical taxonomy and multivariate analysis system, version 1.21. Applied Biostatistics, Setauket, NY
Skvarla JJ, Nowicke JW (1982) Pollen fine structure and relationships of *Achatocarpus* Triana and *Phaulothamnus* A. Gray. Taxon 31:244–249
Sneath PHA, Sokal RR (1973) Numerical taxonomy. Freeman, San Francisco
Sokal RR (1986) Phenetic taxonomy: theory and methods. Ann Rev Ecol Syst 17:423–442
Sperling CR (1987) Systematics of the Basellaceae. Thesis, Harvard University, Cambridge
Swofford DL (1985) PAUP. Phylogenetic analysis using parsimony, version 2.4. Illinois Natural History Survey, Champaign
Swofford DL, Maddison WP (1987) Reconstructing ancestral character states under Wagner parsimony. Math Biosci 87:199–229
Takhtajan A (1969) Flowering plants: Origin and dispersal. Smithsonian Institution Press, Washington, DC
Takhtajan A (1986) Floristic regions of the world. University of California Press, Berkeley
Teeri JA (1982) Carbon isotopes and the evolution of C_4 photosynthesis and crassulacean acid metabolism. In: Nitecki MH (ed) Biochemical aspects of evolutionary biology. University of Chicago Press, Chicago, pp 93–130
Thorne RF (1968) Synopsis of a putatively phylogenetic classification of the flowering plants. Aliso 6:57–66
Thorne RF (1992) An updated phylogenetic classification of the flowering plants. Aliso 13:365–389

Valant-Vetschera KM (1985) C-glycosylflavones as an accumulation tendency: A critical review. Bot Rev 51:1–52

Young DA (1981) The usefulness of flavonoids in angiosperm phylogeny: Some selected examples. In: Young DA, Seigler DS (eds) Phytochemistry and angiosperm phylogeny. Praeger, New York, pp 205–232

Zuckerkandl E, Pauling L (1965) Evolutionary divergence and convergence in proteins. In: Bryson V, Vogel HJ (eds) Evolving genes and proteins. Academic Press, New York, pp 97–166

13 Putative Origin and Relationships of the Order from the Viewpoint of Developmental Flower Morphology

PETER LEINS and CLAUDIA ERBAR

13.1 Introduction

From the viewpoint of developmental flower morphology the most archaic families within the order Caryophyllales are the Cactaceae and Aizoaceae. These families are linked to paeonioid ancestors. The Paeoniales presumably occupy a place near the point of divergence of Dilleniidae and Caryophyllidae from the Magnoliidae. It is a matter of taste whether the Paeoniales are placed in the Dilleniidae or in the Magnoliidae (cf. Leins and Erbar 1991b). The Paeoniaceae agree in some important features with the Magnoliidae and in other ones with the Dilleniidae. With the Magnoliidae they share the spiral sequence of all flower primordia (Plate 13.1,1), the transitions between calyx and corolla leaves, and the epicuticular wax crystalloids of the Aristolochia type (transversely ridged rodlets; Barthlott 1991). With the Dilleniidae they share the centrifugal subdivision of the primary androecial organs (Plate 13.1,1), the persistent calyx, and the large follicular fruit. Even though the Paeoniaceae lack benzylisoquinoline alkaloids, we can still assume that they originated from the Magnoliidae. The most important new character which distinguishes the Paeoniaceae from the Magnoliidae is the fascicled centrifugal androecium. It is, however, conceivable that this structure could be derived from a spiral magnolioid androecium in which the stamens originate directly on the floral apex. The fascicled androecium, in which the primary primordia have a spiral sequence, as in *Paeonia*, may have evolved from a magnolioid androecium via enlargement of the areas occupied by the androecial organ primordia. Consequently, there is only space for five of these enlarged androecial organs in early development. The still active genetic information for a multistaminate androecium then causes a splitting of the five androecial primordia (dédoublement; cf. Leins and Erbar 1991a).

The fascicled centrifugal androecium – if the androecium is multistaminate – is one of the main characters of the subclass Dilleniidae. A spiral sequence of the fascicles is sometimes maintained within the Theales (*Stewartia*, Plate 13.1,2), but the divergence angle has changed from the

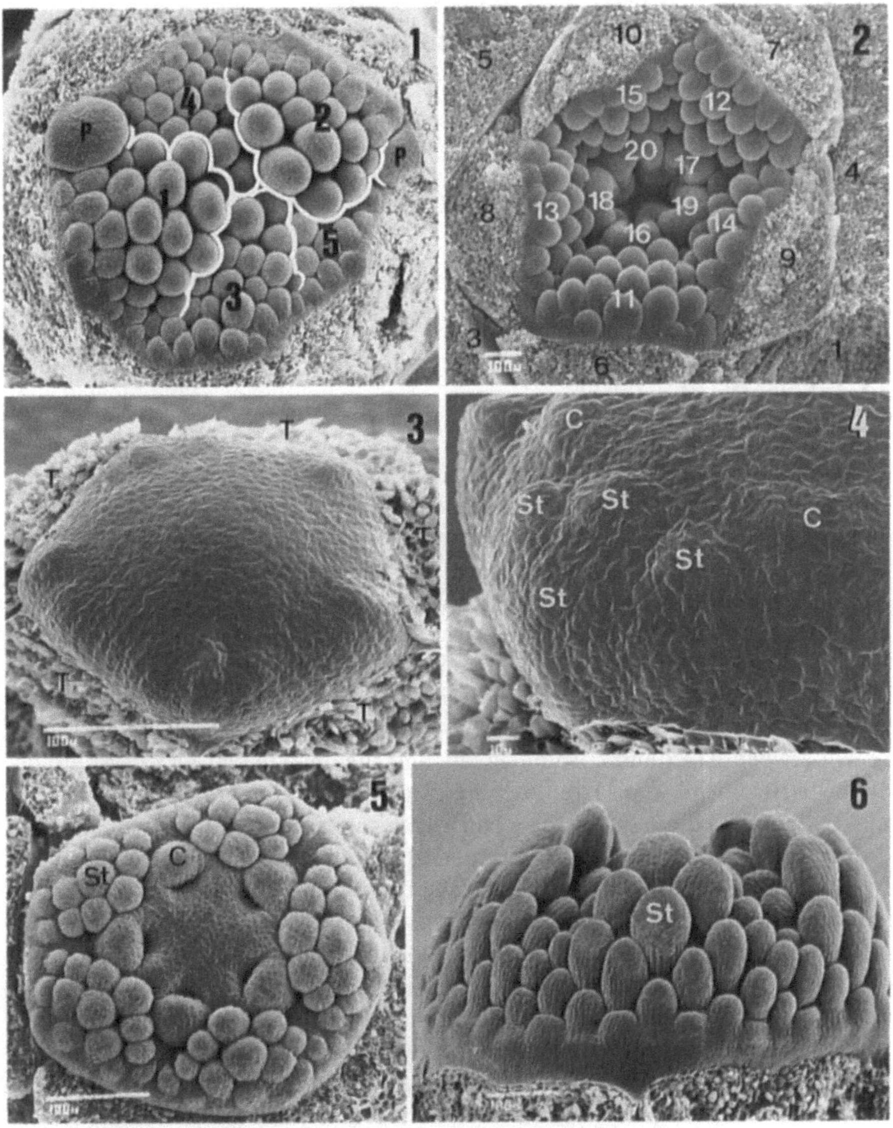

Plate 13.1. *1*, *Paeonia officinalis* L. (Paeoniaceae). Initiation of the androecium; the five stamen fascicles are encircled and numbered according to the spiral sequence; perianth members removed, except the two youngest (*P*). *2*, *Stewartia pseudocamellia* Maxim. (Theaceae). Spiral sequence of organs throughout the flower: *1–5*, sepals (removed, *2* not visible); *6–10*, petals; *11–15*, fascicle primordia; *16–20*, carpels. *3–6*, *Lampranthus promontorii* N.E.Br. (Aizoaceae). Development of the complex centrifugal androecium, starting with five formation centres alternating with five perianth members (*T*). *St*, stamen; *C*, carpel

"golden" divergence angle to 2/5 divergence (Erbar 1986, 1988). In several taxa within the Dilleniidae, the primary primordia do not arise separately but form a common ring wall. The authors have pointed out in several papers (Leins 1979; Leins and Erbar 1991a,b) that simple diplostemonous or haplostemonous androecia in the Dilleniidae probably were derived from fascicled androecia repeatedly during evolution. The formation of a ring primordium instead of separate fascicle primordia, followed by a reduction in size of this ring wall until it disappears completely, are the main steps in one of the evolutionary lines leading to simple androecia. Alternatively, the primary fascicle primordia may fail to subdivide, and thus develop directly to single stamens.

13.2 The Fascicled Centrifugal Androecium as a Basis of Argumentation Concerning the Origin of the Caryophyllales

In the Caryophyllales multistaminate androecia basically correspond to those of the Paeoniales and the Dilleniidae. In the Aizoaceae investigated by us (*Aptenia cordifolia*, *Lampranthus promontorii*, *Sceletium anatomicum*, and *Sphalmanthus* sp.), the primary primordia of the fascicled androecium are not strictly separate from each other (there is more of a tendency to form a ring wall). When the centrifugal subdivision starts, however, five (four) formation centres become distinct, alternating with the five (four) spirally arranged perianth members (Plate 13.1,3–5; cf. Payer 1857; Ihlenfeldt 1960). The borderlines between the formation centres later become blurred, at least basally (Plate 13.1,6). In the tetramerous flowers of *Gunniopsis glabra* and *G. zygophylloides* there are four really distinct primary primordia before these become subdivided (Bittrich, pers. comm.). In Aizoaceae, the basal members of the androecium develop into petaloid staminodia. The second family within the Caryophyllales that is characterized by a multistaminate androecium is the cactus family. In this group we usually find a uniform, prominent ring primordium, on which the stamens arise in centrifugal succession (Plate 13.2,7 and 8). Formation centres, as in the Aizoaceae, are not discernible. Only in the most primitive genus of Cactaceae, namely *Pereskia*, are there some species (*Pereskia bleo* and *P. diaz-romeroana*; Leins and Schwitalla 1986) in which we can observe nearly separate androecial primary primordia that follow the spiral sequence of the numerous perianth members (Plate 13.2,9). They become connected, however, before the centrifugal subdivision starts. The spiral sequence can now be inferred only from the different degree of subdivision of the primary primordia (Plate 13.2,10). In this respect the androecium of *Pereskia* is closely related to that of *Paeonia*. We believe that *Pereskia* is not only the most archaic member of Cactaceae, but of the whole order Caryophyllales. From ancestors with *Pereskia*-like

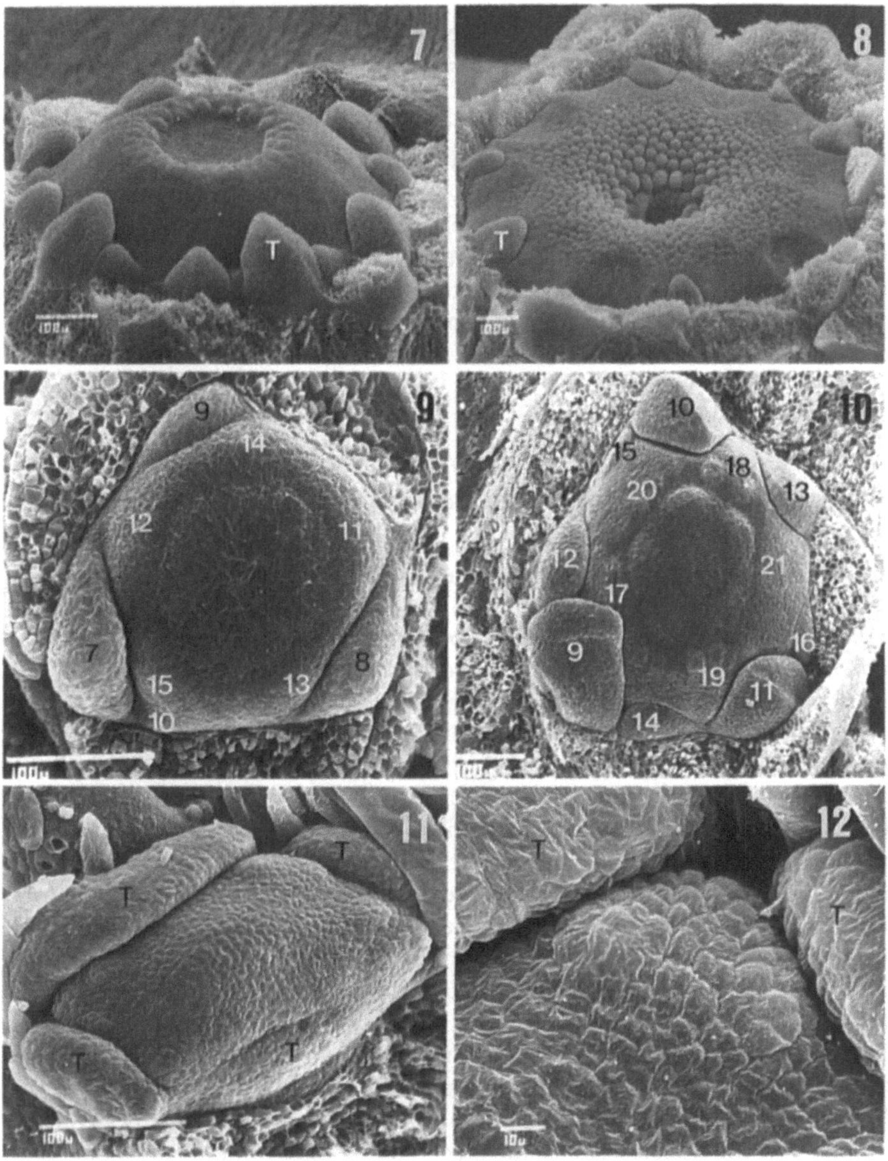

Plate 13.2. *7* and *8*, *Neobuxbaumia polylopha* (DC.) Backeb. (Cactaceae). Centrifugal initiation of stamens on an androecial ring primordium; *T*, perianth member. *9*, *Pereskia diaz-romeroana* Cárdenas (Cactaceae). Young flower bud; numbering according to the spiral sequence of the organs: 7–10, inner perianth members; *11–15*, androecial primary primordia. *10*, *Pereskia bleo* (Kunth) DC. (Cactaceae). Different degree of subdivision of the primary primordia (17–21); 9–16, spirally arranged inner perianth members. *11* and *12*, *Tetragonia tetragonioides* (Pallas) Kuntze (Aizoaceae). Initial stage of the androecium: Alternating with the four perianth members (*T*), four bulges are visible, which start differentiation into stamens. *12*, Part of the flower shown in *11*

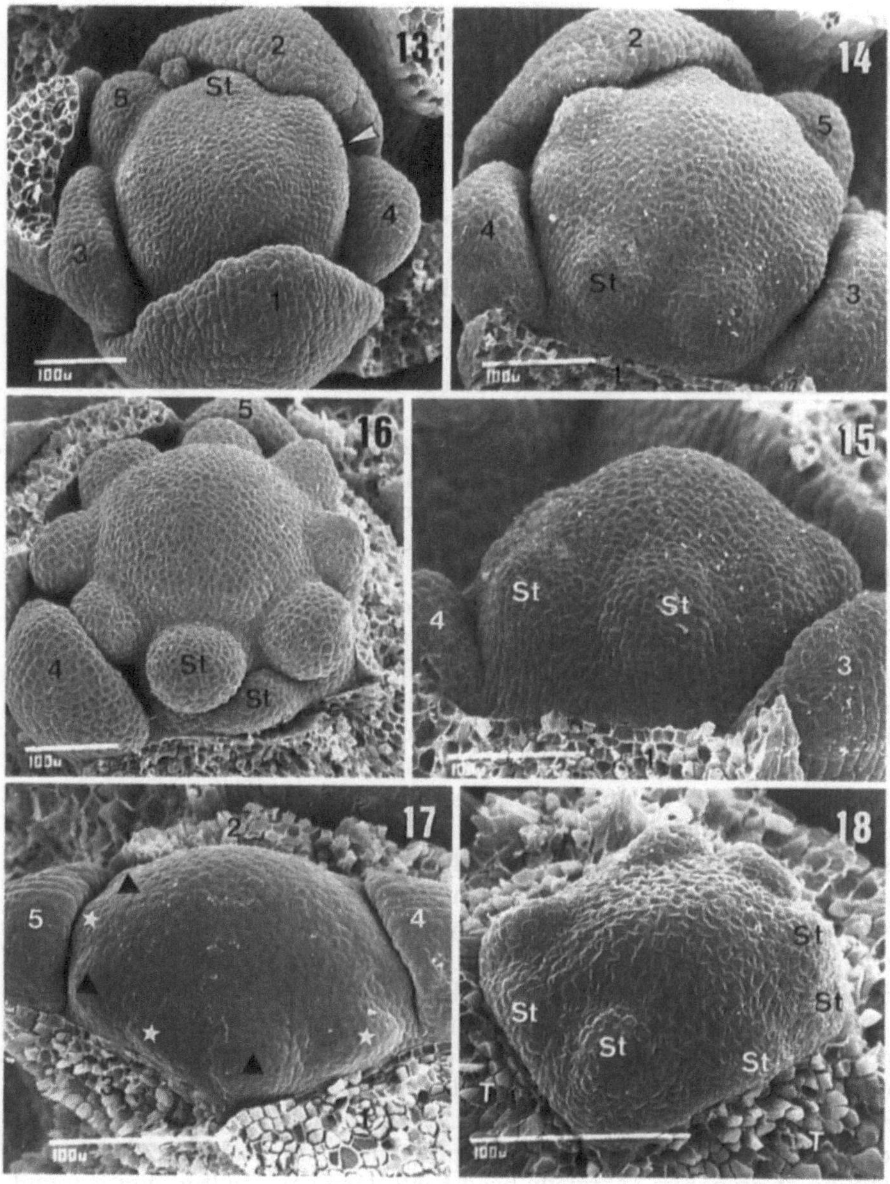

Plate 13.3. *13–16*, *Phytolacca acinosa* Roxb. (Phytolaccaceae). *13–15*, Eight stamen primordia (*St*) originate more or less simultaneously in one whorl, the perianth members (*1–5*) are spirally arranged. The floral apex in *13* shows a broad protuberance on the right (*arrow*), perhaps a remnant of an androecial ring primordium. *15*, Same flower bud as in *14*, side view. *16*, An additional stamen has arisen outside the initial stamen whorl. *17*, *Phytolacca americana* L. (Phytolaccaceae). Ten stamen primordia, arranged almost as in a diplostemonous androecium (also see text). *18*, *Trichostigma peruvianum* (Moq.) H. Walter (Phytolaccaceae). Centrifugal initiation of (twelve) stamens (*St*); the four perianth members (*T*) have been removed

flowers we can perhaps derive the Aizoaceae, in which the stamen fascicles and the formation centres respectively, are usually preserved. As we have pointed out earlier for the Dilleniidae (Leins 1979), also in the Caryophyllales there may have been a tendency in the early evolution towards a reduction of the primary primordia or of the common primary ring primordium, on which consequently only a lower number of stamens could arise. This appears to have happened in the Phytolaccaceae, which we perhaps can link with the Aizoaceae via *Tetragonia*. In young flower buds of *Tetragonia* we can observe more or less distinct bulges alternating with the (five to four) perianth members (Plate 13.2,11 and 12). The relatively few stamen primordia, however, do not follow a strict pattern during initiation (cf. Payer 1857). Nevertheless, a centrifugal development is still hinted at. In *Phytolacca acinosa* and *P. americana* neither primary primordia nor a ring primordium are discernible; at most there can be a more or less broad bulge (arrow in Plate 13.3,13) that may be interpreted as a remnant of an androecial ring primordium. The reduction of this ring primordium corresponds to a more or less fixed stamen number. In *Phytolacca acinosa*, with five spirally arranged perianth members, (mostly) eight stamen primordia are initiated in one whorl simultaneously or in a very rapid sequence (Plate 13.3,14 and 15). Often additional stamens (one or two) arise centrifugally in front of the first and/or second perianth member (Plate 13.3,16). In *Phytolacca americana*, which also has five perianth members, the stamen number is more or less fixed at ten. There seems to be a tendency towards a diplostemonous androecium. This can be concluded from the flower bud shown in Plate 13.3,17, where three stamens can be seen (marked by triangles) alternating with the perianth members 1, 3, 5 and 2, while three others (marked by asterisks) are situated opposite the perianth members. Sometimes an additional stamen arises outside the initial stamen whorl. Our results do not confirm a collateral splitting of initially five uniform primordia in *Phytolacca americana* (*P. decandra*), as has been reported in the literature (Payer 1857; Eichler 1878; Heimerl 1889, 1934; Eckardt 1964).

It wold not be surprising if a change from a centrifugal to a centripetal inception of the stamens were to take place in the Phytolaccaceae, when the primary primordia are completely reduced. However, a centripetal stamen initiation in *Phytolacca icosandra*, as reported by Payer (1857), could not be confirmed by Eckert (1965, Fig. 146). She also did not observe a collateral splitting of primordia. A totally fixed haplostemonous androecium is found in *Rivina humilis*, with four stamens alternating with the perianth (the four perianth members are situated in the median and transversal plane of the flower). In *Trichostigma peruvianum* (Plate 13.3,18) the first four stamens arise in the same position as in *Rivina*. Eight additional stamen primordia originate centrifugally: two in the median plane of the flower and two in the transversal plane, followed by four further stamen primordia arising more or less simultaneously behind the first four stamens (see also Chap. 6 and Ronse Decraene and Smets 1991).

13.3 The Gynoecium as a Basis of Argumentation Concerning the Relationships of the Caryophyllales

In no other group of flowering plants is the placentation so much influenced by intercalary growth in the floral axis as in the Aizoaceae and Cactaceae.

In the Aizoaceae investigated in this study the carpel primordia (mostly five or four) are initiated on a slightly convex floral apex (Plate 13.1,4). The carpel primordia soon become connected through radial septa which arise only at the periphery of the gynoecium, i.e. they do not meet at the centre (Plates 13.1,5 and 13.4,19 and 20). Through intercalary growth beneath the dorsal parts of the carpel primordia the ovary becomes semi-inferior to inferior. At the same time, longitudinal growth of the floral axis leads to an elongation of the insertion area of the septa, displacing them to a steeper position. Between the septa long placentae arise on the floral axis. The placentae are uniform (e.g. *Lampranthus promontorii*, Plate 13.4,20) or have a shallow longitudinal furrow which divides them into two parts (*Sceletium anatomicum* and *Sphalmanthus* sp.).

It has been known for a long time (Payer 1857; Straka 1955; Ihlenfeldt 1960) that in many members of the family intercalary growth in the floral axis during the inception of the ovules is responsible for a shifting of the placentae to the periphery of the ovary (e.g. *Lampranthus*, Plate 13.4,21–26). Hypogynous flowers (Aizooideae) and central placentation (e.g. in *Aptenia*, *Sceletium* and *Sphalmanthus*) can be interpreted as an archaic condition as well as the result of a secondary loss of the corresponding intercalary growth.

In the Cactaceae the placentae are mostly formed on the wall of a floral cup (Plate 13.2,8). As in the Aizoaceae, the placentae alternate with the septa and mostly show no connection with these. In the Cactaceae, however, the septa are only low ledges in the upper part of the inferior unilocular ovary (Plate 13.5,28), which sometimes is more or less superior during its initiation. In an earlier paper (Leins and Schwitalla 1988) this arrangement of the placentae in Cactaceae was called hypanthial placentation. This is the condition in most Cactaceae; in *Rhipsalis* the independent placentae even fuse at the bottom of the inferior ovary. The hypanthial placentation can be understood phylogenetically if one looks at the gynoecium of those primitive members of the cactus family whose archaic androecial pattern is discussed above, namely, *Pereskia*. In *Pereskia* the placentae are clearly connected with the septa. In some species of this genus (e.g. *P. bleo*, *P. diaz-romeroana* and *P. grandifolia*) V-shaped placentae are formed along the bases of the carpel flanks (septa) where these are inserted on the weakly concave floral axis (Plate 13.5,27). We can call it a basal-laminal placentation. The hypanthial placentation can be derived from the basal-laminal placentation via broadening and flattening of the placentae, correlated with an intercalary growth in the floral cup and resulting in the displacement of the ovules away

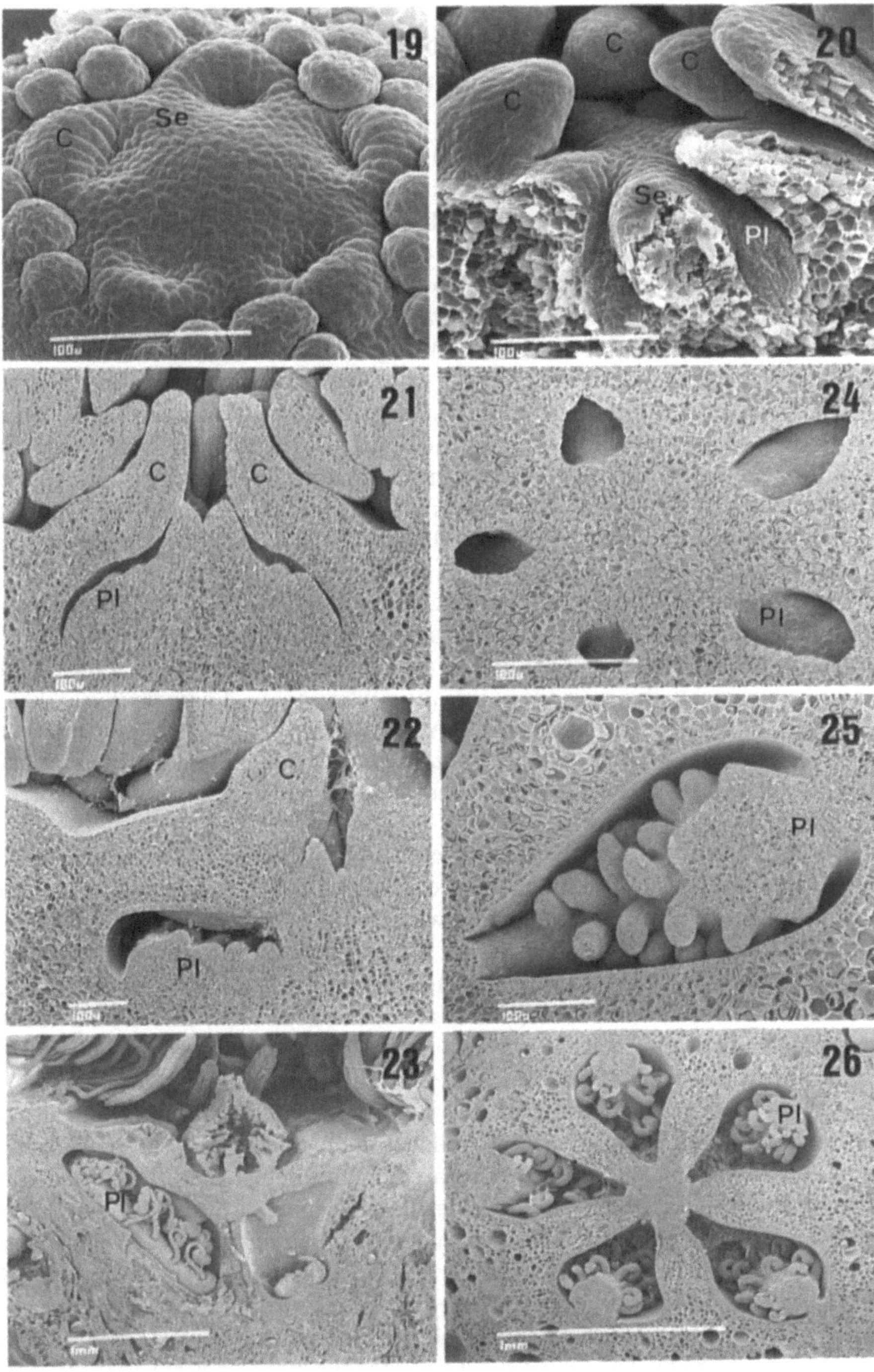

Plate 13.4. *19–26*, *Lampranthus promontorii* N.E.Br. (Aizoaceae). *19*, Young gynoecium. *20*, Dorsal parts of two carpels removed to show the position of the placentae on the floral axis between the septa. *21–26*, Shifting of the placentae to the periphery of the ovary; *21–23*, longitudinal sections, *24–26*, cross sections. *C*, carpel; *Pl*, placenta; *Se*, septum

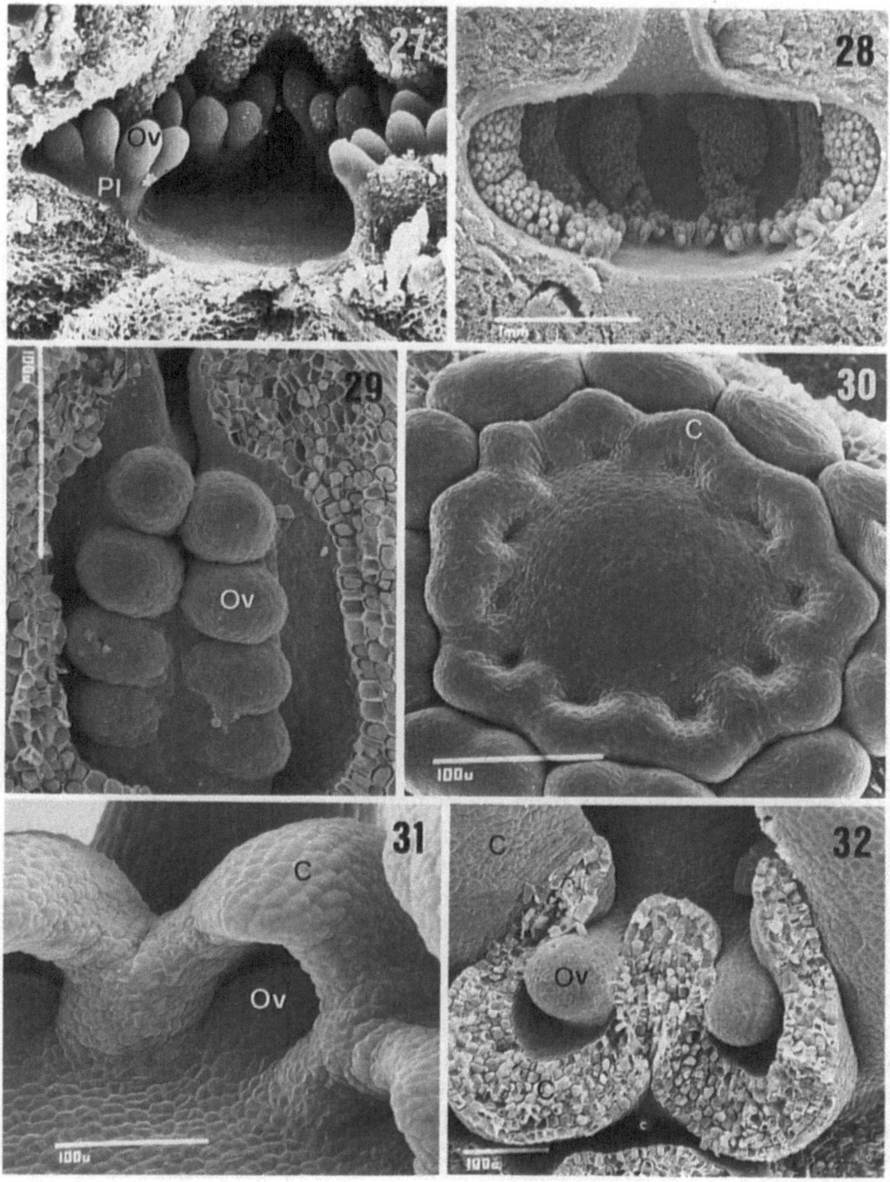

Plate 13.5. *27*, *Pereskia grandifolia* Haw. (Cactaceae). View into a young ovary, longitudinal section, showing the basal-laminal placentation. *28*, *Browningia hertlingiana* (Backeb.) F.Buxb. (Cactaceae). View into a young ovary, longitudinal section, showing the hypanthial placentation. *29*, *Gypsophila repens* L. (Caryophyllaceae). Lateral view into one of the two locules of the ovary. *30*, *Phytolacca americana* L. (Phytolaccaceae). Young coenocarpous gynoecium. *31* and *32*, *Phytolacca acinosa* Roxb. (Phytolaccaceae). Initiation of the single ovule per carpel from the floral axis. In *32* two carpels are cut open. *C*, carpel; Ov, ovule; Pl, placenta; *Se*, septum

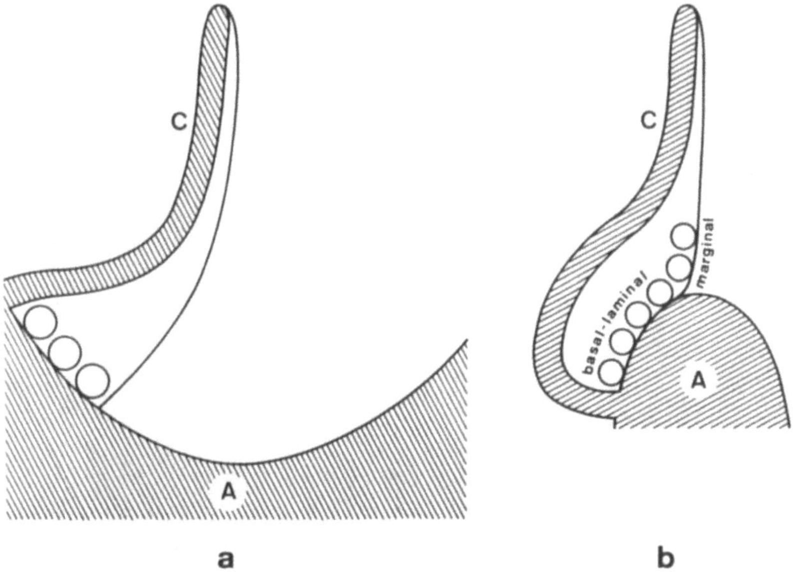

Fig. 13.1a,b. Placentation in *Pereskia*: **a** where the carpel is inserted on a concave floral axis, and Caryophyllaceae **b** where the carpel is inserted on a cone-like floral axis. *A*, Floral axis; *C*, carpel

from the septa. This process can be observed during ontogeny, e.g. in *Opuntia*. The next phylogenetic step is the formation of placentae, which are separate from the septa from the beginning.

The basal-laminal placentation seems to be a basic model for the variety of placentation forms in the Caryophyllales. We can transfer our model from *Pereskia* even to the Caryophyllaceae, where the bases of the septa are in connection with a steep, cone-like floral axis. In principle, there are only differences in the proportions of the carpel parts (Fig. 13.1). In Caryophyllaceae often a marginal placentation apically adjoins the basal-laminal placentation (Plate 13.5,29). The change of the position of the placentae and ovules from the bases of the septa to the floral axis may have occurred repeatedly within the Caryophyllales during early evolution.[1] Therefore it is not surprising if, after a reduction in the number of ovules, the single ovule per carpel, as in *Phytolacca* (Plates 13.5,30–32 and 13.6,33) or *Tetragonia*, originates from the floral axis between the carpel flanks (septa). In the monocarpellate gynoecia of *Trichostigma peruvianum* (Phytolaccaceae; Plate 13.6,34–36), *Rivina humilis* (Phytolaccaceae; Plate 13.6,37) and *Mirabilis viscosa* (Nyctaginaceae; Plate 13.6,38), which show a very uniform ontogeny, the single ovule is formed from a gradual upgrowth of the floral apex, en-

[1] In many cases it is impossible to decide whether the placenta is part of the septum or of the floral axis. This question, however, is insignificant for our phylogenetic considerations.

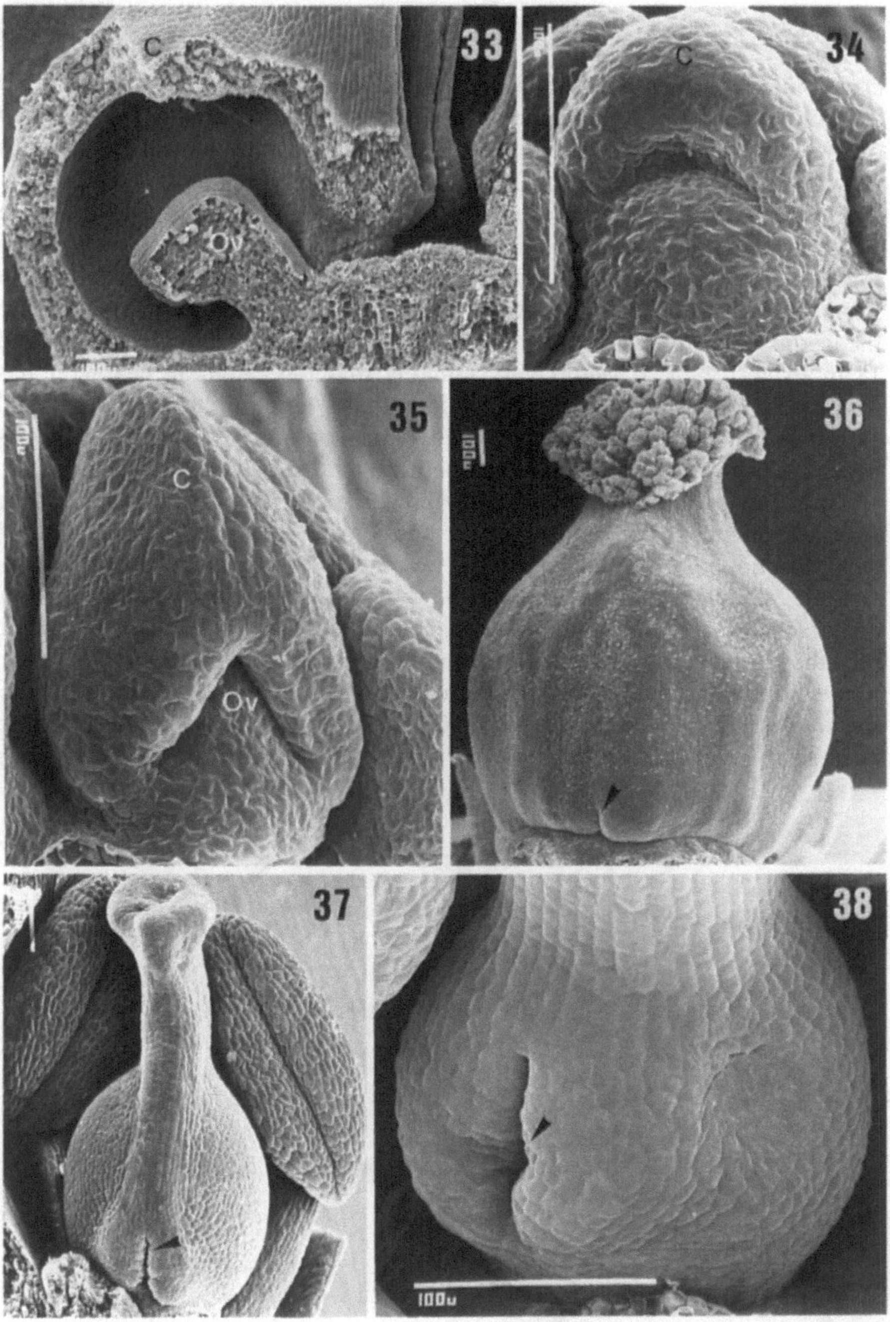

Plate 13.6. *33*, *Phytolacca acinosa* L. (Phytolaccaceae). Carpel flank partly removed to show the position of the single ovule. *34–36*, *Trichostigma peruvianum* (Moq.) H. Walter (Phytolaccaceae). Development of the monocarpellate gynoecium. *37*, *Rivina humilis* L. (Phytolacaceae). Monocarpellate pistil. *38*, *Mirabilis viscosa* Cav. (Nyctaginaceae). Basal part of the monocarpellate pistil. *Arrowheads* in *36–38* indicate the very short ventral slit at the carpel base. *C*, carpel; Ov, ovule

closed in a lateral carpel (cf. Sattler and Perlin 1982 for Nyctaginaceae). An affinity between the Nyctaginaceae and the genera *Rivina* and *Trichostigma* of the Phytolaccaceae is supported by the fact that the ventral slit remains very short at the very base of the carpel in all taxa (Payer 1857).

Let us go back to the presumably most archaic placentation form within the Caryophyllales, namely, the basal-laminal placentation. It is exciting that this unique feature (in addition with a concave floral receptacle, like in *Pereskia*) can also be found at the base of the Dilleniidae, namely in the theaceous genus *Stewartia* (see Figs. 13–19 in Erbar 1986 and Figs. 17 and 18 in Erbar 1988). It seems very unlikely that this common character has developed in parallel. We may assume that *Pereskia*-like Cactaceae and *Stewartia*-like Theaceae have evolved from common paeonioid ancestors which had, however, already developed the tendency towards syncarpy.

In response to the objection of Kubitzki (see Chap. 14), who states that the Cactaceae cannot be placed at the base of the Caryophyllales because of their "pseudoparietal placentation", we would like to make the following remark: When judging which might be the archaic condition in placentation we attach more value to the way the placentae and ovules are connected with the carpels than to the form of the floral axis during anthesis. The clear connection of the placentae with the carpels, i.e. the fact that the placentae and ovules are part of the carpel flanks, is precisely the character in which we regard *Pereskia* as archaic. We assume that the form of the floral axis,

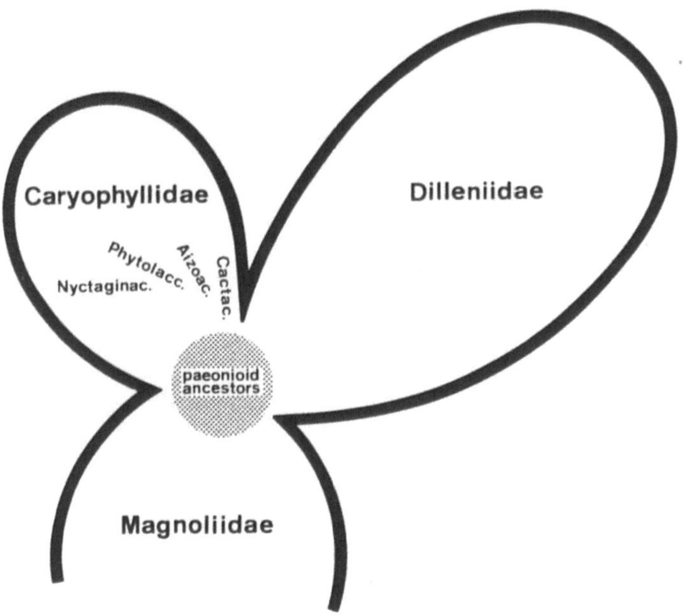

Fig. 13.2. Putative evolutionary relationships between Magnoliidae, Dilleniidae and Caryophyllidae

which is influenced by intercalary growth, was subjected to a high plasticity at the beginning of centrosperm evolution. We should like to emphasize once more that the basal-laminal placentation, i.e. the formation of placentae and ovules on the septa where these are connected to the floral axis (regardless of whether the latter is steeply conical, as in, e.g., Caryophyllaceae and some Aizoaceae, or more or less concave, as in *Pereskia*), is an unusual phenomenon within the angiosperms. Within the Caryophyllales, however, it is a basic model, as described above. The occurrence of marginal placentation in addition to basal-laminal placentation in the Caryophyllaceae can be regarded as a primitive feature in the sense of heterobathmy, just like the production of anthocyanins, whereas in other characters the family appears to be more advanced.

13.4 Conclusion

The Caryophyllales (Caryophyllidae) have often been connected with the Magnoliidae/Ranales (e.g. Takhtajan 1959, 1980, 1987; Cronquist 1981, 1988; Ehrendorfer 1983), the Dilleniidae (Leins 1975; Erbar 1988) or a "Guttiferalean complex" (Cronquist 1957, 1965). However, Cronquist (1957, 1965) refers the "Guttiferalean complex" to the "Ranalian complex". Ehrendorfer (1991) remarks that the Caryophyllales have Ranunculidae- (or ?Dilleniidae-) like ancestors. Summarizing the results of our flower developmental investigations, the Caryophyllales are, at their base, closely connected with the Dilleniidae, and also related to the Magnoliidae via paeonioid ancestors (Fig. 13.2). The most primitive group within the Caryophyllales is the family Cactaceae, particularly the subfamily Pereskioideae, with the Aizoaceae being an early offshoot. From the Aizoaceae, the Phytolaccaceae can be derived and from these, in turn, the Nyctaginaceae. The other groups of Caryophyllales, which are more advanced, are not taken into consideration.

Acknowledgement. We would like to express our sincere thanks to Jens Rohwer for carefully reading and correcting the English text.

References

Barthlott W (1991) Epicuticulare Wachse der Tracheophyten: Systematische, mikromorphologische und ökologische Aspekte. 10th Symp Morphol Anat Syst Göttingen Abstr: 2

Cronquist A (1957) Outline of a new system of families and orders of dicotyledons. Bull Jard Bot Etat Brux 27:13–40

Cronquist A (1965) The status of the general system of classification of flowering plants. Ann Mo Bot Gard 52:281–303
Cronquist A (1981) An integrated system of classification of flowering plants. Columbia University Press, New York
Cronquist A (1988) The evolution and classification of flowering plants, 2nd edn. New York Botanical Garden, New York
Eckardt T (1964) 13.Reihe: Centrospermae. In: Melchior A (ed) A. Engler's Syllabus der Pflanzenfamilien II. Borntraeger, Berlin, pp 79–101
Eckert G (1965) Entwicklungsgeschichtliche und blütenanatomische Untersuchungen zum Problem der Obdiplostemonie. Bot Jahrb Syst 85:523–604
Ehrendorfer F (1983) Dritte Abteilung: Spermatophyta, Samenpflanzen. In: Strasburger E (ed) Lehrbuch der Botanik für Hochschulen, 32nd edn. Fischer, Stuttgart, pp 758–915
Ehrendorfer F (1991) Dritte Abteilung: Spermatophyta, Samenpflanzen. In: Strasburger E (ed) Lehrbuch der Botanik für Hochschulen, 33rd edn. Fischer, Stuttgart, pp 699–835
Eichler AW (1878) Blüthendiagramme II. Engelmann, Leipzig
Erbar C (1986) Untersuchungen zur Entwicklung der spiraligen Blüte von *Stewartia pseudocamellia* (Theaceae). Bot Jahrb Syst 106:391–407
Erbar C (1988) Early developmental patterns in flowers and their value for systematics. In: Leins P, Tucker SC, Endress PK (eds), Aspects of floral development. Cramer, Berlin, pp 7–23
Heimerl A (1889) Phytolaccaceae. In: Engler A, Prantl K (eds) Die natürlichen Pflanzenfamilien, 1st edn, vol 3(1b). Engelmann, Leipzig, pp 1–14
Heimerl A (1934) Phytolaccaceae. In: Engler A (ed) Die natürlichen Pflanzenfamilien, 2nd edn, vol 16c. Engelmann, Leipzig, pp 135–164
Ihlenfeldt H-D (1960) Entwicklungsgeschichtliche, morphologische und systematische Untersuchungen an Mesembryanthemen. Feddes Repert 63:1–104
Leins P (1975) Die Beziehungen zwischen einfachen und multistaminaten Androeceen. Bot Jahrb Syst 96:231–237
Leins P (1979) Der Übergang vom zentrifugalen komplexen zum einfachen Androeceum. Ber Dtsch Bot Ges 92:717–719
Leins P, Erbar C (1991a) Entwicklungsmuster in Blüten und ihre mutmaßlichen phylogenetischen Zusammenhänge. Biol Unserer Zeit 21:196–204
Leins P, Erbar C (1991b) Fascicled androecia in Dilleniidae and some remarks on the *Garcinia* androecium. Bot Acta 104:336–344
Leins P, Schwitalla S (1986) Studien an Cacteen-Blüten. I. Einige Bemerkungen zur Blütenentwicklung von *Pereskia*. Beitr Biol Pflanz 60:313–323
Leins P, Schwitalla S (1988) Placentation in Cactaceae. In: Leins P, Tucker SC, Endress PK (eds) Aspects of floral development. Cramer, Berlin, pp 57–68
Payer J-B (1857) Traité d'organogénie compareé de la fleur. Masson, Paris
Ronse Decraene LP, Smets EF (1991) The floral ontogeny of some members of the Phytolaccaceae (subfamily Rivinoideae) with a discussion on the evolution of the androecium on the Rivinoideae. Biol Jahrb Dodonaea 59:77–99
Sattler R, Perlin L (1982) Floral development of *Bougainvillea spectabilis* Willd., *Boerhaavia diffusa* L. and *Mirabilis jalapa* L. (Nyctaginaceae). J Linn Soc Bot 84:161–182
Straka H (1955) Anatomische und entwicklungsgeschichtliche Untersuchungen an Früchten paraspermer Mesembryanthemen. Nova Acta Leopold NF 17(118):127–190
Takhtajan A (1959) Die Evolution der Angiospermen. Fischer, Jena
Takhtajan A (1980) Outline of the classification of flowering plants (Magnoliophyta). Bot Rev 46:225–359
Takhtajan A (1987) Systema Magnoliophytorum. Officina Editoria, Leningrad (in Russian)

14 A Note on the Relationships of the Order Within the Angiosperms

K. Kubitzki

The systematic position of the Centrospermae (excluding Polygonales and Plumbaginales) within the dicotyledons is a matter of ongoing controversy. While conventionally they are considered to be related with the Magnoliidae in the broad sense, or with the Ranunculales or Illiciales more specifically, more recently affinities with dilleniids, or even something like *Paeonia* (see Chap. 13), have been suggested.

Here I wish to stress that from the chemosystematic point of view, the Centrospermae clearly belong in a major group together with the magnoliids (Magnoliidae including Ranunculales sensu Cronquist 1988) and the monocotyledons. All these taxa contain secondary metabolites originating from a fully expanded shikimate pathway, in which intermediary products, such as allyl and propenyl phenols, cinnamic alcohols, cinnamic acids, and phenylalanine/tyrosine, are channelled into secondary metabolites (Fig. 14.1; see also Kubitzki and Gottlieb 1984a,b). In higher evolved dicotyledons (Hamamelidae, Dilleniidae, Rosidae) intermediary products originating from a truncated shikimate pathway, such as chorismic acid and shikimic acid, give rise to secondary products like anthranilate alkaloids and ellagitannins. In the most highly evolved dicotyledons (Asteridae) the attenuation of this development leads to the near exclusion of the shikimate pathway and its substitution by acetate and mevalonate as intermediary products, from which steroid and iridoid alkaloids, acetylenes and sesquiterpene lactones originate.

The Centrospermae fit into the first group, because the betalains are biogenetically closely related with the alkaloids of the Magnoliidae and monocotyledons. The biogenetic scheme (Fig. 14.1) is self-explanatory in this respect.

In dicotyledons, the transition from woody to herbaceous forms is accompanied by two phenomena: first, the diminution and final disappearance of proanthocyanidins; and second, the accumulation of caffeic acid esters (Mølgaard and Ravn 1988), coumarins, and other cinnamate-derived compounds. Obviously, the large amounts of cinnamate, no longer required for lignin synthesis in herbaceous forms, are channelled into such compounds. The appearance of caffeic acid esters and related compounds is observed in

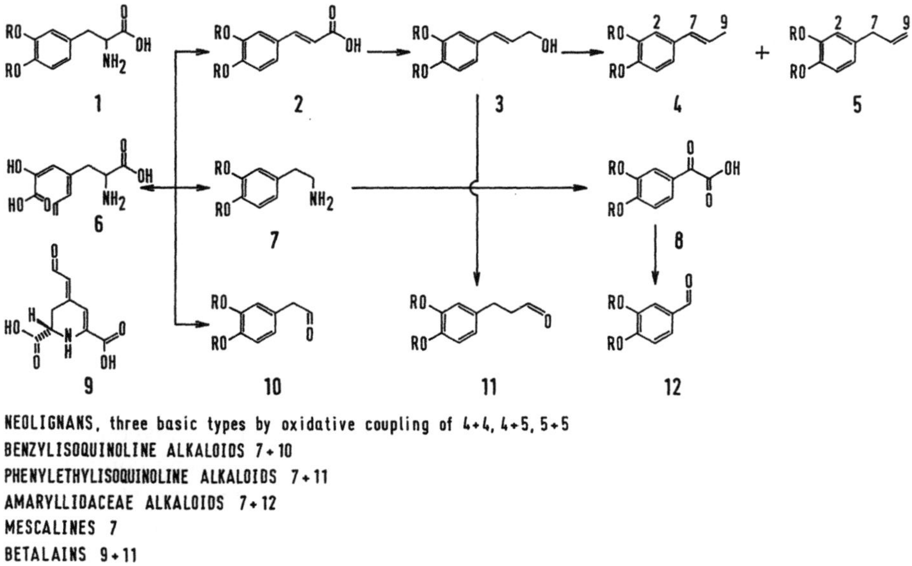

Fig. 14.1. Biogenetic scheme demonstrating the close biogenetic relationships among magnoliids, monocotyledons, and centrosperms. *1*, Phenylalanine; *2*, cinnamic acid; *3*, cinnamylalcohol; *4*, propenylbenzene; *5*, allylbenzene; *6*, 4,5-secophenylalanine; *7*, phenylethylamine; *8*, phenyloxoacetic acid; *9*, betalamic acid; *10*, acetaldehyde; *11*, cinnamaldehyde; *12*, benzaldehyde (all as dihydroxy- forms)

the herbaceous Ranunculales and likewise in many herbaceous members of sympetalous (asterid) orders such as Scrophulariales and Lamiales (Kubitzki 1987; Jensen 1992).

The latter two phenomena apply also to the Centrospermae. The rareness of proanthocyanidins was demonstrated first by Bate-Smith (1962) and has been amply confirmed since (despite the occasional occurrence of these compounds in seed coats, Bittrich and Amaral 1991). Ferulic acid (a methoxylated derivative of caffeic acid) is abundantly present in the cell walls of the Centrospermae (and part of the monocotyledons).

I regard this as circumstantial evidence for a basically herbaceous nature of the Centrospermae, which would exclude any relationship with a basically woody group such as the lower dilleniids. In total, secondary metabolism indicates that the Centrospermae are a basically herbaceous lineage of magnoliid relationship, collateral to the equally herbaceous Ranunculales and monocotyledons. In their possession of tricolpate pollen the Centrospermae have attained the level of the Ranunculales, while their sieve-element plastids are slightly reminiscent of the P-type plastids of lower (monosulcate) magnoliids.

This is in contrast to Leins and Erbar (see Chap. 13), who prefer to consider the Cactaceae and especially *Pereskia* as the most primitive member of all Centrospermae. Leins and Erbar based their opinion on floral morphology, especially androecial development, and on the peculiar pseudoparietal placentation which, among the Centrospermae, is restricted to the Cactaceae and Aizoaceae subfam. Ruschioideae. These floral characters are considered as primitive. This view would imply a derived condition for the well-known, widespread central placentation (with septa) or free-central placentation (with reduced septa) which prevails in all other Centrospermae, an unlikely assumption. Putting Cactaceae at the base of all Centrospermae would also imply a pigment reversal for the anthocyanin families, an assumption for which convincing arguments would be needed.

A dilleniid (or, more specifically, paeonioid) relationship of the Centrospermae is untenable, mainly because the dilleniids are basically woody and have a different secondary chemistry. This is not only a matter of single substances, such as the lack or expected presence of benzylisoquinolines in *Paeonia* (Leins and Erbar 1991), but also a matter of entire biosynthetic pathways, which are fundamentally different in dilleniids and magnoliids plus Centrospermae. For the same reason the Polygonales and Plumbaginales, possessing galloylesters (ellagic acid), are chemically fundamentally different from the Caryophyllales.

The idea of combining dilleniids with Caryopyllales on account of their fascicled stamens is unsatisfactory for the following reason: If stamen fascicles are plesiomorphic, they are of little use for establishing sister group relationships; if they are apomorphic, the prima-facie evidence for their multiple origin has to be invalidated, because it is extremely unlikely that stamen fascicles in the dicotyledons originated only once. The burden of proof is on any conjecture which assumes that they did.

Acknowledgements. Thanks are due to O.R. Gottlieb, Rio de Janeiro, and H. Wyler, Lausanne, for their advice in relation to the biogenetic scheme.

References

Bate-Smith EC (1962) The phenolic constituents of plants and their taxonomic significance. I. Dicotyledons. J Linn Soc Bot 58:95–173

Bittrich V, Amaral MCE (1991) Proanthocyanidins in the testa of centrospermous seeds. Biochem Syst Ecol 19:319–321

Cronquist A (1988) The evolution and classification of flowering plants, 2nd edn. New York Botanical Garden, Bronx

Jensen SR (1992) Systematic implications of the distribution of iridoids and other chemical compounds in the Loganiaceae and other families of the Asteridae. Ann Mo Bot Gard 79:284–302

Kubitzki K (1987) Phenylpropanoid metabolism in relation to land plant origin and diversification. J Plant Physiol 131:17–24

Kubitzki K, Gottlieb OR (1984a) Micromolecular patterns and the evolution and major differentiation of angiosperms. Taxon 33:375–391

Kubitzki K, Gottlieb OR (1984b) Phytochemical aspects of angiosperm origin and evolution. Acta Bot Neerl 33:457–468
Leins P, Erbar C (1991) Fascicled androecia in Dilleniidae and some remarks on the *Garcinia* androecium. Bot Acta 104:336–344
Mølgaard P, Ravn H (1988) Evolutionary aspects of caffeoyl ester distribution in dicotyledons. Phytochemistry 27:2411–2421

15 *Lyallia kerguelensis* Hook. f. and Its Artificial Propagation

ALICIA LOURTEIG

15.1 Introduction

In 1847, J.D. Hooker published the genus *Lyallia* and its unique species, which he had found at Christmas Bay in the north of the Archipelago of Kerguelen during his famous voyage to the Antarctic. He wrote: "I have placed it provisionally amongst Portulaceae, knowing no other order with which it has any equally direct affinity" (p. 549). Hooker was not in possession of any flowers, only fruits; the original description was, therefore, completed by Oliver who obtained flowers from Moseley (see Moseley 1875).

Bentham (1862; see also Bentham and Hooker 1862) placed the genus in the Caryophyllaceae next to *Pycnophyllum* Rémy. A. Gray (1876) retained it in the Portulacaceae, but the following authors (Pax 1989; Engler 1890; Pax and Hoffmann 1934; Mattfeld 1934) conserved it in the Caryophyllaceae. In 1961, Philipson and Skipworth created a new family, Hectorellaceae, for *Hectorella caespitosa* Hook. f., a monospecific genus from New Zealand, and the closely related *Lyallia* (see also Philipson 1987). The family Hectorellaceae was subsequently accepted by Takhtajan (1969), Dahlgren (1975) and Thorne (see Chap. 1), its incorporation into the Caryophyllaceae was made unlikely by the detection of betalains (Mabry et al. 1978) and the specific form of sieve-element plastids in *Hectorella* (Behnke 1975). Finally, *Hectorella* Hook. f. was proposed to be reduced to the synonymy of *Lyallia* by Nyananyo and Heywood (1987). However, contemporary botanists (e.g. Skipworth 1961; Ng Siew Yoong et al. 1975; Richardson 1981; Rodman et al. 1984), investigating the Caryophyllales with modern methods, do not cover *Lyallia* or fail to mention this taxon in either the Caryophyllaceae or the Portulacaceae.

15.2 Taxonomic Description

Genus *Lyallia* Hook. f.

General literature: Hooker (1847: pp. 548–549, Plate 122); Bentham (1862: p. 73); Bentham and Hooker (1862: p. 153); Pax (1889: p. 88, Fig. 31); Ewart (1907); Pax and Hoffmann (1934: p. 329, Fig. 135); Mattfeld (1934).

Calyx 4-partite, persistent. Petals 0. Stamens 3. Ovary elliptic, 2- to 3-ovulate. Style the same length. Stigma 2-fide, papillose. Flowers subsessile. Peduncle 2-bracteate; bracts connate at the base. Fruit 1-locular, 1-seeded (from abortion). A seed arises from the base of the fruit; funicle ascendent, fleshy; embryo peripheral. Suffrutex caespitose, glabrous, much branched; leaves tightly imbricate. Penduncles solitary. Flower inconspicuous, hidden among the foliage.

Type: *L. kerguelensis* Hook. f.
Dedicated to Dr. D. Lyall, friend of J.D. Hooker and who participated in the expedition.

Lyallia kerguelensis Hook. f.

Additional literature: Oliver in Moseley (1875); Dyer (1874); Gray (1876); Hooker (1879: pp. 1–15, Plates 1 and 2); Engler (1890); Werth (1911); Cour (1958); Philipson and Skipworth (1961); Skipworth (1961); Cranwell (1961); Lourteig and Cour (1963); Young and Schopfield (1973). Tufts or cushions (Fig. 15.1a–c) usually 15–45 cm but up to 60 cm and exceptionally to 1 m in diameter. Roots and branches ligneous; densely branched, branches 0.5–1.0 cm in diameter. Leaves green (Fig. 15.1d), densely imbricate, alternate, coriaceous or subchartaceous, ascendent, concave, oblong to linear, three to four times longer than wide, sessile, obtuse or irregularly dentate, one nerve 2-4-5-branched, ±prominent (2.5–3.0 mm long). Flowers solitary (Fig. 15.1e), hidden among the foliage, shorter than the leaves. Peduncle erect, enlarged toward the base, 2-bracteate. Bracts generally long, connate at the base, linear, concave, apex ± acute. Sepals 4, almost free, oblong, elliptic, subacute (2.5–3.0 mm), subequal. Petals 0. Stamens 3, the relative position of which is not constant; filaments filiform and about the length of the petals; anthers roundish, small, 2-thecate. Pistil (Fig. 15.1f) about the same length of the other floral cycles; ovary ellipsoid, glabrous, 1-locular, 1-, 2- or 3-ovulate; style straight; stigma 2-fide, lobules linear-oblong, papillose. Fruit obovoid (Fig. 15.1g), subglobose, retaining at the top part of the base of the style, 1-seeded (usually other ovules are aborted). Seed irregularly shaped (1–1.5 mm), generally reniform (Figs. 15.1h and 15.2). Testa blackish; testa cells isodiametric to elongate, tabular to slightly tuberculate at the dorsal region towards the micropyle. Cell walls straight. No obvious microsculpturing of testa surface (Fig. 15.3). Sieve-element plastids

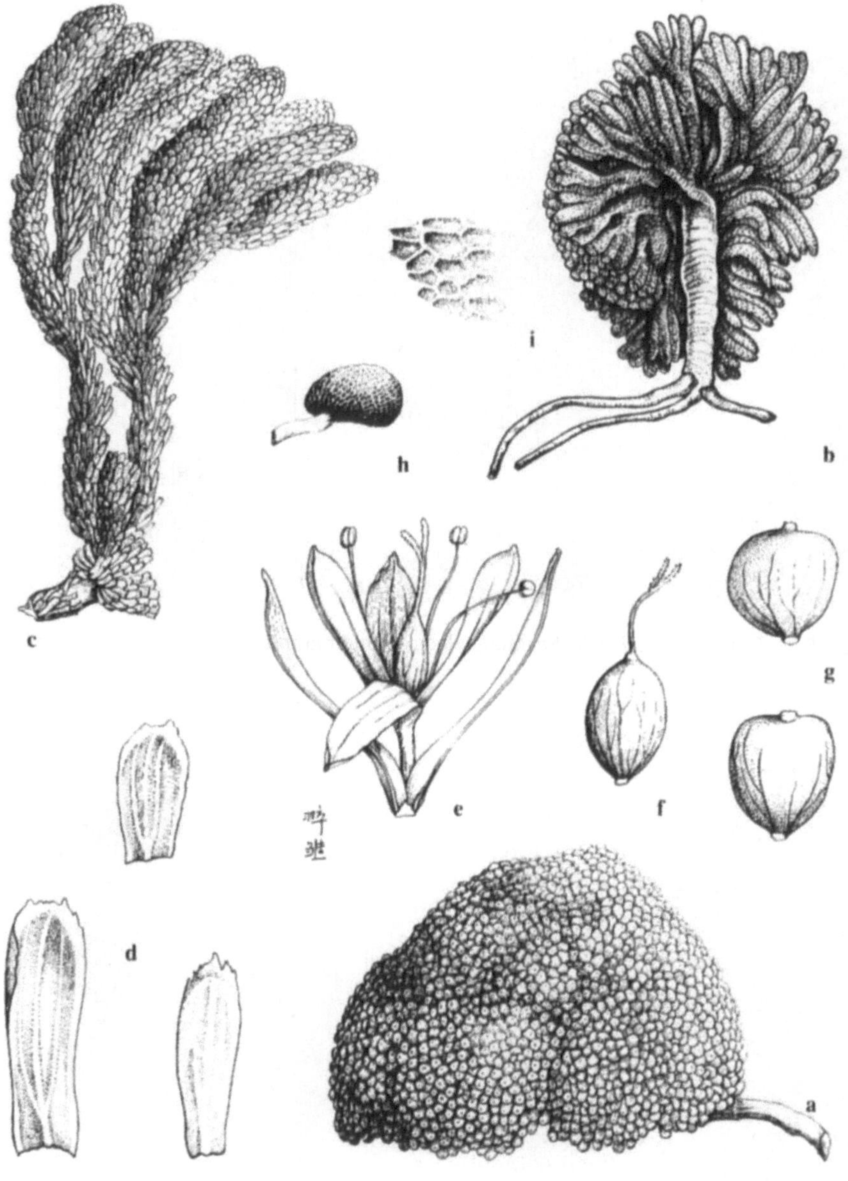

Fig. 15.1a–h. *Lyallia kerguelensis*. **a** Tuft, ×0.25; **b** underside of tuft; **c** part of a plant, ×1.5; **d** leaves, ×10; **e** flower with bracts, ×10; **f** pistil, ×15; **g** and **g′** fruits, ×10; **h** seed, ×7.5; **i** detail of the same, ×±50. (Tuft after Aubert de la Rüe, P; **c** Cour 319, P; **e–h** Coulaud 56, P. Drawings by R. Tang)

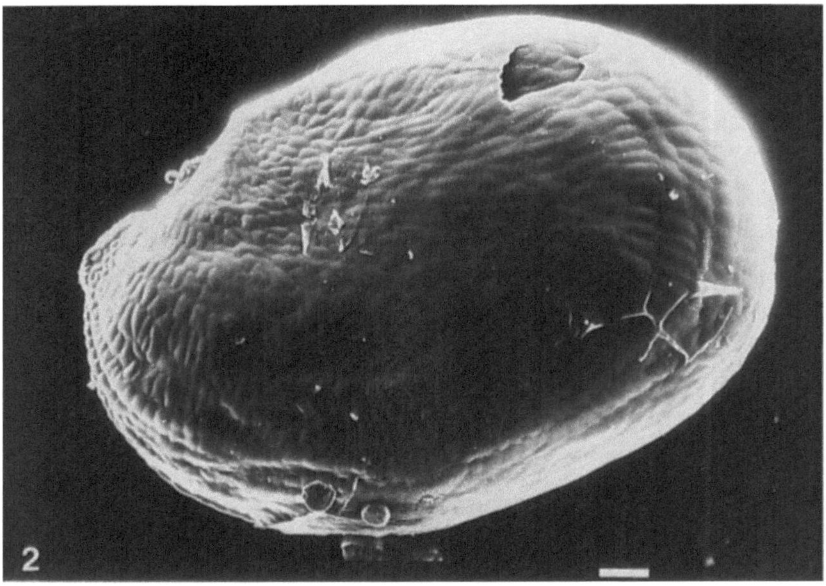

Fig. 15.2. *Lyallia kerguelensis*. Scanning electron micrograph of seed; *bar* denotes 100 μm

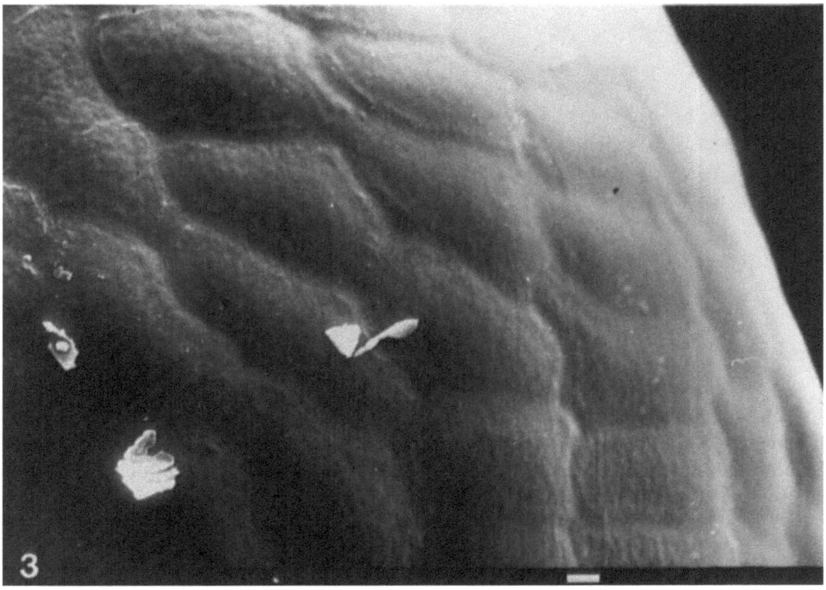

Fig. 15.3. *Lyallia kerguelensis*. Scanning electron micrograph of testa surface; *bar* denotes 10 μm

of form P3cfs (see Chap. 5). Pollen 3-colpate with a spinulose and sparsely punctate ektexine (Nowicke and Skvarla 1977).

Type: Kerguelen, Christmas Harbour, on dry gravelly soil, 400 ft above sea level, very rare, in fruit, Hooker (766) May 1840 Holotype K. Isotype BM, P.

15.3 Geographical Distribution and Ecology

Endemic to the Archipelago of Kerguelen. Sparsely distributed, not continuously but not frequent; at sea level receiving sea spray, on very humid, well-drained (not waterlogged) soils, also on plateaus. Always found in very exposed situations, windy places, such as depots of moraines, syenites, originated by glaciers but carrying rain water. Often it grows supported on cushions of *Azorella Selago* Hook. f.; we have observed deteriorated tufts. It is only in regions outside man's area of activity that the largest cushion (up to 1 m) has been found, on the western side of the island, in a syenite soil under the influence of sea spray. At Christmas Harbour, near Port-aux-Français, and at other spots where there is much human traffic or where cattle have been introduced, the species has disappeared. We consider its area to be in regression and the species to be threatened with extinction.

15.4 Material Examined

Kerguelen. Kerguelen's Land: Kidder (Venus Exp.) GH, K, P. Antart. Exp. 1839–1843 s. col. GH, K, S. UPS. Moseley I 1874 BM, K, P, S, C. Christmas Harbour, Hooker (766) 1839–1843 BM, K, P, Ib., Mc Cormick a. 1840 BM. Pointe Molloy, P. Cour (320) XII 1958 P. Peninsule Courbet, plateau près du Haut du Château, Aubert de la Rüe 14 XII 1949 P. Vallée entre Le Pouce et le Mont W. Thompson, Pointe Poincaré, ipse 25 XII 1949 P. Pied du Dôme Rouge, ipse III 1931 P. Port-aux-Français, J.-C. Jolinon (9) I 1959 P. Royal Sound, Eaton 26 II 1875 BM, K. Ib., Antarctic Research Expedition (A.R.E.) (B 125a) 20 XI 1929 BM. Ile Longue (Long Island): A.R.E. (B 171 b) 11 II 1930 BM. Ib., ipse 2 XI 1929 BM. Ib., Plateau W-N-W, P. Coulaud (24) 19 IX 1961 P. Ile Antares: Passe Royale, Aubert de la Rüe 16 I 1950 P. Ile Australia: P. Cour (318) II 1957 P. Ib., ipse (821) II 1958 P. Murray Island: A.R.E. (B 215 b) 23 II 1930 BM. Ile Mussel: P. Coulaud (56 et 57) 31 I 1962 P. Ib., Cour (319) II 1957 P. Peninsule Rallier du Baty: Vallée de Larmor, vers Deux Frères, J.-C. Jolinon (746) 11 II 1984 P. Anse Syénite, ipse (757) 14 II 1984 P. Près de la plage, sur syènite, ipse (759) 14 II 1984 P. Baie Syénite, ipse (762) 14 II 1984 P. Les Deux Frères, ipse (772) 20 II 1984 P. Vallée de la Plage Jaune, ipse (777) 22 II 1984 P.

15.5 Artificial Propagation

With the aim of saving this strictly endemic plant, doubtless threatened, we attempted to bring it into cultivation. The first problem was to get seeds. My efforts were devoted to obtaining the admission of a botanist in the expedition of Terres Australes and Antarctiques Françaises of 1984, for a study of the geology of the western side of the archipelago (also badly known in botanical terms), from Christmas Harbour to Peninsule Rallier du Baty. Jean-Claude Jolinon was accepted to carry out botanical work, especially on *Lyallia*, in this memorable performance of 45 days by foot, under conditions of a very severe climate and difficult terrain. He obtained seeds that we shared, for cultivation, between the gardens of Paris and Kew.

The seeds cultivated in Paris, without receiving any special treatment, did not germinate. On 26 March 1986, the 12 seeds sent to Kew Gardens were placed in a solution of 5% sodium hypochlorite for 30 minutes and rinsed twice in sterile water. The seed coats showed signs of contamination and, therefore, the embryos were dissected out and sown on a range of media: Knudson, C., Murashige and Skoog (M & S), half strength M & S, and Vacin and Went.

Germination, on a cool shelf, started within 2 weeks and 6 of the 12 seeds germinated. There was no apparent difference between seedlings on different media. The plants were between 10 and 15 mm tall. In November 1986, three of them were sent to the Alpine department in an attempt to grow them in compost, using more natural methods. The other three were retained to try to multiply them. The plants in the herbaceous Alpine section grew reasonably well at first but, even on a cool bench, did not survive the 1989–1990 summers. The last material died early in 1991 (personal report from J.L.S. Keesing, Kew Gardens).

References

Behnke H-D (1975) *Hectorella caespitosa*: Ultrastructural evidence against its inclusion into Caryophyllaceae. Plant Syst Evol 124:31–34

Bentham G (1862) Notes on Caryophylleae, Portulaceae, and some allied orders. J Linn Soc Bot 6:55–77

Bentham G, Hooker JD (1862) Genera plantarum, vol 1. Reeve, London

Cour P (1958) A propos de la flora de l'Archipel de Kerguélen (Mission 1957–58). Rev Terr Aust Antarct Fr 4(5):10–22

Cranwell LM (1961) Subantarctic pollen and spores. 1. *Lyallia* of Kerguelen. Pollen Spores 3:11–20

Cranwell LM (1963) The Hectorellaceae: Pollen type and taxonomic speculation. Grana Palynol 4:195–202

Dahlgren R (1975) A system of classification of the angiosperms to be used to demonstrate the distribution of characters. Bot Not 128:119–147

Dyer T (1874) Proc Linn Soc 4(June): xxxiv

Engler A (1890) Übersicht über die botanischen Ergebnisse der Expedition. Die Forschungsreise S.M.S. Gazelle 1874–1876. Botanik 4(1): xv

Ewart AJ (1907) The systematic position of *Hectorella caespitosa* Hook. f. J Linn Soc Bot 38:1–3

Gray A (1876) Botany. In: Kidder JH (ed) Contributions to the natural history of Kerguélen Island, made in connection with the United States Transit of Venus Expedition, 1874–75. II. Bull U S Nat Mus 1(3):20–30

Hooker JD (1847) The botany of the Antarctic voyage of H.M.S. Discovery ship Erebus and Terror, in the years 1839–43, vol 2. Reeve, London

Hooker JD (1864) Portulacaceae. In: Handbook of the New Zealand flora: a systematic description of the native plants of New Zealand. Reeve, London, pp 26–27

Hooker JD (1879) An account of the botanical collections made in Kerguelen's Land during the transit of Venus Expedition in the years 1874–75. Philos Trans R Soc 168:1–15

Lourteig A, Cour P (1963) Essai de distribution géographique des plantes vasculaires de l'archipel de Kerguélen. CNFRA Publ 3:65–70

Mabry TJ, Neuman P, Philipson WR (1978) *Hectorella*: A member of the betalain suborder Chenopodiineae of the order Centrospermae. Plant Syst Evol 130:163–165

Mattfeld J (1934) Nachtrag zu den Carryophyllaceae. In: Engler A (ed) Die natürlichen Pflanzenfamilien, 2nd edn, vol 16c. Engelmann, Leipzig, pp 365–367

Moseley HN (1875) On the botany of Marion Island, Kerguelen's Land, and Yong Island of the Heard Group. J Linn Soc 14:389–390

Ng Siew Yoong, Philipson WR, Walker JRL (1975) Hectorellaceae – a member of the Centrospermae. N Z J Bot 13:567–570

Nowicke JW, Skvarla JJ (1977) Pollen Morphology and the relationship of the Plumbaginaceae, Polygonaceae, and Primulaceae to the order Centrospermae. Smithson Contrib Bot 37:1–64

Nyananyo BL, Heywood VH (1987) A new combination in *Lyallia* (Portulacaceae). Taxon 36:640–641

Pax F (1889) Caryophyllaceae. In: Engler A, Prantl K (eds) Die natürlichen Pflanzenfamilien, 1st edn, vol 3(1b). Engelmann, Leipzig, pp 61–94

Pax F, Hoffmann K (1934) Caryophyllaceae. In: Engler A (ed) Die natürlichen Pflanzenfamilien, 2nd end, vol 16c. Engelmann, Leipzig, pp 275–364

Philipson WR (1987) The treatment of isolated genera. Bot J Linn Soc 95:19–25

Philipson WR, Skipworth JP (1961) Hectorellaceae: A new family in dicotyledons. Trans R Soc N Z Bot 1(4):31

Richardson PM (1981) Flavonoids of some controversial members of the Caryophyllales (Centrospermae). Plant Syst Evol 138:227–233

Rodman JE, Oliver MK, Nakamura RR, McClammer JV Jr, Bledsoe AH (1984) A taxonomic analysis and revised classification of Centrospermae. Syst Bot 9:297–323

Skipworth JP (1961) The taxonomic position of *Hectorella caespitosa* Hook. f. Trans R Soc N Z Bot 1:17–30

Takhtajan A (1969) Flowering plants, origin and dispersal. Oliver & Boyd, Edinburgh

Werth E (1911) Die Vegetation der subantarctischen Inseln, Kerguélen, Possession- und Heard-Eiland. II. In: Drygalski EV (ed) Deutsche Südpolar-Expedition 1901–1903. 8. Botanik 2:221–371

Young SB, Schopfield EK (1973) Palynological evidence for the late glacial occurrence of *Pringlea* and *Lyallia* on Kerguelen Islands. Rhodora 75:239–248

Genera Index

MIX
Papier aus verantwortungsvollen Quellen
Paper from responsible sources
FSC® C105338

If you have any concerns about our products,
you can contact us on
ProductSafety@springernature.com

In case Publisher is established outside the EU,
the EU authorized representative is:
Springer Nature Customer Service Center GmbH
Europaplatz 3, 69115 Heidelberg, Germany

Printed by Libri Plureos GmbH
in Hamburg, Germany